Advances in Insect Pollination Technology in Sustainable Agriculture

Advances in Insect Pollination Technology in Sustainable Agriculture

Editor
Dr. Amarjit S. Tanda
School of Horticulture
University of Western Sydney
Richmond, Australia.

I.K. International Pvt. Ltd.
NEW DELHI

Published by

I.K. International Pvt. Ltd.

4435–36/7, Ansari Road, Daryaganj

New Delhi-110 002 (India)

E-mail: info@ikinternational.com

Website: www.techsarworld.com

ISBN 978-93-90620-71-5

Published/Printed by Krishan Makhijani for I.K. International Pvt. Ltd., 4435–36/7, Ansari Road, Daryaganj, New Delhi-110 002.

Dedicated to my family

Mother Mrs Bhagwanti

Father S. Tara Singh Tanda

My wife Mrs Paramjit Kaur Tanda

Geni Tanda **Daughter**

Dr Gurminder Singh Bharee **Son-in-Law**

Mr Amargurjot Singh Tanda **Son**

Foreword

Insect pollination plays a significant role in the agroindustry and serves as a basic pillar for crop production. Crop plants depend on vectors to transfer pollen, which can include water, wind, and animal pollinators like bats, moths, hover flies, birds, bees, butterflies, wasps, thrips, and beetles. Cultivated field crops are typically pollinated by animals. Animal-based pollination contributes to 30% of global food production, and bee-pollinated crops contribute to approximately one-third of the total human dietary supply. Bees are considered important pollinators due to their effectiveness and wide availability. Bee pollination provides excellent value to crop quality and quantity, improving global economic and dietary outcomes. Taking a systematic crop by crop new approach, the present book depicts advances in insect pollination technology in sustainable agriculture. On this newly designed subject of Entpollinatology, Dr. Amarjit Singh Tanda has attempted to highlight the theme of insect pollination technology which offers a new insect pollination management technology as a vital tool to modern agroindustry. The first five chapters cover the world's food sustainability without insect pollinators, the historical outlook of insect pollinators, pollination concepts, supplementary pollinator services, climate change, and conservational aspects of pollinators. The next few chapters include chemical ecology in cross-pollination, seasonal management for improved foraging of honeybees, and their economic significance. The last five chapters describe the importance of wildbees, natural and artificial feeding of domesticated honeybees, biochemical in insect pollination, urbanization and its impact, fitting pollinators in integrated pest and management technology, and challenges and threats for pollinator conservation.

This book is an uncommon blend of pure and applied entpollinatological or insect pollination research. The discussions include using minimum pesticides and the possibility of eliminating toxic chemicals from farm practices. I congratulate Dr. Amarjit S. Tanda, the editor of this book, for bringing out a valuable collection of chapters touching on the most advanced and important aspects of insect pollination technology in sustainable agriculture. This book will serve as a reference tool for scientists, students, policymakers, and other agroindustry professionals.

Gurdev S. Khush FRS

Member US National Academy of Sciences
Adjunct Professor Emeritus, University of California,
Davis, CA 95616 USA
Former Head, Plant Breeding Genetics and Biotechnology,
IRRI Los Baños, Philippines

Preface

Pollination is the procedure by which pollen is transferred from the male part of a flower to the female part. This process enables fertilization and the production of seeds and fruits. Mostly, the pollination mechanism occurs within a plant species, but in cases where it occurs between two different plant species, a hybrid offspring is developed. Self-pollination happens when pollen from the same plant arrives at the stigma of a flower or ovule. Pollination is often helped by animals such as birds, insects, and bats as well as by the wind, rain, and water. Almost of the global food supply is obtained from crop plants that need pollination. The financial and ecological benefit of natural pollination by native bee pollinators in farm crops can be observed in their improved quality and yield. However, there is a recent decline in bee pollinator populations owing to habitat destruction, pesticide use, climate change, and diseases. Entpollinatology or insect pollination management technology as proposed by us is thus crucial to modern agroindustry. Attempts to conserve and encourage bee pollinators, and add new pollinators to monoculture habitats, as in commercial orchards, are significant for sustainable agricultural production. Featuring fourteen chapters contributed by leading experts on pollination and pollinating agents, this book covers a broad and comprehensive range of topics on advances in insect pollination technology in sustainable agriculture. The book aims to update the information on the global scenario of entpollinatology, a newly proposed term by us. Chapter 1 gives an overview of the world's food sustainability without insect pollinators. To know the novel historical outlook of insect pollinators the comprehensive studies are mentioned in chapter 2. In chapter 3, a summarized picture of pollination concepts and crop production is elucidated. Supplementary pollinator services in yield optimization of crop plants through management of bees are portrayed in chapter 4. Chapter 5 describes the climate change and pollinators and conservational aspects of pollinators. Additionally, the potential of emerging technologies including chemical ecology in cross-pollination is elaborated in chapter 6. In chapter 7, seasonal management for improved foraging of honeybees and enriched honey production has been presented. Chapter 8 aims to inform about insect pollination technology and its economic significance. Wild bees are important in sustainable agriculture are well-enumerated in chapter 9. Chapter 10 focuses on the natural foraging and artificial feeding of domesticated honeybees. The use of plant biochemicals in insect pollination for crop improvement has been highlighted in chapter 11. Chapter 12 describes urbanization and its impact on native pollinators. Pollinators are equally important in integrated pest and management technology as narrated in chapter 13. Finally, chapter 14 describes the challenges and threats for pollinator conservation. The book will help new researchers by foregrounding their knowledge in this area.

I am greatly thankful to all the contributors for the conscientious and diligent work they have done in compiling their great, grandiose, and majestic chapters. Without their determined efforts, this book project would not have been possible.

I am indebted to Professor Gurdev S. Khush, for showering his blessings on me by writing the Foreword of this book.

I must make mention of my immediate family members, and friends without whose involvement and support, completion of this book would not have been possible. No one can do any creative work, such as the writing of a scientific book, without the cooperation of his/her spouse. My wife Paramjit showed utmost patience and cooperation during the preparation of this book. My son Amargurjot again made sure that I was kept away from the routine type of tiring and monotonous household duties. My daughter Geni and son-in-law Dr. Gurminder frequently motivated me by calling me on the progress of the book.

Amarjit S. Tanda

The Editor

AMARJIT S. TANDA, obtained BSc (Agri) (1974), MSc (1976), PhD (1980), degrees in Entomology with Merit Scholarship and Fellowship from Punjab Agricultural University, Ludhiana. He specializes in insect pest and nematode management of crops, integrated pest management technology (IPMT), biological control, entpollinatology (or insect pollination), and gnotobiology. He is a peer reviewer of Springer Nature books and refereed journals of national and international repute. Dr. Tanda crafted insect damage indexing technique, investigated vacuuming and brushing methodology for IPMT. Established role of honeybees in Asiatic cotton hybrid seed production. Proposed "Entpollinatogy" for insect pollination. Designed a new and simple technique for root exudates collection in vitro. Standardized gnotobiotechniques for root culture screening, and testing nematicides. Served alma mater for 15 years. Worked at School of the Horticulture University of Western Sydney, Richmond, Australia and CSIRO, North Ryde NSW, Australia. Awarded Charted Biologist degree by the Institute of Biology, UK. Presented Guest Speaker Certification Award by Norwegian Plant Protection Institute, Ås, Norway. Springer Nature published his first book *Molecular Advances in Insect Resistance of Field Crops: Modern and Applied Approaches*. Recipient of Best Poem Award, Bhasa Vibhag Punjab, and PAU Best Poet Award. Presented with Life Achievements Award, Punjabi Association Brisbane and Melbourne, Australia. Published 9 books in literature, co-edited *Bibliography of Entomology*. Published 135 research papers, 7 book reviews, 7 book chapters, and 117 science articles. Elected Fellow of the Linnean Society London. Honorary Member, American Association of Professional Apiculturists, Australian Society of Plant Scientists, Australian Native Bee Association Inc. Life Fellow of Entomological Society of India, Member, Insect Science Society of India, Ecological Society of India, and Plant Pathological Society of India.

Contributors

1. **Abu Bakar Muhammad Raza**, Department of Entomology, University of Sargodha, Pakistan
2. AICRP on Biological Control, AAU, Jorhat centre
3. **Amarjit S. Tanda,** School of Horticulture, University of Western Sydney, Richmond, Australia
4. **Arijit Kundu**, Department of Botany & Forestry, Vidyasagar University, Midnapore- 721102, India
5. **C. Patel**, PhD Scholar, GB Pant University of Agriculture and Technology, Pantnagar, Uttrakhand-263145, India
6. **Deeksha M. G.**, Research Scholar, DST Inspire Fellow, Indian Agricultural Research Institute, New Delhi-110012, India
7. **G. S Giri**, Assistant Professor, Department of Entomology, Tirhut College of Agriculture, Dholi, RPCAU, Pusa, Bihar-843121, India
8. **H. Khurram Shurjeel**, Department of Entomology, University of Sargodha, Pakistan
9. **Irfan Akram**, Department of Entomology, The Islamia University of Bahawalpur, Pakistan
10. **K. Kishor**, Assistant Professor, Department of Agronomy, PG College of Agriculture, RPCAU, Pusa, Bihar-848125, India
11. **K. Nishanthini**, Department of Entomology, Faculty of Agriculture, Annamalai University, Tamil Nadu-608 002, India
12. **Mubasshir Sohail,** Plant Protection Division, Nuclear Institute of Agriculture, NIA, Pakistan
13. **Muhammad Abu Bakar**, Department of Entomology, University of Sargodha, Pakistan
14. **Muhammad Anjum Aqueel**, Department of Entomology, The Islamia University of Bahawalpur, Pakistan
15. **Muhammad Haseeb Ahsan,** Department of Entomology, The Islamia University of Bahawalpur, Pakistan
16. **Muhammad Yasin**, Department of Entomology, The Islamia University of Bahawalpur, Pakistan
17. **Prabhu Prasanna Pradhan**, Department of Entomology, Odisha University of Agriculture and Technology, Bhubaneswar, Odisha, India

18. **Prakash Karmakar**, Department of Botany & Forestry, Vidyasagar University, Midnapore- 721102, India
19. **R. Kanagarajan**, Department of Entomology, Faculty of Agriculture, Annamalai University, Tamil Nadu-608002, India
20. **Rudra Narayan Borkakati**, AICRP on Biological Control, AAU, Jorhat Centre. Department of Entomology, Assam Agricultural University, Jorhat-785013, India
21. **Sajjad Ali,** Department of Entomology, The Islamia University of Bahawalpur, Pakistan
22. **Sourabh Bisui**, Department of Botany & Forestry, Vidyasagar University, Midnapore- 721102, India
23. **Swagatika Sahoo**, Department of Entomology, Odisha University of Agriculture and Technology, Bhubaneswar, Odisha, India
24. **Ujjwal Layek**, Department of Botany, Rampurhat College, Birbhum- 731224, India
25. **Ziyad Abdul Qadir**, HBRI, NARC, Islamabad, Pakistan

Contents

Foreword *vii*

Preface *ix*

The Editor *xi*

Contributors *xiii*

1. **World Food Sustainability without Insect Pollinators is Impossible** **1**
   – *Amarjit S. Tanda*
2. **Global Agroindustry in the Absence of Insect Pollinators: Historical Outlook** **29**
   – *Amarjit S. Tanda*
3. **Pollination Concepts and Crop Production** **44**
   – *R. Kanagarajan and K. Nishanthini*
4. **Supplementary Pollinator Services in Yield Optimization of Crop Plants through Management of Bees** **64**
   – *Ujjwal Layek, Sourabh Bisui, Arijit Kundu and Prakash Karmakar*
5. **Climate Change and Pollinators and Conservational Aspects of Pollinators** **80**
   – *Gouri Shankar Giri, Chenesh Patel, Deeksha M.G. and Kaushal Kishor*
6. **Chemical Ecology in Cross-Pollination** **106**
   – *K. Nishanthini and R. Kanagarajan*
7. **Seasonal Management of Honeybees for Their Improved Honey Production** **115**
   – *Muhammad Anjum Aqueel, Hafiz Khurram Shurjeel, Abu Bakar Muhammad Raza, Muhammad Yasin, Irfan Akram, and Muhammad Haseeb Ahsan*
8. **Insect Pollination Technology and Its Economic Significance** **132**
   – *Amarjit S. Tanda*
9. **Wild Bees are Important in Sustainable Agriculture** **163**
   – *Amarjit S. Tanda*
10. **Natural Foraging and Artificial Feeding of Domesticated Honeybees** **184**
    – *Muhammad Anjum Aqueel, Hafiz Khurram Shurjeel, Sajjad Ali, Muhammad Abu Bakar, Mubasshir Sohail, Ziyad Abdul Qadir, Amir Manzoor, Shams ul Islam and Imran Khalid*
11. **Plant Biochemicals in Insect Pollination for Crop Improvement** **203**
    – *Swagatika Sahoo, Prabhu Prasanna Pradhan and Rudra Narayan Borkakati*

12. Urbanization and Its Impact on Native Pollinators **231**

- *Amarjit S. Tanda*

13. Pollinators are Equally Important in Integrated Pest and Management Technology **261**

- *Amarjit S. Tanda*

14. Challenges and Threats for Pollinator Conservation **278**

- *Usman Ghani, Ahmad Nawaz, Tariq Mustafa, Saiqa Muneer, Ahmad Ghafar, Shanza Akbar, Sadia Bibi, Marium Aziz and Aqsa Aslam*

Index **301**

1

World Food Sustainability without Insect Pollinators is Impossible

Amarjit S. Tanda*

ABSTRACT

In the sexual reproduction of flowering plants, pollination, and is a crucial step in the transfer of pollen to the stigma of the flower gynoecium. The majority of flowering plants rely on animals for this transfer mechanism of pollen. As flower visitors gain no direct benefits by flower pollination a few rewards must lure them. To visit flowers is by offering food is nectar, pollen, or oils the best way plants allure animal pollinators. While searching for nectar, pollen from the flower reproductive parts pollen from anthers stick to the body of the pollinator. Based on reports in the literature, we conclude that to the body of the pollinators new pollen may adhere on foraging subsequent blooms in search of more food, and the pollen from its body may get deposited to the stigma of crop flowers. Involving animal pollinators and plants in mutualism pollination is fundamental to the sustainability of agro-natural ecosystems. To a global economy of nature is inestimable and the value to agriculture is huge. From highly managed agriculture to remote wilderness nevertheless, pollination is a threatened system. Reducing pollinator forage is better otherwise insecticides will start hitting pollinators directly and herbicides indirectly? In croplands and natural areas' habitats, destruction has reduced pollination. By demographic change in apiculture and pollination, availability diseases attack honeybees. Using native wild bee pollinators, management of potential crop pollinators, and exploitation of the sensitive climatic system are required for agricultural protection processes, tree culture, and a sustainable crop environment. In this chapter, after a description of insect pollinators, we highlight selected works on why insect pollinators matter in crop improvement needed applications published over the last two decades.

Keywords: *Sexual reproduction • Pollination • Pollen grains • Fowering plants • Nectar • Pollen • Mutualism • Agricultural • Natural • Productivity • Beekeeping • Honeybee • Diseases • Conservation • Sustainable • Wild • Pollinators • Human • World*

* School of Horticulture, University of Western Sydney, Richmond, Australia and CSIRO North Ryde NSW, Australia.

E-mail: drtanda193@gmail.com

1.1 INTRODUCTION

Crop pollination is an essential mechanism and an important ecological function. Without pollinators, the human race and all of earth's terrestrial ecosystems would not survive. Over 80 percent of the world's flowering plants require a pollinator to reproduce. Regarding the biodiversity in agricultural systems and ecological services of insect pollinators, intensive agriculture has adverse effects. To transfer pollen as they forage, many crop plant species depend on many animals, and great champions are the insects. By offering the natural food gifts of pollen or nectar and they give directions to the crop flowers using their scents and visual cues, plants attract pollinators by their vibrant colors and alluring fragrances. This floral biotechnology has brought about strong relationships between flowers and the animals that pollinate them. Modern agribusinesses depend on insect pollinator services globally. Honeybees, native wild bees, and flies offer billions of dollars (about $235 and $577 billion U.S.) in economic value in every pollination season and serve a significant role in the world economy (FAO, 2018; Tanda, 2019a, 2020 a-c, 2021a-e; Tanda, 2022 a, b).

1.1.1 Insect Pollinators

Many species of bees, beetles, flies, wasps, thrips, butterflies, and moths are all successful pollinators. Such insects become effective plant pollinators because they share two important features: they fly, and so are capable of foraging several flowering plants in a very short time.

1.1.2 Entpollinatology (or Insect Pollination)

Entpollinatology (from Ancient Greek ἔντομον (entomon or ent), meaning 'insect', + pollination or pollinate, the act of transferring pollen grains and -λογία (-logia), meaning 'study of') is the first time proposed technical term by the author as the scientific study, management, and utilization of insect pollinators, a branch of zoology.

Pollinatechnology is described as Pollina = pollen + technology = the techniques or methods used to transfer pollen grains from the male anther of a flower to the female stigma. Entpollinatology is the first time proposed technical scientific term describing as the branch of biology in which insects are involved or used to transfer pollen grains to the stigma of the flower for pollination which is a very critical process in the fruit production of crop plants (Tanda, 2019a, 2020 a-c,2021a; Tanda, 2022a, b; Tanda, 2022 a, b).

Because of the high level and uniformity of modern agroindustry, numerous plants requiring pollination are relying upon honeybee services (Aizen et al., 2008). There is no other insect more beneficial to the globe than honeybees. The world's staple nourishments, including wheat, corn, and rice reproduce without insect pollination. Every one of these yields represents 65% of global food production, still leaving as much as 35% depending on pollinating insects (Klein et al., 2007). To set seeds and fruits, about one-third of the world's crops need flower pollination and animal husbandry relies on bee pollination for fodder production (Dias et al., 1999). Numerous crops and wild plants rely on insects, birds, bats, and other flower pollinators with insects as the chief players for crop production. In a variety of crop pollination, wild solitary and social bees aid in flower pollination in managed and natural environments. Many economic crops such as apple, citrus, tomato, melon, strawberry, apricot, peach, cherry, mango, grape, olive, carrot, potato, onion, pumpkin, bean, cucumber, sunflower, various nuts, a range of herbs, cotton, alfalfa, and lavender rely upon insect bee pollination services (Tanda, 2019, 2020 a, b, 2021a-e). No doubt, honeybee (*Apis mellifera*) is in front line of insect pollination globally, the role of native wild bee cannot be ignored. Crop pollination system is a biological management, food security and key to sustainable environment. Crops that produce fiber such as cotton or wood also need flower pollination for lifetime produce more prominent yields when flower pollinators are accessible (Tanda and Goyal, 1979; Tanda, 2019, 2020 a, b, c, 2020a-d,2021a-e; Tanda, 2022a, b). Bees are the key pollinators for numerous fruit and vegetable plants. Especially in pollen limited plants, advance pollination technology is a tool for expanding profitability without using expensive plant protection processes. Bee pollination bolster efficiency in numerous crops even without growers salute, so long as environment and supplementary bee pollinators are accessible. For agriculture, designing larger fields and landscapes plant pollinators may become limited or finished. Expanding over use of pesticides for pest and disease management is more harmful to crop pollinators, except if arranged and embraced with outrageous consideration.

People bother less about their conservation, even knowing the significance of insect pollination in food production. Agro-environments are being altered to deal with the great demand of world food security, we should know which pollinators are of great importance for more food production, and their protection in our agro environment. Insect flower visitor varies from crops and cultivars. Numerous crops do not require insect pollinators, however, may give rise to excellent fruit quality and seed production if cross-pollinated, and many flowers are surely self-pollinated or cross-pollinated. Some flowers need particular pollinators while many are cross-pollinated by different visitors, and still, other plants are pollinated by water currents or wind blowing. Efficient pollinators differ from one location to another of the same plant. The solo crop pollination services of honeybees offer over \$14 billion in economic value to the agroindustry. Honeybees are indispensable for the production of apple, cherry, melon, alfalfa, almond, sunflower, and many berries. Only a few bee species can be used for crop pollination at the commercial level. Further, apiary strength and advanced management system are greatly important for the sustainable agroindustry globally.

1.2 NO INSECT POLLINATORS NO WORLD FOOD SUSTAINABILITY

Insect pollinators are indispensable to the production of many of the micronutrient-rich fruits, vegetables, nuts, seeds, and oils we consume. About, 75 percent of the world's crops producing fruits and seeds for human consumption depend, at least in part, on pollinators for sustained production, yield, and quality. The beekeeping industry is significant because it offers honeybees a shielded place to work and make a colony to live in. It's important to maintain the bee population healthy, as bees pollinate many of our food crops. Apiculture also gives an environment in which to study bee habitat and their behavior.

To assess the relationship between bees and the landscape, US Department of Agriculture (2019) provided land managers and scientists with different methods. It offers a basis for making decisions about where to put their apiaries for the summer and fall after crop pollination ends so that the colonies will be in a position to build up strong healthy numbers in time for the migration to California for almond pollination. As it produces honey, a unique product, that is, the honeybee is the most versatile commercial pollinator. Honeybee reduction worldwide, and their impertinence for agroecosystem was seriously questioned in the history (Aizen and Harder, 2009; vanEngelsdorp et al., 2007). By persistent exposure to acaricides required to suppress the parasitic mite Varroa and bee's health is damaged negatively by the over applications of pesticides in cropping areas (Desneux et al., 2007; Barnett et al., 2007; Karise, 2007). Any decline of supervised honeybees and the loss of wild pollinators are of increasing concern as there was a relationship between insect pollinators and the world food security. Doubtlessly, the health of honeybees is at great risk globally. Major damages to honeybees and other plant pollinators are thereby the demolition and segmentation of wild and semi-wild environments as well as land use escalation in the agro-environment (Rathcke and Jules, 1993; Tscharntke et al., 2005; Kremen et al., 2004, 2007; Steffan-Dewenter and Westphal, 2008;). During different life stages of honeybees the most important matter is that they are damaged by parasitic mites (*Varroa destructor*, *Acarapis woodi*, *Tropilaelaps* spp.), bacteria (*Paenibacillus larvae*, *Melissococcus plutonius*), fungi (*Nosema* spp., *Ascosphaera apis*), many viruses, and scavengers from beetles, mice, and bears. While for others they remain elusive, but few parasites for bees are well described. Non-Apis bees are needed in a particular crop species or can be manipulated in the greenhouse agroindustry. Various practices and techniques that support the commercialization of wild non-*Apis* pollinators are also in high demand.

For native bee's unmanaged pollinators, the most important requirements are the conservation and restoration of their environment. Including appropriately managed farms, urban parks, and golf courses, many bees can nest in small habitat patches and in the natural areas, wildlands, bottom of form and human-dominated areas. Bees are challenged by contracting nesting sites, reproductive habitat, and bio-needs, for example, their parasites and brood diseases. Wild native and other pollinators could be effectively used to enhance the pollination in plants in case these could be multiplied in big populations and utilized for crop pollination jobs. For more competently pollinated crops there is a necessity to comprehend the pollination mechanism and pest management system. Pollination is a pivotal role in the setting of seeds in angiosperms. Numerous plants with blooms depend upon bees for the movement of male sex cells (Renner, 1988; Nabhan and Buchmann, 1997). As guests of blooms receive no direct advantages by pollinating flowers, there should be some gifts for them rendering this valuable service. Most plants offer them sufficient amounts of pollen, nectar, or different oils for their visiting job. During this visit to the plants, the dust of male sex cells from the anthers gets adhered to the hairy parts of the visitors. At the point when the creature visits consequent flowers looking for more food rewards, dust from its hairy parts gets deposited on the receptive stigma and new dust may get adhered to the hairs of the forager.

In a natural ecosystem, the entpollinatological procedure, is a rudimentary need for sexual proliferation in seed delivering plants (spermatophytes), taking into consideration hereditary recombination and the development of a hereditarily exceptional seed. This mingling mechanism of genetic material expands the capacity of probably a portion of a plant's posterity to get by in a universe of erratic ecological changes. Maintenance of this genetic variability in a population is necessary for evolution by natural selection process to occur, and therefore is the ability of a plant population to adapt to changing environmental influences. By cross-pollination, the gene variation mechanism in the next age is greater, where pollen transfer from the flower of one plant is moved to the stigma of new blooming plant. In meiosis, gene reshuffling also happens, to produce gametes. Thus, even in self-pollinated plants, they permit that genetic variations of the cultivar is maintained.

It has become invaluable for some crops to depend on pollination using insects, birds, or bats. For pollination, numerous plants depend on wind or water currents but must develop amounts of pollen dust to be surely received by the receptive stigmas. The plants using insects as pollinators waste fewer amounts of pollen as compared to the wind or water pollination process. Also, the plants may exhaust extra vitality using these creatures as pollinators, showing the generation of nectar a model example to remunerate pollinators. Likewise, insect visitors can move disease pathogens from one flower to another during the pollen transportation process. Colossally blooms contrast in shading, aroma, shape, and size; and are foraged by creatures having various morphological and taxonomic variations. These characteristics are available in insect orders of Hymenoptera, Lepidoptera, Coleoptera, and Diptera including different species of birds and bats (Tanda, 2019a, 2020 a-c, 2021a-e; Tanda, 2022a, b).

For effective pollination, in pollination biology, crops should particularize on the little subcategory of flower foragers. Nevertheless, many interaction partners and their morphological and taxonomical variations furthermore, undoubtedly, blooms exhibit combinations of traits showing morphological, behavioral, and physiological characteristics of pollinators (Faegri and van der Pijl, 1979). Deeply hidden with diluted nectar red odorless flowering plants are modified to humming or perching birds whereas blue blossoms with less hidden and concentrated nectar with pleasant aroma are bee's choice (Baker, 1975). Pollination syndromes are reported in taxonomic diverse plants and appear to be a result of specialization and evolution convergence. Since the Cretaceous time frame, pollinators and plants have developed together millions of years ago. Crop pollination relies on the animal behavior of different species, which move pollen dust from male parts to the female parts, the main procedure in the setting of seed in angiosperms. Bee pollinators offer a crucial service in the natural environment that gives rise to flower crossing and crop sexual multiplication and provides livelihoods. This natural mechanism further thus helps in crop and natural bio-diverse system protection. Lowered crop yields and malformed fruit quality often develop from ineffective pollination process rather than from a lack of other agrochemicals. In biological systems, the eye clues of inadequate pollination are more unobtrusive than in crops, however, the results can be as acute as the vanishing of species of a plant. A perceptible reduction in animals for consuming fruits and seeds, vegetation deterioration, and eventually, if foundation species are involved, the death of healthy environment and their services. Natural environments and agricultural systems rely on plant-pollinator diversity for sustainable bio-system. Dry wood, sand, bare land, mud, resins, embankments without vegetation, carrion, caves, and host plants for bees, moths, and beetles help in pollinator's abundance and sustainable bio-ecosystem. Pollinator biodiversity is huge and gigantic. In the world, over 20,000 bee species as well as many other insects and vertebrates are flower pollinators. Crop pollinators differ from many other sources of essential eco environment because they are often part of highly specific plant-pollinator mutualistic relationships. In specific niche requirements for the plants and their insect pollinators, pollinator losses can have falling effects on the entire agroecosystem. Some wild bees that pollinate herbaceous plants depend on dry wood nest openings, plant fruitfulness is reduced when the wood nesting sites are demolished.

1.3 ROLE OF HONEYBEES IN CROP PRODUCTION

As far as the agricultural financial returns matters honeybees are the most important, crucial and profitable pollinators. These hyper-efficient insects can offer pollination to virtually any crop plant. Almonds almost entirely depend upon honeybee cross-pollination for better seed setting and maximum yield. Driving up prices and disrupting the marketplace, without honeybees, yield for blueberries, squash, watermelon, and other fruits would be greatly decreased. A honeybee colony is worth 100 times more to the community than to the beekeeper—meaning the value they deliver extends well beyond their actual price (USDA). Honeybee pollination has helped make fruits, nuts, and vegetables more accessible to consumers (USDA, 2018; Tanda, 2019a, 2020 a, 2021a-e; Tanda, 2022a, b).

In the agroindustry, the significance of pollination has been recognized for centuries (Kevan and Phillips, 2001). Temple carvings of Ancient Assyrian show pollinating female date palm winged deities with male reproductive flower parts ensuring that dates would develop on the trees (Buchmann and Nabhan, 1996). In logs to manage and multiply captive colonies of stingless bees, is described in Maya of Mesoamerica. The flower pollination technology in the history was established by Koelreuter (1733–1806) and described in pollination ecology (Sprengel, 1750–1815).

Though the significance of crop pollination and pollinators conservation in the natural environment has been established for a long, however still it is trusted that somehow angiosperms get cross-pollinated naturally set seeds, form fruits and keep it on to the next age. So the pollination bioecological technology has not much progress, and there is a sufficient space for new students and known apiculturists to add to the entpollinatlogical studies and the crops they cross-pollinate, may be in the natural or agricultural environment. Amazingly, for several main crops globally, even the names of key and insignificant plant pollinators are not still not known.

The mechanism of flower pollination includes the transfer of pollen dust from the stamens to the receptive stigmas of the flower. Especially this biofloral-mechanism is pivotal in crop farming where males and females are available in separate flowers. Cross-pollination is a big asset that is essential in setting seeds and food production. In fruit and vegetable producers, bee pollinators play a great role. The pollinators are of great importance because they add to the fruit quality by effective fertilization process and ameliorate the progeny genetically by enhancing plant seed setting. For fruit and vegetable seed production, bee pollination becomes more essential when other agro-inputs such as pesticides, water, fertilizer, and pest management practices are considered.

Currently, the bee pollinators are declining due to the changes in different agricultural practices, segmentation of habitat, and pesticide overuse, nesting site abolition, and attack due to brood diseases and predators (Fig. 1.1). Damages to honeybees due to pesticides only are about $14.3 million in the USA annually. Owing to a serious decline of bee pollinators, mostly decision makers are badly enlightened about the huge bio-ecological catastrophe worldwide. To eradicate utmost poverty in developing countries, honoring the significance of bee pollinators will be an important ethical and practical drive in the world.

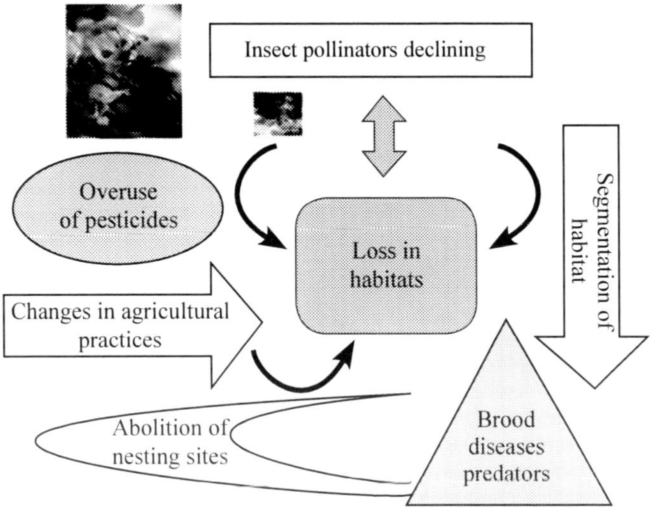

Fig. 1.1 Schematic representation of decline in insect pollinator populations and diversity

Honeybees contribute fairly in rural development to secure and sustainable livelihoods, further offering a critical part in crop pollination and thus enhancing the fruit quality and crop production. Many don't know the profitability of plant pollinators is manifolds than the importance of honey production globally. The contribution of wild bumblebees, solitary and stingless bees is of prominent importance besides managed honeybees. Numerous flowering plants depend on bee pollination. These creatures transfer the pollen grains, and most of these are the members of orders Diptera such as flies, Coleopterwa as beetles, Lepidoptera including butterflies and Hymenoptera: Apoidea, the most important to which bees are related. Honeybees are contemplated to be the most effective crop pollinator because of their morphological modifications for the maximum pollen gathering capacity.

1.4 SONICATION MECHANISM

Sonication or buzz pollination is a mechanism of vibrations performed by solitary bees to release pollen which is firmly held by the anthers. Pollen dust can only be free when vibrations are given to the stamens by the movement of bees in many flowering plants. Not by the honeybees, pollination by buzzing is carried out by bumblebees, carpenter bees, and by the Melipona stingless bees. Here implicit a new proportion for using the bees for crop pollination that need to be pollinated in greenhouse environment without wild pollinating players. For seed production in crops, commercial honeybees are used as an important pollinator worldwide. For the pollination of greenhouse crops and ornamental plants, bumblebees and solitary bees are being utilized (Estes et al., 1983). In spite of their significant part in crop production, the honeybee's numbers and their bio-environment is deteriorating as the result of damages to habitat, more monocultural practices, and ill effects of agrochemicals (Fig.1.1).

Due to decline of honeybee and native bee populations due to predators and bee diseases along with more crop acreage depending on bee pollinators, there have been a bee pollination disaster coming near soon in few years all over the universe. The services of managed bee pollinators are decreasing, however, the demand for bee pollination is increasing in many countries and the bee pollination catastrophe is making concerned in bee management, culture and conservation (Fig. 1.2). Advance agricultural practices and environment management systems may have significant influence on the biodiversity and the bee population protection.

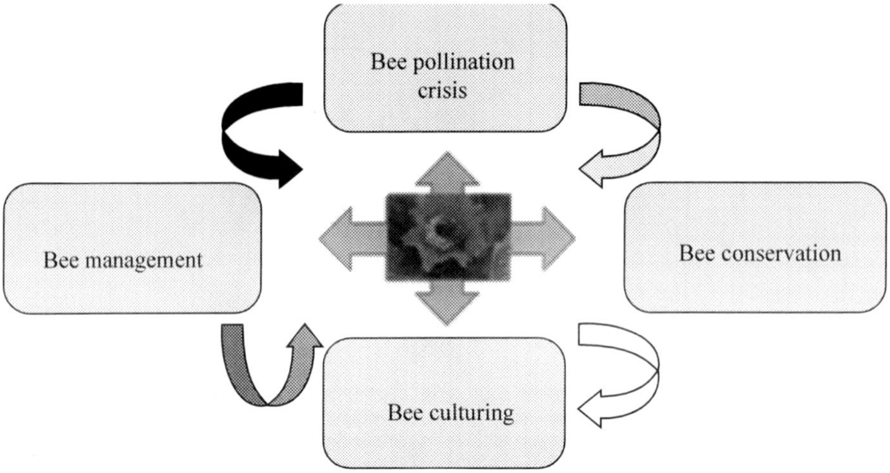

Fig. 1.2 Schematic representation of bee pollination crisis solution (Tanda, 2021a)

Crop pollination is an endangered process from more technical agriculture to an uncultivated wild environment. Insecticidal sprays harming bees directly and indirect effects were observed by herbicides minimizing bee populations. Bees contribute significantly in the modern cropping system. In reality, it is in crop pollination not honey production, the honeybees take participation. Certain improvements, pesticides and cultivars designed by gene technology are considered to be detrimental for beekeeping. In research, honeybees are highly deliberated insect after Drosophila due to many reasons. The value of crops that require pollination by honeybees, in the United States alone, is estimated to be around $24 billion each year and commercial bee pollination was valued around $10 billion annually. There is a tendency to consume more bee-pollinated crops, more honeybees multiplication and use in crop pollination. Truly honeybees are not managed well but growers should be educated to receive maximum benefits from the honeyBeehives and their hardships. Contribution of beekeeping to agroindustry is enormous. In India, honeybees cross-pollinate over 50 million hectares of cropping area. In oilseeds, pulses, fruits and vegetables bees enhanced the yields by cross-pollination. Considering the recurring shortages of pulses, cooking oils, fruit and vegetables, the significance of bee pollination can be acknowledged. An old co-developed practice using pollinators and crops in a mutulistic relationship, bee pollination is basic to agricultural crop production.

Crop pollinators and bee culture demographic alteration strategies of honeybee diseases are a big threat to the industry. Agroforestry, agricultural sustainability and development in conservation, significant assistance of wild natural pollinators, domestication of unused potential pollinators, and more environmentally sensitive human exploitation of the biodiversity are urgently required. As reciprocal selective factors, pollinators and plants have been kindred and interlinked in mutualism for 200 million years. To produce some amazing inventive pollination mechanisms, they have developed together. There several entomophilous species of crops are adapted to certain insects for seed setting such as Phlox to a diurnal butterfly, Hemoris;Yucca to a tineid moth, Pronuba; figs to a wasp, Blastophaga; red clover to bumblebees; Trollius to a blade fly, Chiastochaeta. Some female wasp species of *Scolia*, look like the flowers of *Ophrys insectifera* in some orchids. As the look of flowers alike females, incidentally pollination occurs, male wasps forage the blossoms not for nectar or pollen. In organic evolutionary development, this mutualism in pollinators and plants is one of the best important occurrences. To the agriculture world about 200 billion US dollars per year is the added potential economic value of insect crop pollination. On our planet to cross-pollinate 250,000 wild plants, about one lakh various pollinator species of animals are found. About 1,500 birds and mammals species give out the services as the flower pollinators joining with bees, wasps, moths, butterflies, flies, and beetles. Globally, birds, fruit bats, possums, lemurs snails, slugs, flying foxes, and gecko serve as efficient crop pollinators, however, hummingbirds are the effective flower pollinators in many states of the US. At a very alarming rate, the population level of wild and domesticated bee pollinators is decreasing worldwide. Due to change in their food preferences and habitats, nesting sites, ecosystem shrinkage, forests and grassland bio-environments, insecticidal damages, diseases and predators are a danger to many pollinators. Specificity of biophysical needs, quality of being rare or ability to offer humans desirable resources, few ecosystems are a greater risk. In the universe, about 500 insects including bees, wasps, butterflies, moths, ants, beetles are an important energy source and protein source as natural supplements. In the U.S.A. at Cornell University, about 20% of the caterpillars, it was observed that monarch butterfly caterpillars eating Bt corn toxic pollen blown on to milkweed plants near Bt corn fields had suffered significant adverse effects. For moths and butterflies, and deplete nesting materials for bees, these toxins can finish nectar sources for flower pollination and host plants affect larvae disapprovingly. To minimize pollinator's exposures to poisoning chemicals, farmers can move to more insect pollinator-friendly practice of agriculture. In India, due to the indiscriminate application of chemicals as well as environmentotal pollution, the abundance of approximately 1,500 butterfly species, is alarmingly fluctuating. By the eradication of immature and adult food sources, habitats and sites for nesting, other human alterations such as deforestation, and farm land extension and unlimited urbanization are also a challenge to butterfly species to extinction. The deliberate habitat eradication in several regions of the Malabar Tree Nymph, Travancore Evening Brown, Bhutan Glory and Kaiser-I-Hind Butterfly are almost near elimination. Under local law, most glorious and species in danger have protection and safety measures levels. For the survival of human and animal bio-ecology, after bees and moths only butterflies are efficient pollinators for seed setting, food, fruits and fiber production. To protect butterflies and their environment, based upon wildlife agricultural practices can help a lot. Over the past decade, the growers of the Himalayan region have been complaining about the reduction in apple fruit production and quality due to bee pollination difficulties. During the apple flowering season they have all disappeared now, there used to be a lot of wild bees, butterflies and moths in the past. Insect pollinator's scarcity has now become an evaluative component in insufficient crop pollination services. Augmenting the populations of honeybees, sting less bees, bumblebees, and solitary bees, the crop pollination problem can be solved. In China, Maoxian County of Sichuan practice of pollination by hand in apples is a common procedure. Among policy planners, beekeepers and farmers, the importance of bee pollinator's awareness has to be build. For pollination of different crops growers are already utilizing honeybees and solitary bees in many western countries. From crop pollination to standard honey production methods, the center of attention of beekeeping requires to change (Tanda, 2019a, 2020 a-c; 2021a-e; Tanda, 2022a, b).

1.5 INSECT POLLINATORS AND THEIR IMPORTANCE IN FOOD SUSTAINABILITY

Pollination by insect's supplies basic support of the form and operation of a huge natural ecosystem, and it increases ornamental, recreational, and ethnic aspects of people living, and its importance to the community. Taking into consideration that profitable and ecological significance, it is important to know the reason of the decline in species, its analysis, and their potential results, and monitoring requirements, protection measures. Most successful lifeforms are among angiosperms in this universe that develop seeds and fruit. Flower reproductive systems differ considerably among insect species, but two mechanisms are significant for breeding sexually in all angiosperms: firstly the movement of pollen from the anthers to the stigma of a flower, secondly the seed setting after effective fertilization, the blending of the sperm nuclei with egg nucleus to develop a fertile embryo and seed (Figs. 1.3, 1.4).

Fig. 1.3 Schematic scientific representation of sexual reproduction involved in raspberry flowers (Tanda, 2020b).

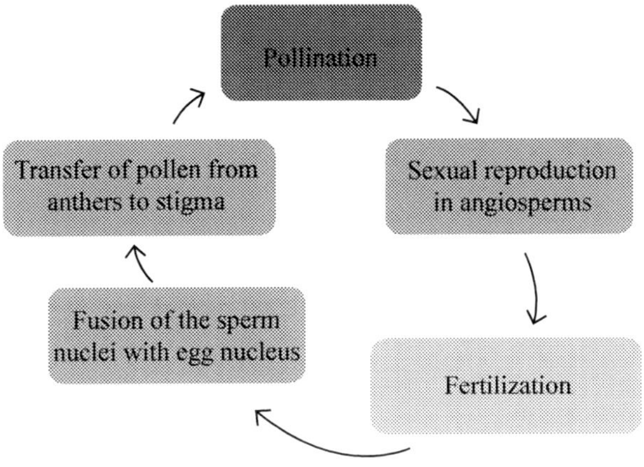

Fig. 1.4 Simplified representation of sexual reproduction in angiosperms

Many crops self-pollinated, implies that pollen dust movement on a single plant among flowers, as the anthers tap the adjoining receptive stigma. However, most of the flowers, rely on the pollen from other flowers called cross-pollinated flowers. To fulfill their sexual reproductive requirements more than 200,000 insect pollinator species are offering this service. Of the world's angiosperms more than three-fourths rely on wind and water for cross-pollination. About 130 million years ago, the fossil records are the evidences where the angiosperms have shown an extraordinary heterogeneity.

1.6 POTENTIAL GENETIC COMPLICATIONS IN INBREEDING AND SELF-SEED FERTILIZATION

Having both male and female reproductive organs having only male reproductive organs as staminated, blossoms can be perfect, or with only female reproductive parts as pistillated flowers. Flowers may be monoecious (staminate and pistillate flowers), dioecious (staminate and pistillate flowers present on separate crop plants), or even trioecious (staminate, pistillate, and perfect blossoms occur on different plant species). In different species, many modifications used in the breeding system of a species may be found within dioecy such as gynodioecy, in which flowers have either female or hermaphrodite blooms (Richards, 1997). Of all crop plant species about three-quarters develop perfect blossoms. About 5% flowers are dioecious, and a little more than 5% are monoecious blooms (Molnar, 2004). Among flowers of different plants cross pollination can take place even within the flowers of a single plant. It is called autogamous flowers, if pollination happen within the same flower, a plant may be self-fertile and self-pollinating. It is known as geitonogamous, if pollination occurs between flowers of the same plant. A plant is called xenogamous if it is cross-pollinated. They are perfect or hermaphrodite if the male and female parts are in the same flower, receive mixed self and outcross pollen dust happen in many field crops. Because of the many potential genetic complications associated with inbreeding in plants, perpetual self-fertilization could be a problem (Charlesworth and Charlesworth, 1987). In different taxa, adaptations that decrease the likelihood of selfing is found. The mechanisms of dioecy and monoecy encourage outcrossing, and that has attained superiority in the environment (Davies et al., 2004). For the dispersal of pollen dust, main among the various reports provided for their eye-catching ascendancy is the evolution of mutualistic relationships in animal population (Baker and Hurd, 1968; Faegri and van der Pijl, 1979; Labandeira et al., 1994; Stebbins, 1950, 1974) and seed formation (Herrera, 1989; Kevan, 1984; van der Pijl, 1982).

Mutualistic associations among creatures offer transfer of reproductive cells which permits for big genetic variations in plants reaching to a bioecological chances with the methods of seed spread. Insect pollinator enhances the field in which potential mates can be obtai ned and encourages mechanism of outcrossing in angiosperms, the meeting of gametes from different genetically flowers. Outcrossing is crucial though monoecious plants enhancing genetic variability by associated recombinations getting self-pollen on the same plant. Monoecious plants give rise to male and female blossoms at various stages, reducing self-pollination. In hermaphrodite flowers, when the male and female flower parts mature at different stages selfing in blooms is unsuccessful. Here the male and female organs of the same flowers are isolated, the probability of self-pollination in some plants, is minimized. Allowing self-pollination before the flower is too old to set fruit, the male and female parts of the flower come closer together as the flower ages in those species. Several angiosperms are self-incompatible, means pollen deterrent to selfing, and cannot do seed setting. In complex and variable ways, self-incompatibility is under jurisdiction and it includes the interaction of incompatible alleles and their results of two different mother plants (Matton et al., 1994). Mechanisms to stop up self-fertilization can suddenly cease to function as a result of maturity or effect of temperature, the efficacy of self-incompatibility becomes low. Even when cross-pollination is not possible, shattering barricades down assures sexual reproduction in plants. Several species persevere victoriously surviving with self-pollinating and self-fertile flowers in spite of the omnipresence of outbreeding. In agriculture, few self-fertile plants that can self-pollinate are of great significance. Where their native bee pollinators are not present many can set-up themselves in non-indigenous crop regions. In detail, the nature and evolutionary biology of crop-breeding systems are demonstrated by Richards (1997). In plants gene differences could ease the development of insect or disease resistance, by permitting animals to adjust spatially and temporally in variable environments. Many ovules give rise to numerous seeds in kiwi fruit, tomato, watermelon, cucumber, or squash; however, a single ovule develop in to a fruit as in coconut, plum, almond, avocado, and cherry. In some plants require pollens to impregnate all the egg cells to develop seed from a pollen grain and an egg cell resulting from the sperm cell unification. Few eggs will not be inseminated without seed if a flower gets insufficient amount of pollen grains. Undersized fruits which have less value in the market are developed due to the result of incomplete pollination and fertilization. Single blossom may be visited by some different pollinators, the sufficient pollination often requires this. That would not be sustainable in natural environment, as few fruits are the result of selective crossing techniques and gene engineering (Schery, 1972). Bananas without seeds are the results of clean plant triploids due to sudden or via diploid hybridization and tetraploids and are vegetative proliferation. Natural parthenocarpic consequences, as seedless tangerines, are those in which organic products create without effective fertilization treatment could fall flat in light of the fact that these self-incompatible varieties are developed in single crop farming. Interestingly, grapes without seeds, are stenospermocarpic; fertilization happens, however the product is seedless due to development of immature embryos (Schery, 1972).

1.7 CROP INSECT POLLINATORS AND BIOHABITATS

Deguines et al. (2016) indicated that benefits from agricultural intensification may be offset by depletion in pollination services, and supports the need for an ecological intensification of agriculture through optimization of ecosystem. Flower foragers and plant interactions have been approximated to 400,000 species, exceptionally the nature of connection between flowers and pollinators differs. All pollinators never offer pollination, though some visit flowers for nectar or pollen. Efficient bee pollinators often have behavioral and anatomical traits that greatly enhance the effectiveness and precision of pollen transfer (Barth, 1991; Faegri and van der Pijl, 1979; Proctor et al., 1996; Lewinsohn et al., 2006).

Crop pollination, in general, is a beneficial relationship for each other, during visitation pollinating insects get some form of nutritional food and pollen transfer. For some flies, butterflies, birds, and bats, pollen itself is a gift, helping as the basic food need for brood as a protein-rich source (Roulston and Cane, 2000). Many plants vary in nectar, pollen, resins, oils, fragrances and other gifts to encourage visiting and pollen spread (Buchmann, 1987; Barth, 1991; Roulston et al., 2000; Dafni et al., 2005; Roulston and Cane, 2000). Degree of interdependence differ in pollinators and flowers. Some crops rely on a single insect pollinator or genus, which in turn, has restricted pollen or nectar sources. The mutualism between pollinators and in the Yucca (Agavaceae), the yucca moths of Tegeticula, is an important example of a close relationship (Pellmyr, 2003). In developing yucca seeds, yucca moth is the chief pollinator since 40 million years. To gather and compact big amount of pollen grains from yucca blossoms, up to even moth's 10% body weight, female moths have tentacles, utilized for this activity. After collecting a pollen load, the moth flies off and forage at other plant, in which eggs are laid by female moth. Female transfer little of pollen dust on the stigma for crossing and setting seed, also as a nutritious food for the offsprings, showing a unique biofloral behavioral activity. In plant-pollinator mutulism, such sterotyped associations are unusual. Similar relationships exploiting immediate opportunities are found in many cases, if not most. About in 5 insect orders 45 species visited on *Geranium thunbergii* plants in a natural environment; of these, 11 insect species are known as key flower pollinators (Kandori, 2002). In rocky mountains, the alpine sky pilot, *Polemonium viscosum*, is fertilized mainly by bumblebees at heights and by flies at low altitudes as key pollinators of a specific species which can differ temporally or spatially (Galen et al., 1987).

Plants need pollen deposit on receptive stigma to set seed and develop fruits for the last 3,500 years, it is well known now in the agricultural industry for pollination process. Till the 17th century (Camerarius, 1694) and after that mechanism was recognized slowly, that the seeds are developed from the deposit of pollen on stigma, was not well eloquent. Plants reproductive parts are significant factors of categorization, Carolus Linnaeus, the systematist recognized. It was not well communicated until the end of a century (Kölreuter, 1761; Sprengel, 1793), the procedure that animals take part in cross-pollination of plants, the transfer of pollen dust from one flower to another plant. Into the utilization of insects by new blossoms and flowers by insect pollinators, Kolreuter's studies, in natural environment watching these processes throughout the days, provided him a wonderful intuition and a unique perception. These natural niche associations design a platform for the plant hybridization among closely associated crop species, Kolreuter acknowledged. Insect foraging was necessary for seed development in many economic vegetables, fruits and flowers and manipulated his ideas to applied form by evolving a new methodology of artificial pollination and developed the first cross-hybrid from two flower species (Mayr, 1986; Sinnott, 1946) documented at the University of Karlsruhe, Germany by Joseph Gottlieb Kölreuter. To convice the fellow scientists, about the great perceptions and innovative ideas of Kölreuter and Sprengel were unsuccessful for a half-century, however, later on their findings had an intensed influence on the mind of Charles Darwin, a great biologist. A new era of experimental bee pollination biology was documented in 'The Origin of Species by Means of Natural Selection or the Preservation of Favoured Races in the Struggle for Life in 1859. What permits living creatures to perform their duties as discriminating representatives, in the same way as abiotic agents and exactly described pollination process, Darwin elaborated the mechanism of co-living and survival. On plant-pollinator associations, in The Various Contrivances the Origin of Species, by which British and Foreign Orchids are bee crossed and the intercrossing effects on fruit quality described by Darwin at a large-scale in 1862. Which placed the basis for agricultural practices and applied research, description of the developing process to explain the adaptive mutualism between pollinators and flowers presented the motivation for biological studies in crop pollination technology (Müller, 1869; Delpino, 1868–1875; Knuth, 1906, 1908, 1912; Müller and Delpino, 1869).

1.8 MANAGEMENT OF INSECT POLLINATORS

Prasifka et al. (2018) reported related difference in bee foraging with specific floral traits, measured advantages of pollinators to hybrid crops, and used genetic assets in sunflower and other plants to find markers connected with principal floral characteristics. A model for using nectar-related traits to increase flower pollinator interactions, future work to enhance pollinator rewards should enable sunflower. Domestication of bees for pollination allowed for increasing yield and for financial gains and exhibition of biotic pollination techniques showed direction to significant agricultural technology, with extensive economic results (Fig. 1.5). Worldwide the western honeybee, *Apis mellifera* L., is the leading domesticated pollinator, for its great efficiency as a pollinator, and wax, honey producer (Free, 1993; Kearns et al., 1998; McGregor, 1976; Delaplane and Mayer, 2000). For 6,000 years, for its honey and wax, from Eurasia, the honeybee has been searched (Dobbs, 1750; Crane, 1983, Crane, 1990 and hive domestication is to ancient Egypt time (Crane, 1999).

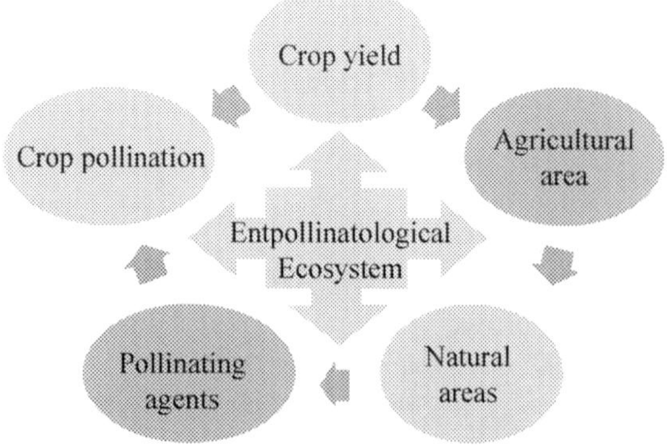

Fig. 1.5 Schematic representation of pollinating agent's role in crop entpollinatological ecosystem

In the 1600s, first in North America with European colonists, *A. mellifera* rapidly became the key pollinator for advance agroindustry, and managed hives were transfered worldwide (Sheppard, 1989). In North America, new beekeeping dates to 1862, when L. L. Langstroth, a Philadelphia minister domesticated honeybees, exploited the bee space to make portable frame technology. To extract more honey, he exploited top-bar hives and enhanced the bee efficiency without injuring the bee workers (Langstroth, 1862). Langstroth's new technology ensued high level commercial beekeeping and honey production and now bees serve to more than 100 economic crops providing pollination (Free, 1993; Kearns et al., 1998; McGregor, 1976; Delaplane Mayer, 2000;Tanda, 2019, 2020a-c,2021a-e). Flower pollinators for which effective management process have been evolved constitute many bumblebee species (Bombus), for tomato pollination in greenhouses mostly (de Ruijter, 1997; Hughes, 1996; Kevan et al., 1991; Macfarlane et al., 1994; Plowright, 1996; van Heemert et al., 1990), and leafcutting bees (Megachile rotundata) (Bohart, 1972a; Frank, 2003), and in arid Pacific Northwest for alfalfa crop pollination. For alfalfa pollination, Alkali bees (*Nomia melanderi*) (Bohart, 1972a) are also used (Stephen, 2003). In the eastern United States, the Japanese horn-faced bee, *Osmia cornifrons*, mostly Mason bees, are domesticated in apple orchards for pollination (Batra, 1982; Bohart, 1972b), though in Japan these bees are utilized for pollinating the entire apple crop on a larger area. In the northwestern United States, *O. lignaria* is employed to pollinate apple plantations (Bosch et al., 2000; Bosch and Kemp, 2002) and in eastern Canada for cherry fruit fertilization (Bosch and Kemp, 1999, 2000, 2001; Sheffield, 2006). Advance procedures are available for the reproduction of this species (Griffin, 1993; Torchio, 2003). For multiplication and management of various crop pollinators methods are accessible (Free, 1970; Kevan et al., 1990; Batra, 1994a, b; Bosch and Kemp, 2001; Shepherd et al., 2003; Torchio, 1990, 2003). Except honeybees, for many crop pollination, more effective pollinators such as bumblebees, megachilids, and other wild bees are available (Cane, 2002; Javorek et al., 2002; Maeta and Kitamura, 1981; Tepedino, 1997). As an alternative flower pollinators *Osmia* species work in almond orchards (Bosch et al., 2000; Bosch and Kemp, 2000; Torchio, 2003), pears (Maeta et al., 1993), blueberries (MacKenzie et al., 1997) for sweet clovers (Richards, 2003) and blackberries and red raspberries (Cane, 2005). To save field and greenhouse crops against fungal pathogens and pests, there are many biocontrol forces found (Kevan et al., 2005). In blueberry crop to control *Monilinia vaccinicorymbosi*, or for a destructive mummy berry disease, bees are utilized to provide *Bacillus subtilis* as a biocontrol agent (Dedej et al., 2004). To supply *Trichoderma harzianum* 1295-22, they also have been exploited in strawberries as commercially produced bio-agent to suppress *Botrytis cinerea*, the pathogenic fungus (Kovach et al., 2000). As biological detectors for atmospheric polluting or poisonous substances (Bromenshenk et al., 1995) and land mines (Bromenshenk et al., 2003), it is an ongoing attempt to explore the possibility of honeybees in Pomegranate (*Punica granatum*) raspberry, *Rubus parvifolius* in improving fruit quality and yield (Tanda, 2020b 2021a) (Figs. 1.6 and 1.7).

Their services can be used by changing ecological conditions to encourage their biodiversity and population abundance or to impact the functioning, bee pollinators can also be manipulated passively (Shepherd et al., 2003; Vaughn et al., 2004). Growing to encourage the growth of flower food assets, offering artificial nesting matter, sites, and securing nesting habitats, acquiescent management constitute several activities. To encourage the biodiversity of the crop pollinators and the populations of a special species, many new techniques are developed (Kevan et al., 1990).

Fig. 1.6 Effect of pollination by honeybees in pomegranate (*Punica granatum*) in improving fruit quality and yield (Tanda, 2021a).

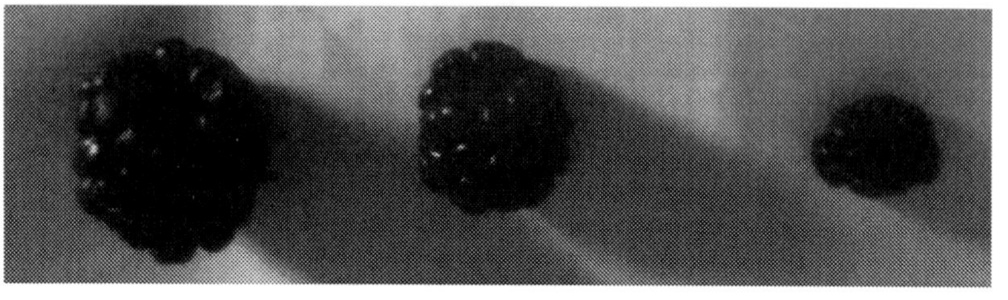

BP NP SP

Fig. 1.7 Bee-pollinated fruit (BP), naturally-pollinated fruit (NP) and not-pollinated, self-pollinated) (SP) fruits of raspberry, *Rubus parvifolius* (Tanda, 2020b).

1.9 CROP INSECT POLLINATORS AND THEIR ECONOMIC IMPORTANCE

Agroindustry comprises one of the most important economic sectors. The yield of most crop species is boosted by bee pollination services. Globally, more than 75% of the 115 leading crop species rely on or at least benefit from insect pollination, whereas wind and self-pollination are enough for only 28 crops. So, to an estimated 35% of global crop production, is achieved by animal pollination (Klatt, et al., 2013; Tanda, 2019, 2020 a, b, c; Tanda and Goyal, 1979). In crop production cross-pollination has great significance for financial and bio-ecological benefits. Bee pollination supplies many advantages in agroindustry by offering many bio-products globally. Fruit production itself is derived directly from the bee action in pollination. In other cases, though pollination does not result in productivity of the produce itself, the procedures contribute to quality carrot and good tomato size production is due to bee pollination services. Through food chain associations there are indirect advantages there as well. Directly, an annual value of $109 million, alfalfa seed crop is produced, along with hay valued at $4.6 billion per year for livestock forage indirectly by bee pollination services (Morse and Calderone, 2000). Though indirect results overemphasize the economic value of crop pollination, the findings have been utilized in many research programs. Annual value of honeybee pollination, to the U.S. agriculture, has been calculated at $18.9 billion (Levin, 1983), $1.6–5.7 billion (Southwick and Southwick, 1992), $150 million (Rucker et al., 2005), $9 billion (Robinson et al., 1989a, b), and $14.6 billion (Morse and Calderone, 2000). In Canada, honeybee pollination annual profits has been reported as $443 million (Scott-Dupree et al., 1995). Alfalfa leafcutting bees and bumblebees play an important role in pollination however, the part of *A. mellifera* is not distinctive. In annual bee pollination, about $2 billion to $3 billion benefits may be added to the role of wild and other bee species (Prescott-Allen and Prescott-Allen, 1986; Southwick and Southwick, 1992; Losey and Vaughan, 2006). As plant pollinators of economically significant crops, many vertebrates also work. Species utilized for cotton, timber, silk, and other valuable products, and depend on bats mainly for pollination in Bombacaceae family of trees (Bawa, 1990; Watson and Dallwitz, 1992). Which are main origin of alcoholic beverages (tequila, mescal) cacti and agaves, and several other valuable products such as sisal fibers, also rely on bats and birds for seed fertilization-related activities (Grant and Grant, 1979; Valiente-Banuet et al., 1996; Arizaga and Ezcurra, 2002; Slauson, 2000, 2001; Arizaga et al., 2002; Fleming et al., 2001a, b; Rocha et al., 2005). For about 30% of edible oils, fibres, nutraceuticals, alcoholic beverages, and medicines utilized by the people, developed from plants, and bee pollinators are basically predominant (McGregor, 1976; Roubik, 1995; Buchmann and Nabhan, 1996) worldwide. Estimating their economic significance in agriculture sector, the contribution of bees and natural pollinators, and forecasting the results of their deprivation are more demanding. Species number and the scarcity details obtainable for species, such approximates are more intricated. In their role to estimate the economic importance of their eco-services given by pollinators, Losey and Vaughan (2006) during studies did not estimate a dollar value of the plant pollination. Further for maintenance of natural crop ecosystems, it is practicable to presume that an important number of flowers in uncultivated terrestrial environment rely on bee pollinators. Valualuable to humans, for example, water filtration, carbon sequestration, floods and soil erosion control, all these cropping systems offer a lot of environmental jobs (Daily et al., 1997). Other than flower pollination in their immature stages, the difficulty is that bee pollinators may offer many environmental jobs. The significance of these jobs is equally complicated to estimate, especially without a full knowledge of pollination mechanism and biological systems. Honeybee is the main bee pollinator that applies a strong effect on its population through their enlarged phenotypes that connects to disease resistance traits (Easton-Calabria et al. 2019). Honeybees confirm resistance to pathogens and pests in its enlarged phenotypes such as honey, propolis, venom, beeswax, bee bread and royal jelly, etc. To show antipathogenic properties and to play as a hive level defensive tool against diseases, each of these enlarged phenotypes have proved.

1.10 THREATS TO INSECT POLLINATORS

According to a UN report, Howard and Johnson (2019) mentioned that about one million species now face disappearance. Apiculturists warn that this is not only a crisis for nature, but for humanity all over the world. Why they so feared are and what does it mean for our life systems? Localized bee population diminishing and a worldwide decrease in the number and survival of flower pollinators, a threat to pollinators contributing to trophic collapse (Dobson et al., 2006) was noted in 1990s. About this crucial crop pollinator, big damages to hives was found in the USA (Watanabe, 1994). Regarding the forgotten crop pollinators, Buchmann and Nabhan, 1996 warned even nonmanaged species and put down the matter related to big devastation in biodiversity and ecological systems. On pollination and bee pollinators, an immediate U.S. plan was asked by Buchmann and Nabhan. As a fundamental eco-service system a stress on insect pollination (Daily et al., 1997) brought an interest at the international level (Costanza et al., 1997; Allen-Wardell et al., 1998; Eardley et al., 2006). In conservation of biodiversity, the entpollinatology, bio-ecology and floral biology sciences have now been well established. Worldwide scientists and agriculturists were worried about a decline of bee pollinators and ill-effects on biodiversity globally occurring in the mid-1990s. In 2000, at the Conference of Parties (COP) of the Convention on Biological Diversity (CBD) to establish an International Initiative for the Conservation and Sustainable use of pollinators (the International Pollinators Initiative, or IPI), concerned the policymakers. On agricultural biodiversity within the work of programme to encourage program coordination globally 5th Meeting of the Conference of Parties (COP) contemplated this to be an alternate enterprise, and appealed for the development of an Action Plan for the IPI. In 2002, Action Plan developed on the São Paulo Declaration on Pollinators recommendations was embraced at COP 6 (decision VI/5). All conservationists, growers, planners, architects in landscape and stakeholders are required to follow policy designers and perform for the significance of crop pollinators and sustainability of biodiversity around the world.

1.11 CONCLUSION

This chapter presents that the world food cannot sustain without insect pollinators and their role in agroecosystem. No doubt commercial honeybees are well established as effective crop pollinators, domesticated wild bumble and solitary bees are also important for the prosperous fertilization of many crop plants, and native bees offer voluntary job. For conservation of wild plantations, insect pollination is significant for wild ecosystem sustainability. It is very complicated to understand history of insect pollination, bee behavior and their management for pollination services. Confirming that a non-wild bee will not distort the environment, moving hives from native to new fields for plant pollination jobs only then the action should be taken. Bees perform a big service in the agroecosystem, though they are not very conspicuous and little in size. Of a big mutual connection between humans and the natural environment, the relationship between bees and humankind is figurative. This book integrates diverse, several related animals mutually for the production of a qualitative and quantitative crop. It strives to offer the evaluative analysis of various bee management plans in crop improvement, food and environmental sustainability and bee pollination requirements of various field crops enhancing fruit quality and production. This new book on entpollinatology scores over other similar books published globally by providing several innovative biotechnological reports. No such comprehensive book found on the subject which could amalgamate production inhibiting and production promoting elements for the sustainable and boosted food productivity in agroindustrial sector. The compilation of this book is unique in the sense that it does not deal with the conventional way of discussing crop pollinators but also takes into consideration the role of pollinators and their better utilization in the larger context of ecologically management and safety of pollinators on the other. The recent research details presented in this beautiful volume will be beneficial to plant pollination biologists, apiculturists, students, and teachers, agriculture scientists, botanists, bio-ecologists, entomologists, researcher's in genetics, plant breeding, horticulture, seed growers and seed agencies.

The book contains various chapters to provide information on all subjects of insect pollination or entpollinatology. It considers issues such as why crop insect pollinators matter, global agroindustry in the absence of pollinators, pollination technology in crop improvement programs, insect pollination in hybrid seed production system, entomophilous crops, sustainability of natural ecosystem: essentials and challenges, improving pollinator density and diversity boosting crop yields, current threats to pollinator populations and food productivity, insecticides: challenges and solutions to pollinators, crop pollinators: conservation and strategies, pollination in cultivated crops, importance of non-insect pollinators in crop production, important issues and future of crop pollination, new advances in pollination materially engineered artificial pollinators, the assessment of pollination techniques, and management of plants and wild bee life for pollination strategies for the applied agricultural procedures.

Interestingly, the chapters on crop pollinator's protection and preservation, melittophilogical studies under confinement, entpollinatology in crop improvement program and biochemistry of entpollinatology in crops provide a base for future perceptions in entpollinatology. At the end of each chapter important issues and future of crop pollination along with new advances in pollination have been beautifully described, the future aspects to consider for research studies.

REFERENCES

- Aizen M, Harder LD (2009) The global stock of domesticated honeybees is growing slower than agricultural demand for pollination. Curr Biol 19: 915–918
- Aizen M, Garibaldi L, Cunningham S, Klein A (2008) Long-term global trends in crop yield and production reveal no current pollination shortage but increasing pollinator dependency. Curr Biol 18: 1572–1575
- Arizaga S, Ezcurra E (2002) Propagation mechanisms in *Agave macroacantha* (Agavaceae), a tropical arid-land succulent rosette. Am J Bot 89: 632–641
- Arizaga S, Ezcurra E, Peters E, de Arellano FR, Vega E (2002) Pollination ecology of *Agave macroacantha* (Agavaceae) in a Mexican tropical desert. I. Floral biology and pollination mechanisms. Am J Bot 87: 1004–1010
- Baker HG (1975) Sugar concentrations in nectars from hummingbird flowers. Biotropica 7: 37–41
- Baker HG, Hurd PD Jr (1968) Intrafloral ecology. Annu Rev Entomol 13: 385–414
- Barnett EA, Charlton AJ, Fletcher MR (2007) Incidents of bee poisoning with pesticides in the United Kingdom, 1994–2003. Pest Manag Sci 63: 1051–1057
- Barth FG (1991) Insects and flowers: the biology of a partnership (trans: Biederman-Thorson MA) Princeton University Press, Princeton, 408 pp
- Batra SWT (1982) The hornfaced bee for efficient pollination of small farm orchards. Research for small farms. In: Kerr WH and Knutson LV (eds.) Research for small farm, USDA Miscellaneous Publication 1422, pp 116–120

- Batra SWT (1994a) *Anthophora plumipes villosula Sm.* (Hymenoptera: Anthophoridae), a man- ageable Japanese bee that visits blueberries and apples during cool, rainy, spring weather. Proc Entomol Soc Wash 96: 98–119
- Batra SWT (1994b) Diversify with pollen bees. Am Bee J 134: 591–593
- Bawa KS (1990) Plant-pollinator interactions in tropical rainforests. Annu Rev Ecol Syst 21: 399–422
- Bohart GE (1972a) Management of wild bees for the pollination of crops. Annu Rev Entomol 17: 287–312
- Bohart GE (1972b) Management of habitats for wild bees. In: Proceedings tall timbers conference on ecological animal control by habitat management no. 3, Tallahassee, pp 253–266
- Bosch J, Kemp WP (1999) Exceptional cherry production in an orchard pollinated with blue orchard bees. Bee World 80(4): 163–173
- Bosch J, Kemp WP (2000) Development and emergence of the orchard pollinator *Osmia lignaria* (Hymenoptera: Megachilidae) Environ Entomol 29: 8–13
- Bosch J, Kemp WP (2001) How to manage the blue orchard bee as an orchard pollinator. Sustainable agriculture network handbook series book 5. National Agricultural Library, Beltsville, 88 pp
- Bosch J, Kemp WP (2002) Developing and establishing bee species as crop pollinators: the example of *Osmia* spp. (Hymenoptera: Megachilidae) and fruit trees. Bull Entomol Res 92(1): 3
- Bosch J, Kemp WP, Peterson SS (2000) Management of *Osmia lignaria* (Hymenoptera: Megachilidae) populations for almond pollination: methods to advance bee emergence. Environ Entomol 29(5): 874–883
- Bromenshenk JJ, Smith GC, Smith GC, Watson VJ (1995) Assessing ecological risks in terrestrial systems with honeybees. In: Butterworth FM (ed.) Biomonitors and biomarkers as indicators of environmental change. Plenum Press, New York, pp 9–30
- Bromenshenk JJ, Henderson CB, Seccomb RA, Rice SD, Etter RT, Bender SFA, Rodacy PJ, Shaw JA, Seldomridge NL, Spangler LH, Wilson JJ (2003) Can honeybees assist in area reduction and landmine detection? J Mine Action 7(3): 24–27
- Buchmann SL (1987) The ecology of oil flowers and their bees. Annu Rev Ecol Syst 18: 343–369
- Buchmann SL, Nabhan GP (1996) The forgotten pollinators. Island Press, Washington, DC
- Camerarius RJ (1694) Epistola ad M.B. Valentini de sexu Plantarum [Letter on the Sexuality of plants]. In: Ostwald's Klassiker der exakten *Naturwissenschaften*, No. 105, 1899. Verlag von Wilhelm Engelmann, Leipzig
- Cane JH (2002) Pollinating bees (Hymenoptera: Apiformes) of U.S. alfalfa compared for rates of pod and seed set. J Econ Entomol 95(1): 22–27
- Cane JH (2005) Pollination potential of the bee *Osmia aglaia* for cultivated red raspberries and blackberries (Rubus: Rosaceae) HortScience 40(6): 1705–1708

- Charlesworth D, Charlesworth B (1987) Inbreeding depression and its *Evolutionary* consequences. Annu Rev Ecol Syst 18: 237–268
- Costanza R, d'Arge R, de Groot R, Farber S, Grasso M, Hannon B, Limburg K, Naeem S, O'Neill RV, Paruelo J, Raskin RG, Sutton P, van den Belt M (1997) The value of the world's ecosystem services and natural capital. Nature 387: 253–260
- Crane E (1983) The archaeology of beekeeping. Duckworth, London, 260 pp
- Crane E (1990) Bees and beekeeping. Comstock, Ithaca
- Crane E (1999) The world history of beekeeping and honey hunting. Routledge, London
- Dafni A, Kevan PG, Husband B (eds.) (2005) Practical pollination biology. Enviroquest Ltd., Cambridge
- Daily GC, Matson PA, Vitousek PM (1997) Ecosystem services supplied by soil. In: Daily G (ed.) Nature's services: societal dependence on natural ecosystems. Island Press, Washington, DC, pp 113–132
- Davies TJ, Barraclough TG, Chase MW, Soltis PS, Soltis DE, Savolainen V (2004) Darwin's abominable mystery: insights from a supertree of the angiosperms. Proc Natl Acad Sci USA 101(7): 1904–1909
- de Ruijter A (1997) Commercial bumblebee rearing and its implications. In: Richards KW (ed.) Proceedings of the 7th international symposium on pollination. ISHS Acta Horticulturae 437, Belgium, pp 261–269
- Dedej S, Delaplane KS, Scherm H (2004) Effectiveness of honeybees in delivering the biocontrol agent. *Bacillus sutilis* to blueberry flowers to suppress mummy berry disease. Biol Control 31: 422–427
- Delaplane KS, Mayer DF (2000) Crop pollination by bees. CAB International, Oxon
- Deguines, N., Jono, C., Baude, M., Henry, M., Julliard, R., Fontaine, C. (2016) Large-scale trade-off between agricultural intensification and crop pollination services. Front. Ecol. Environ., 12, 212–217
- Delpino F (1868–1875) Ulteriori osservazione sulla dicogamia nel regno vegetale. Atti della Societa Italiana di Scienze Naturali Milano 1 and 2: 265–332
- Desneux N, Decourtye A, Delpuech JM (2007) The sublethal effects of pesticides on beneficial arthropods. Annu Rev Entomol 52: 81–106
- Dias BSF, Raw A, Imperatri-Fonseca VL (1999) International pollinators initiative: the Sao Paulo declaration on pollinators. Report on the recommendations of the workshop on the conservation and sustainable use of pollinators in agriculture with emphasis on bees, Brazilian Ministry of the Environment, Brazil, 79 pp
- Dobbs A (1750) Concerning bees and their method of gathering wax and honey. Phil Trans R Soc 46: 536–549
- Dobson A, David Lodge D, Alder J, Cumming GS, Keymer J, McGlade J, Mooney H, Rusak JA, Sala O, Wolters V, Wall D, Winfree R, Xenopoulos MA (2006) Habitat loss, trophic collapse, and the decline of ecosystem services. Ecology 87(8): 1915–1924

- Eardley C, Roth D, Clarke J, Buchman S, Gemmill B (eds.) (2006) Pollinators and pollination: a resource book for policy and practice. African Pollinator Initiative, Pretoria, 77 pp
- Easton-Calabria August, Demary Kristian C, Oner Nola J (2019) Beyond pollination: Honeybees (*Apis mellifera*) as zootherapy keystone species. Frontiers in Ecology and Evolution 6 161 https: //www.frontiersin.org/article/10.3389/fevo.2018.00161
- Estes J, Bonnie B, Amos J, Sullivan R (1983) Pollination from two perspectives: the agricultural and biological sciences. In: Jones CE, Little RJ (eds.) The handbook of experimental pollination biology. van Nostrand Reinhold, New York, pp 173–183
- Faegri K, van der Pijl L (1979) The principles of pollination ecology, 3rd edn. Pergamon Press, Oxford
- FAO (2018) The Beeconomy: Economics and Insect Pollination. Biodiversity. April 27, https: //modernag.org/biodiversity/beeconomy-economic-value-pollination/
- Fleming TH, Sahley CT, Holland JN, Nason JD, Hamrick JL (2001a) Sonoran Desert columnar cacti and the Evolution of generalized pollination systems. Ecol Monogr 71: 511–530
- Fleming TH, Tibbitts T, Petryszyn Y, Dalton V (2001b) Current status of pollinating bats in the southwestern United States. In: O'Shea T, Bogan M (eds.) Monitoring bat populations in the United States. U.S. Geological Survey, Fort Collins
- Frank G (ed.) (2003) Alfalfa seed and leafcutter bee production and marketing manual. brooks. Irrigated Alfalfa Seed Producers Association, Alberta, 160 pp
- Free JB (1970) Insect pollination of crops. Academic Press, New York
- Free JB (1993) Insect pollination of crops, 2nd edn. Academic Press, San Diego
- Galen C, Zimmer KA, Newport ME (1987) Pollination in floral scent morphs of *Polemonium viscosuma* mechanism for disruptive selection on flower size. Evolution 41: 599–606
- Grant V, Grant KA (1979) The pollination spectrum in the southwestern American cactus flora. Plant Syst Evol 133: 29–37
- Griffin BL (1993) The orchard mason bee. Knox Cellars Publishing, Bellingham, Washington, DC
- Herrera CM (1989) Seed dispersal by animals: a role in angiosperm diversification? Am Nat 133: 309–322
- Howard Emma and Georgie Johnson (2019) We are losing the web of life': why the global nature crisis is as dangerous as climate change. https: //unearthed.greenpeace.org/2019/05/06/nature-crisis-biodiversity-dangerous-climate-change-extinction/
- Hughes MJ (1996) Commercial rearing of bumblebees. In: Matheson A (ed.) Bumblebees for pleasure and profit. International Beekeeper Research Association, Cardiff, pp 40–47
- Javorek SK, Mackenzie KE, Vander-Kloet SP (2002) Comparative pollination effectiveness among bees (Hymenoptera: Apoidea) on lowbush blueberry (Ericaceae: *Vaccinium angustifolium*) Ann Entomol Soc Am 95(3): 345–351
- Kandori I (2002) Diverse visitors with various pollinator importance and temporal change in the important pollinators of *Geranium thunbergii* (Geraniaceae) Ecol Res 17(3): 283–294

- Karise R (2007) Foraging behaviour and physiology of bees: impact of insecticides. Ph.D. thesis, Estonian University of Life Sciences, Tartu, Estonia, pp 1–123
- Kearns CA, Inouye DW, Waser NM (1998) Endangered mutualisms: the conservation of plant-pollinator interactions. Annu Rev Ecol Syst 29: 83–112
- Kevan PG (1984) Pollination by animals and angiosperm biosystematics. In: Grant WF (ed.) Plant biosystematics. Academic, Toronto, pp 271–292
- Kevan PG, Phillips TP (2001) The economic impacts of pollinator declines: an approach to assessing the consequences. Conserv Ecol 5(1): 8
- Kevan PG, Clark EA, Thomas VG (1990) Insect pollinators and sustainable agriculture. Am J Alternative Agr 5: 13–22
- Kevan PG, Straver WA, Offer M, Laverty TM (1991) Pollination of greenhouse tomatoes by bumblebees in Ontario. P Entomol Soc Ont 122: 15–19
- Kevan PG, Shipp VL, Kapongo JP, Al-mazra'awi MS (2005) Bee pollinators vector biological control agents against insect pests of horticultural crops. In: Guerra-Sanz M, Roldán Serrano A, Mena Granero A (eds.) First short course on pollination of horticulture plants. CIFA, La Mojonera, Spain, pp 77–95
- Klatt BK, Holzschuh A, Westphal C, et al. (2013) Bee pollination improves crop quality, shelf life and commercial value. Proc Biol Sci. 2013; 281(1775): 20132440. Published 2013 Dec 4. doi: 10.1098/rspb.2013.2440
- Klein AM, Vaissiere BE, Cane JH, Steffan-Dewenter I, Cunningham SA, Kremen C, Tscharntke T (2007) Importance of pollinators in changing landscapes for world crops. Proc Royal Soc B 274: 303–313
- Knuth P (1906) Handbook of flower pollination. Oxford University Press, Oxford
- Knuth P (1908) Handbook of flower pollination, vol 2 (trans: Ainsworth Davis JR) Clarendon Press, Oxford, 705 pp
- Knuth P (1912) Handbook of flower pollination, vol 2 (trans: Ainsworth Davis JR) Clarendon Press, Oxford
- Kölreuter JG (1761) Vorläufige Nachricht von einigen das Geschlecht der Pflanzen betreffenden Versuchen und Beobachtungen. Leipzig
- Kovach J, Petzoldt R, Harman GE (2000) Use of honeybees and bumblebees to disseminate *Trichoderma harzianum* 1295–22 to strawberries for *Botrytis* control. Biol Control 18: 235–242
- Kremen C, Williams NM, Aizen MA, Gemmill-Herren B, LeBuhn G et al. (2004) The area requirements of an ecosystem service: crop pollination by native bee communities in California. Ecol Lett 7: 1109–1119
- Kremen C, Williams NM, Aizen MA, Gemmill-Herren B, LeBuhn G et al. (2007) Pollination and other ecosystem services produced by mobile organisms: a conceptual framework for the effects of land-use change. Ecol Lett 10: 299–314
- Labandeira CC, Dilcher DL, Davis DR, Wagner DL (1994) Ninety-seven million years of angiosperm insect association: paleobiological insights into the meaning of coevolution. Proc Natl Acad Sci USA 91: 12278–12282

- Langstroth LL (1862) The hive and the honeybee. C.M. Saxton, New York
- Levin MD (1983) Value of bee pollination to U.S. agriculture. Bull Entomol Soc Am 29: 59–51
- Lewinsohn TM, Prado PI, Jordano P, Bascompte J, Olesen JM (2006) Structure in plant-animal interaction assemblages. Oikos 113: 174–184
- Losey JE, Vaughan M (2006) The economic value of ecological services provided by insects. Bioscience 56: 311–323
- Macfarlane RP, Patten KD, Royce LA, Wyatt BKW, Mayer DF (1994) Management potential of sixteen North American bumblebee species. Melanderia 50: 1–12
- MacKenzie KE, Javorek S, Richards KW (1997) The potential of alfalfa leafcutter bees (*Megachile rotundata* L.) as pollinators of cranberry (*Vaccinium macrocarpon* Aiton) Acta Hortic 437: 345–351
- Maeta Y, Kitamura T (1981) Pollinating efficiency of *Osmia cornifrons* in relation to the number of bees required for economic fruit production. Honeybee Sci 2(2): 65–72
- Maeta Y, Goukon K, Tezuka T (1993) Utilization of *Osmia cornifrons* as a pollinator of Japanese pears (Hymenoptera, Megachilidae) Chugoku Kontyu 7: 1–12
- Matton DP, Nass N, Clarke AE, Newbigin E (1994) Self-incompatibility: how plants avoid illegitimate offspring. Proc Natl Acad Sci USA 91: 1992–1997
- Mayr E (1986) Joseph Gottlieb Köreuter's contributions to biology. Osiris 2(2): 135–176
- McGregor SE (1976) Insect pollination of cultivated crop plants. USDA handbook 496. Department of Agriculture, Agricultural Research Service, Washington, DC, 411 pp
- Molnar S (2004) Plant reproductive systems. http: //www.geocities.com/we_evolve/ Plants/ breeding_sys.html. Accessed 10 Mar 2006
- Morse RA, Calderone NW (2000) The value of honeybees as pollinators of U.S. crops in 2000. Bee Cult 128(3): 1–15
- Müller H (1869) DeCandolle. Prodromus systematis naturalis regni vegetabilis 16: 16
- Müller H, Delpino F (1869) Application of the Darwinian theory to flowers and the insects which visit them. Am Nat 5: 271–297 (trans Packard RL, 1871)
- Nabhan GP, Buchmann SL (1997) Pollination services: biodiversity's direct link to world food stability. In: Daily GC (ed.) Nature's services: societal dependence on natural ecosystems. Island Press, Washington, DC, pp 133–150
- Prasifka JR, Mallinger RE, Portlas ZM, et al. (2018) Using Nectar-Related Traits to Enhance Crop-Pollinator Interactions. *Front Plant Sci*. 2018;9: 812. Jun 18. doi: 10.3389/ fpls.2018.00812
- Pellmyr O (2003) Yuccas, yucca moths, and coevolution: a review. Ann Mo Bot Gard 90: 35–55
- Plowright C (1996) Bumblebee rearing. Melissa 9: 10–12
- Prescott-Allen C, Prescott-Allen R (1986) The first resource: wild species in the North American economy. Yale University Press, New Haven
- Proctor M, Yeo P, Lack A (1996) The natural history of pollination. Timber Press, Portland

- Rathcke BJ, Jules ES (1993) Habitat fragmentation and plant-pollinator interactions: from specialization to generalization. In Waser N, Ollerton J (eds.) The University of Chicago Press, Chicago and London, pp 173–199
- Renner SS (1988) Effects of habitat fragmentation on plant-pollinator interactions in the tropics. In: Newbery DM, Prins HHT, Brown ND (eds.) Dynamics of tropical communities. Blackwell Science, Oxford, pp 339–360
- Richards AJ (1997) Plant breeding systems, 2nd edn. Garland Science, London
- Richards KW (2003) Potential use of the alfalfa leafcutter bee *Megachile rotundata* to pollinate sweet clover. J Apicult Res 42(1/2): 21–24
- Robinson WS, Nowogrodzki R, Morse RA (1989a) The value of honeybees as pollinators of U.S. crops. Part I of a two-part series. Am Bee J 129: 411–423
- Robinson WS, Nowogrodzki R, Morse RA (1989b) The value of honeybees as pollinators of U.S. crops. Part II of a two-part series. Am Bee J 129: 477–487
- Rocha M, Valera A, Eguiarte LE (2005) Reproductive ecology of five sympatric *Agave littaea* (Agavaceae) species in Central Mexico. Am J Bot 92: 330–1341
- Roubik DW (ed.) (1995) Pollination of cultivated plants in the tropics. Food and Agriculture Organization Agricultural Services Bulletin 118, Rome, 196 pp
- Roulston TH, Cane JH (2000) Pollen nutritional content and digestibility for animals. Plant Syst Evol 222(1–4): 187–209
- Roulston TH, Cane JH, Buchmann SL (2000) What governs the protein content of pollen grains: pollinator preferences, pollen-pistil interactions, or phylogeny? Ecol Monogr 70: 617–643
- Rucker RR, Thurman WN, Burgett M (2005) Internalizing reciprocal benefits: the economics of honeybee pollination markets. North Carolina State University, Department of Agricultural and Resource Economics, Raleigh, NC, Unpublished manuscript http: // legacy.ncsu.Âedu: 8020/classes/ecg701001/personal/Internalization_of_Reciprocal_ Benefits-full_March_2005.pdf. Accessed 6 Mar 2005
- Schery RW (1972) Plants for man, 2nd edn. Prentice-Hall, Englewood Cliffs NJ
- Scott-Dupree C, Winston M, Hergert G, Jay SC, Nelson D, Gates J, Termeer, B and Otis G (1995) A guide to managing bees for crop pollination. Canadian Association of Professionals Apiculturists. Canadian Department of Agriculture, Aylesford, 41 pp
- Sheffield CS (2006) Diversity and management of bees for the pollination of apples in the annapolis valley of Nova Scotia. Ph.D. thesis University of Guelph, Guelph, Ontario
- Shepherd MD, Buchmann SL, Vaughan M, Black SH (2003) Pollinator conservation handbook: a guide to understanding, protecting, and providing habitat for native pollinator insects. The Xerces Society in association with the bee works. The Xerces Society, Portland, 45 pp
- Sheppard WS (1989) A history of the introduction of honeybee races into the United States. Am Bee J 129: 617–619
- Sinnott EW (1946) Botany principles and problems, 4th edn. McGraw-Hill, New York, p 726

- Slauson LA (2000) Pollination biology of two chiropterophilous agaves in Arizona. Am J Bot 87: 825–836
- Slauson LA (2001) Insights on the pollination biology of *Agave* (Agavaceae) Haseltonia 8: 10–23
- Southwick EE, Southwick L Jr (1992) Estimating the economic value of honeybees (Hymenoptera: Apidae) as agricultural pollinators in the United States. J Econ Entomol 85: 621–633
- Sprengel CK (1793) Das entdeckte Geheimniss der Natur im Bau und in der Befruchtung der Blumen [The secret of Nature in the form and fertilization of flowers discovered]. Friedrich Vieweg der Ältere, Berlin
- Stebbins GL (1950) Variation and evolution in plants. Columbia University Press, New York
- Stebbins GL (1974) Flowering plants evolution above the species level. Harvard University Press, Cambridge, MA
- Steffan-Dewenter I, Westphal C (2008) The interplay of pollinator diversity, pollination services and landscape change. J Appl Ecol 45: 737–741
- Stephen WP (2003) Solitary bees in North American agriculture: a perspective. In: Strickler KV, Cane JH (eds.) For non-native crops, whence pollinators of the future? Thomas Say Publications in Entomology, Entomological Society of America, Lanham, pp 41–66
- Tanda AS (2022a) Chapter 11: Pest Management Technology and Bee Pollinators' Integration. In: Tanda A S (ed.) Advances in Integrated Pest Management Technology: Innovative and Applied Aspects. Springer Nature Switzerland (in press)
- Tanda AS (2022b) Chapter 1: Mutulistic Plant Associations related to Insect Resistance. In: Tanda A S (ed.) Molecular Advances in Insect Resistance of Field Crops. Modern and Applied Approaches. Springer Nature Switzerland (in press)
- Tanda AS (2021a) Biofloralphenology, Foraging behavior and entpollinatological effect of honeybees in pomegranate (*Punica granatum*) fruit quality and yield. Journal of Horticulture 08 2
- Tanda AS (2021b) Bee pollinators matter in crop improvement need immediate conservation strategies. Journal of Insect Science (Accepted)
- Tanda AS (2021c) Comparative pollination efficacies of *Apis mellifera* L. and *Tetragonula carbonaria* (Smith) on peach (*Prunus persica* L.) Indian Journal Entomology (Accepted)
- Tanda AS (2021d) Why insect pollinators important in crop improvement?. Indian Journal Entomology (Accepted)
- Tanda A S (2021e) Insect pollinators matter in sustainable world food production. Indian Journal Entomology (Accepted)
- Tanda AS (2020a) Biogenetic engineering in developing insect resistant crops: constraints and applications. *5th Edition of Global Congress on Plant Biology and Biotechnology* (*GPB 2020*) Valencia, Spain

- Tanda AS (2020b) An Assessment of honeybee (*Apis mellifera*) foraging activity and pollination efficacy in Australian raspberry *Rubus parvifolius* at Rosehill. Australian Journal of Entomology (Submitted)
- Tanda AS (2020c) Entpollinatology is for sustainable agroecosystems and global agroindustry. Indian Journal of Entomology. (Submitted)
- Tanda AS (2020d) Entpollinatology: A strong relationship between plants and insects for crop improvement. *6th Edition of Global Conference on Plant Sciences and Molecular Biology* (*GPMB 2020*) Paris, France
- Tanda AS (2019a) Entomofauna enhance the quality and quantity in Okra. Indian Journal Entomology 81(1): 16-17
- Tanda AS (2019b) Entomophilous crops get better fruit quality and yield: An appraisal. Indian Journal Entomology 81(2): 227-234
- Tanda AS (2019c) Floral biology, foraging behavior and efficiency of European honeybee (*Apis mellifera*) in bitter gourd (*Momordica charantia* L.) pollination at Sydney Australia. Bee World. (Submitted)
- Tanda AS (1984) Bee pollination increases yield of 2 interplanted varieties of Asiatic cotton (*Gossypium arboreum* L.) The Amer. Bee Jour. 124: 539-540
- Tanda A S (1983) Assessing the role of honeybees in a field of Asiatic cotton (*Gossypium arboreum* L.) American Bee Journal 123: 593-594
- Tanda AS and Goyal NP (1979) Insect pollination in Asiatic cotton (*Gossypium arboreum*) J Apic Res 18(1): 64–72
- Tepedino VJ (1997) A comparison of the alfalfa leafcutting bee (*Megachile rotundata*) and the honeybee (*Apis mellifera*) as pollinators for hybrid carrot seed in field cages. In: Richards KW (ed.) Proceedings of the 7th International symposium on pollination, Lethbridge, Alberta. International Society for Horticultural Science, Leiden, the Netherlands, pp 457–461
- Torchio PF (1990) Diversification of pollination strategies for U.S. crops. Environ Entomol 19(6): 1649–1656
- Torchio PF (2003) The development of *Osmia lignaria* Say (Hymenoptera: Megachilidae) as a managed pollinator of apple and almond crops: a case history. In: Strickler K, Cane JH (eds.) For nonnative crops, whence pollinators of the future? Proceedings, entomological society of America. Thomas Say Publications in Entomology, Lanham, pp 67–84
- Tscharntke T, Klein AM, Kruess A, Steffan-Dewenter I, Thies C (2005) Landscape perspectives on agricultural intensification and biodiversity – ecosystem service management. Ecol Lett 8: 857–874
- US Department of Agriculture (2018) The Beeconomy: Economics and Insect Pollination. Biodiversity. April 27,

  https: //modernag.org/biodiversity/beeconomy-economic-value-pollination/
- US Department of Agriculture (2019) Honeybee colonies more successful by foraging on non-crop fields Public Release: 20-Mar-https: //www.eurekalert.org/pub_releases/2019-03/udoa-hbc031619.php

- Valiente-Banuet A, Arizmendi MC, Rojas-Martínez A, Domínguez-Canseco L (1996) Ecological relationships between columnar cacti and nectar feeding bats in Mexico. J Trop Ecol 12: 103–119
- van der Pijl L (1982) Principles of dispersal in higher plants. Springer, Berlin, p 214
- van Heemert C, de Ruijter A, van den Eijnde J, van der Steen J (1990) Year round production of bumblebee colonies for crop pollination. Bee World 71(2): 54–56
- vanEngelsdorp D, Underwood R, Caron D, Hayes J Jr (2007) An estimate of managed colony losses in the winter of 2006–2007: a report commissioned by the apiary inspectors of America. Am Bee J 147: 599–603
- Vaughan M, Shepard M, Kremen C, Black SH (2004) Farming for bees: guidelines for providing native bee habitat on farms. The Xerces Society for Invertebrate Conservation, Portland
- Watanabe ME (1994) Pollination worries rise as honeybees decline. Science 265: 1170
- Watson L, Dallwitz MJ (1992) The families of flowering plants: descriptions, illustrations, identification, and information retrieval. http: //delta-intkey.com/. Version: 2 June 2006

2

Global Agroindustry in the Absence of Insect Pollinators: Historical Outlook

Amarjit S. Tanda*

ABSTRACT

Approximately 80 percent of all flowering plant species are specialized for pollination by animals, mostly insects, and they affect 35 percent of the world's crop production, increasing the output of 87 of the leading food crops worldwide. This chapter traces the history of pollination. The basic principle of sex differentiation in plants may have been known as early as 1500 B.C. Hebrews learned the value and art of date pollination from Egyptian and Babylonian experts. An Assyrian architectural relief of that period shows two divine creatures, each presumably holding a male date inflorescence over a female inflorescence. The Arabic writer, Kazwini, who died about 682 A.D., as saying that the date is the only tree that is artificially fertilized. Growers of dates today use this method to assure a set of dates in their groves. In 1694, Rudolph Jacob Camerarius stated that there are two different parts of the flower, the stamens and the pistil, and that they must work together to produce ripe seed. Arthur Dobbs (1750) was the first to observe how bees pollinated flowers.

Keywords *Pollen tube • Pollination • Pollen grains • Inflorescence • Mechanism • Flowering Plants • Nectar • Pollen • Female • Insect pollination*

2.1 INTRODUCTION

Globally, insects including animals pollinate about seventy-five percent of all plant species, and about ninety percent of all flowering plants. It is established that it takes a pollinated flower to develop so many of the fruits we depend on in our human diets. So animal pollination being responsible for this mechanism and about one of every three bites we take each day. These pollinators are entirely vital to our life system on this earth as we are aware of this seed setting and fruit production process. The plant and pollinator relationships coevolved over the millennia (Tanda, 2022 a,b). Recent research has updated the insect tree of life to indicate

* School of Horticulture, University of Western Sydney, Richmond, Australia and CSIRO North Ryde NSW, Australia.

E-mail: drtanda193@gmail.com

that insects evolved from ocean-dwelling crustaceans around 480 million years ago and about 450 million years ago, the plants first colonized land. For the first 300 million years, the plant kingdom was dominated by non-flowering plants, such as ferns, conifers, and cycads. The ginko and its early relation, relied entirely on wind pollination to complete their life processes, insects were not part of the equation. The angiosperms produced abundant pollen, which is high in nutrition, and insects began to notice. As insect species started using pollen for food, there was suddenly a perfect avenue for plant pollination to occur on a grand scale. The feeding insects would spread pollen from plant to plant vastly improving the plant kingdom's ability to disperse genetic information across the landscape. Flowering plants that were pollinated by insects had a distinctive advantage over the archaic wind-pollinated species and their numbers exploded across the globe. Competition between plants for attraction of pollinating insects lead to the development of many different size, shapes, colors and fragrances of flowers we see today. Eventually this competition also lead to the addition of nectar as a sweet reward to insect visitors, which further diversified the flower structures (Tanda, 2022 a,b).

About 300,000 to 500,000 species of flowering plants, comprising 9 out of 10 plants on planet. The efficiency of insect pollination has proven to be a mainstay of natural ecosystem and vastly influenced the evolution and proliferation of flowering plants around the globe. Take some time during Pollinator Week to learn about and identify pollinators in your neighborhood so you can better understand how to support them. Consider reducing your home pesticide use since pesticides are a major part of the problem. Or, best of all, plant some pollinator habitat on your property to provide valuable resources for these invaluable members of our global ecosystem.

2.2 INSECT POLLINATION OR ENTPOLLINATOLOGICAL AWARENESS IN HUMANS

More than 5000 years ago, the mechanisms of pollination and fertilization were assumed from the practical utilization of global agroindustry and seed production when Assyrian priests had rituals of dusting pollen from the male date palms on the inflorescences of female receptive stigmas (Fig. 2.1 a, b). Theophrastus (300 B.C.) the Greek philosopher, has also deduceed on the union of the two sexes for the production of seed in plants. However, the idea was not generalized until 1682, when the botanist Nehemias Grew stated that pollen must reach the stigma to insure the development of seeds. In 1694, Rudolph Jacob Camerarius published (De sexu Plantarum epistole), in which he stated that based upon his experiments there are two different parts of the flower, the stamens and the pistil and that they must work together to produce seed. It would be Amici (1824, 1830) after Brown's (1833) detailed descriptions, to firstly observe pollen tubes and its passage down the style and into the ovule (Amici, 1824, 1830, 1847) (Fig. 2.1 a, b). The contact between the tip of the pollen tube and the egg cell remained unanswered until Strasburger proved that the pollen tube does not stay intact, but its apical tip disintegrates upon contact with the embryo sac and that one of its "nuclei fuses with the nucleus of the egg" (Maheshwari 1950; Cresti and Linskens 1999). Darwin (1862) proved conclusively the significance of pollination in perpetuation and vigor maintenance of

(a) (b)

Fig. 2.1 Historical perspective of the discovery of sexual reproduction in plants. (a,b) Ancient representations of the first historical evidence that pollen had to be brushed on the stigma surface as a means to assure seed production. As long as 5000 B.C. Both Assyrian priests (b) and Egyptian gods were pictured on ceremonial fertilization of date palms. (Photo Courtesy, Boavida et al., 2005)

crop plants. Double fertilization was afterwards simultaneously described by Sergius Nawashin and Leon Guignard in the dawn of the nineteenth century and nowadays is believed to be a ubiquitous phenomenon (Friedman, 2001, Boavida, et al., 2005).

Fig. 2.2 The origin of sexual and asexual reproduction. (Photo courtesy by Sanders (2019)

Sanders (2019) stated that evolutionists are completely unable to explain the origin of sexual reproduction—or any type of reproduction for that matter! They have produced a lot of contradictory hypotheses and often appeal to logical fallacies to make them work. Ultimately, none of them have any observable evidence to back them up. To make matters worse, asexual lineages should be going extinct in the evolutionary paradigm, yet there are thousands of obligate asexual kinds in the world, many of which are ancient by evolutionists' own admission. The biblical perspective provides a much more satisfactory answer. Nature made both forms of reproduction in the beginning, commanding the animals and man to reproduce and fill the earth he had made (Fig. 2.2).

Golicz et al. (2018) Sexual reproduction requires cell fate specification and involves epigenetic changes, while lncRNAs are known to be involved in cell differentiation and have the ability to alter chromatin states. Thus, lncRNAs can be particularly suited for fulfilment of roles in sexual reproduction in both plants and animals (Fig. 2.3).

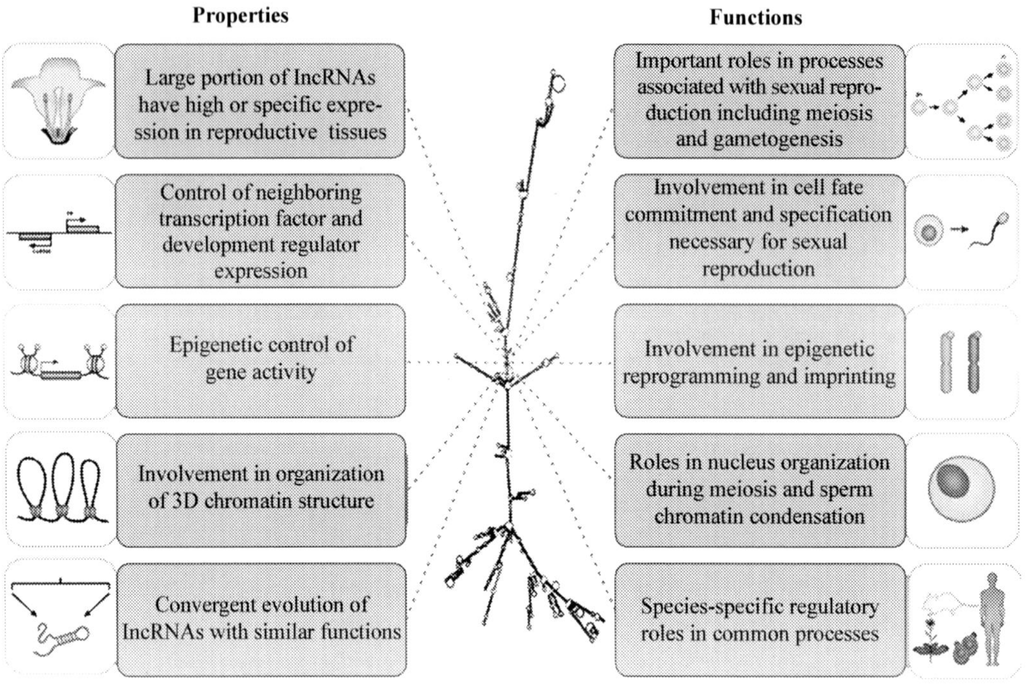

Fig. 2.3 Long noncodingRNAs in sexual reproduction. (Photo courtesy by Golicz et al. (2018).

2.3 EMPIRICAL DEMONSTRATION IN POLLINATION-HISTORICAL PAGES

Rudolf Jakob Camerarius (1665–1721), the German physician and botanist is attributed with the first empirical demonstration in pollination historical pages that plants breed sexually. Camerarius brought to light the game of the different parts of a flower in seed production. He observed that some bisexual (with both male and female reproductive organs) flowers, that a stamen (male pollen-producing organ) and a pistil (female ovule-producing organ) were both needed for seed production. Seed fertilization mechanism was invested by scientists several decades after Camerarius. In the study of flower pollination, many scientists followed Camerarius's studies, one of the most eminent was Charles Darwin. In 1862, Darwin published an important book on pollination: *The Various Contrivances by which Orchids are Fertilized by Insects*. In part, Darwin wrote this book on orchids in support of his theory of Evolution proposed in *The Origin of Species*, published in 1859. Darwin revealed that many orchid flowers had evolved elaborate structures by natural selection in order to facilitate cross-pollination. He proposed that orchids and their insect pollinators evolved by interacting with one another over many generations, a process as co-evolution (Ames, 1946). The proposal of sex in plants was given by Theophrastus the fourth century B.C., however, later Muller (1882–1883) investigated plant reproduction and mechanism of insect pollination. He wrote a book *fertilization of flowers* in German. Knuth published *Fower pollination* in German, later on, in 1906–1908 translated into English. Relationship between bees and flowers, the new proposal presented by Aristotle. However, in 1870 Koelreuter and Dobbs who first reported the role of insects in pollination. Sprengel (1793) stated that "every peculiarity of plant anatomy and physiology is related to the peculiarity of structure and behavior of the insects which visit and pollinate flowers" the research work of Darwin 1883 using both hand and insect pollination techniques, helped greatly in understanding the theories of plant perpetuation and vigor maintenance through cross-pollination.

2.4 PRINCIPLE OF SEX DIFFERENTIATION IN CROPS

The basic principle of sex differentiation in plants may have been known as early as 1500 B.C. Hebrews grasped the significance and art of date pollination from Egyptian and Babylonian experts (Goor, 1967). An Assyrian architectural relief of that period shows two divine creatures, each presumably holding a male date inflorescence over a female inflorescence (Faegri and van der Pijl, 1966). Kerner (1878) demonstrated , "When we consider that from time immemorial, Chinese and Japanese gardeners have produced asters, camellias, chrysanthemums, peonies, pinks, and roses, of which the majority are the results of crossing, we may assume with certainty that the practice of dusting flowers of one species with pollen of another species first came into use in those countries." The Arabic writer, Kazwini, in 682 A.D., Werkenthin (1922) quoted as saying that the date is the only tree that is artificially pollinated to set fruit. To ensure a set of dates in their groves, date growers utilize this procedure. However, if this indicated recognition of sex in plants, the mechanism was not taken to other crops. A botanist, Nehemias Grew, 1682 described that pollen must reach the stigma to ensure the development of seeds it was not sure. He assumed that the stamens of a flower spread their pollen dust directly on flower the stigma of the same plant (Dowden, 1964).

In 1500, Valerius Cordus (1515–1544), German botanist used the word pollen, powder or fine flour, to describe the flower pollen grains. A Venetian named Prospero Alpini (1553–1616 or 1617) in 1592, studied Egyptian flora while serving as physician to his government's consul in Cairo and published observations on the fertilization of date palms disproving classical plant correlations. An Englishman, in the 1600s, John Ray (1627–1705) strived a grand systematic description of all known plants and animals. In 1660, out of this came, a three-volume work that listed and classified the plants in the area surrounding Cambridge. That effort was significant in that it sought to organize plants along lines of similarity and discussed, for the first time, sexuality among plants. Nehemiah Grew (1641–1712), Another Englishman, wrote in 1682 of the possibility of plant sexuality, suspecting that the parts of the flower played separate roles in plant reproduction. At the Botanic Garden in his home town of Tubingen, German botanist, Rudolph Camerarius (1665–1721), studied extensively. In 1694, he observed that the flower anthers, the pollen-bearing parts of the stamen, were the plant's male organs, while the style and stigma were the female organs. He established what Ray, Grew, and others had doubted before, and explained the pollination process. English botanist Stephen Hales (1677–1761) studied both plants and animals to delineate the similarities in their circulation systems: that sap is to a tree what blood is to a horse or dog. His research studies established that water air and the light is important for the growth of plants and he described his findings in his book *Vegetable Staticks* (1727).

Bai et al. (2019) ruled out the hypothesis that monoecism originates from a bud mutation, proposing that sexes were ascertained during floral development and may be influenced by environment. The findings established the critical period of sex differentiation in inflorescence buds and detected the internal and external differences between male and female buds on the same monoecious *P. chinensis*. Additionally, monoecious materials were obtained through graft propagation. These findings and resources can provide the foundation for further exploration of sex determination mechanisms and the utilization of monoecious *P. chinensis*.

2.5 HONEYBEES UTILIZATION IN CROP IMPROVEMENT SYSTEM

Use of honeybees for pollination of cultivated crops and increasing their yield started with the practice of moving honeybee colonies to the crops during flowering period. In Virginia, to examine the crop losses in an orchard of 22,000 Bartlett pear trees, in 1880, Mr. M.B. Waite was appointed by USDA. It was due to the problem of self-sterility in the pear plants, he found. He also suggested the necessity of cross-pollination for a superior fruit set. He further advocated the use of honeybees for pear pollination by moving hives into pear orchards, as pear pollen grains were heavy and sticky to be transported by the wind movements. For apple pollination in the late nineteenth century honeybees were also hired. Apple fruit trees gone sterile as honeybees were not there prior to the turn of present century, in Australia and the banks of Margabi River in Central Asia. After the introduction of beehives to these regions, apple orchards again began setting fruit successfully. For the pollination of self-sterile commercial varieties of apple in USA, Benton (1896) recommended 5–6 colonies of Apis mellifera per 100 trees. The transfer of the male gametes to the female portion of the flower and the fusion of the cells in the ovule is a critical period in the biolife of angiosperms. By the utilization of bee pollinators, man contributes significantly to the effectiveness of cell fusion process and to the insurance that the plant will be productive of fruit or seeds to his profit (Tanda, 2022 a,b).

Koelreuter, the father of plant hybridization, also observed the uniformity and sterility of most hybrid crop plants, and found the contribution of nectar, wind, and insects in the spread pollen dust. Christian Konrad Sprengel (1750–1816) advanced his studies in greater detail flower structures and the role of insect pollination. He established that the capacity of some flowers to self-pollinate was attributed to the different maturation stages of the anther and stigma. Working on ferns and mosses in the late 1800s, Wilhelm Hofmeister (1824–1877) investigated that these plants have alternative generations, sexually reproducing in one generation while reproducing asexually in the next time. In microscopic details he observed the plant cell division and reported that plant ovules grow into embryos. Gregor Mendel's findings of chromosomes and the laws of heredity in the late 1800s and the application of plant cross breeding by Luther Burbank established a new era in agriculture sectors. In the long history of farming, for the first time agronomists could directly manipulate the reproduction of their crops. Plant breeding became an important part of mass crop production during the twentieth century. A new branch of agricultural research known as genetic engineering was unfolded by the discovery of DNA (deoxyribonucleic acid). Biotechnologists expected that quality, quantity, and consistency of crop can be attained in the laboratory studies by editing gene structures and through the use of cloning process. Although much progress has been achieved in these fields of research, many breeders worried about the climatic influence and the effect on production safety using these biotechniques. As people have become more deeply involved with reproductive processes, utmost care must be taken to prevent environmental catastrophes that could affect both man and nature. The pollen was the male part which after falling upon the stigma, was capable of fertilizing the ovary, London Arthur Dobbs, explained to the Royal Society. He further established that the pollen must come from its own species (Grant, 1949a). Watson (1751) carried date pollen 20 miles and cross-pollinated a previously fruitless tree. In 1761, Koelreuter found that bees are the bioagents who transfer the pollen grains from the male parts to the female parts of the flower (Grant, 1949b). He was the first breeder to cross-pollinate and produced a hybrid fruit between two plant species (Sinnott, 1946). In 1763, Arena also wrote on the subject of cross-pollination in plants and noted that it was carried out by insects (Lutz, 1918). Sprengel (1793) was the first scientist to explore sex in crop plants, the important role played by pollinators, and the importance of cross-pollination in crops. His studies encouraged future work on sex in plants and the role played by insect pollinators. Thomas A Knight (1799) established the significance of cross-pollination and hybrid vigor in crop plants: "Nature intended that a sexual intercourse should take place between neighboring plants of the same species." He found that the site of the pollen within the flower was ". . . generally well adapted to place it on the bodies of insects; and the villous coat of the numerous family of bees, is not less well calculated to carry it." Later on Herbert (1837) supported the importance of cross-pollination in plants. On the formation of the pollen tube and its passage down the style and into the ovule, the studies were reported by Amici, in 1830. By recognition of the fact that there is sexual fusion between gametes in the ovule, it was further followed by many (Sinnott, 1946). Darwin (1889) demonstrated the significance of pollination in perpetuation and vigor maintenance of the crop plants. He made observations in many crop species, utilizing hand and insect pollination methods and established the significance of cross-pollination in plants. Since his studies was published, not enough information has been appended to the crop pollination knowledge necessities in plants.

2.6 BOTANY AND PLANT REPRODUCTION PROCEDURES

Botany developed from human's wishes, to know how plants live and reproduce. Functionally, plant animal reproduction systems are similar. Several crop plants grown in agroindustry have their reproductive organs in their flowers. It is essential that fertilization takes place for the production of fruits and seeds, whether the plants are cross-pollinated or self-pollinated.

Colored petals and attractive scents is not necessary and is by no means characteristic of all flowering plants. Mentioned in *Encyclopædia Britannica, Inc.* (Fig. 2.4), the significant distinguishing feature separating flowering plants from gymnosperms is that the ovules of angiosperms are produced within enclosed containers called carpels.

There was an evolution of insect pollination or entpollinatology in parallel with the taxonomy of species description. Proctor, Yeo and Lack reported the studies of Camerarius,

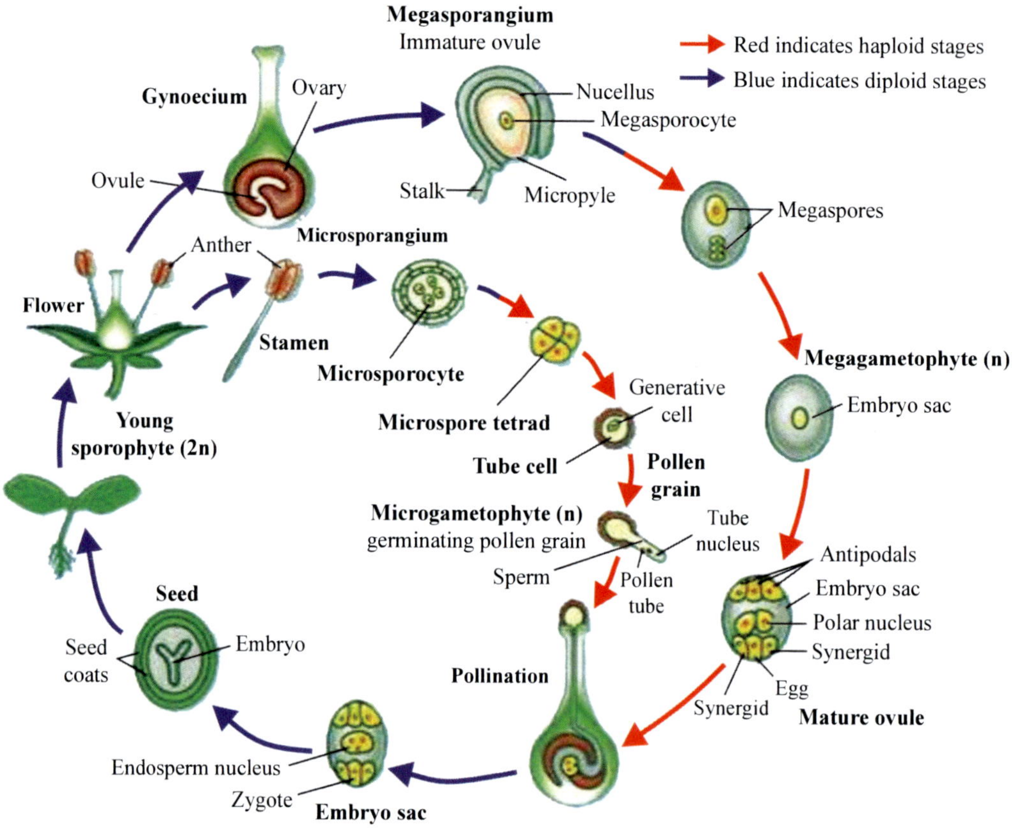

Fig. 2.4 Life cycle of a typical angiosperm. The angiosperm life cycle consists of a sporophyte phase and a gametophyte phase. (Photo courtesy by *Encyclopædia Britannica*, Inc.)

Bradley, Muller, Logan and others in their book 1996, during the first half of the eighteenth century. During same time many botanists found that the "*Farina Fecundens*" produced by the stamens and the stigma were essential to set seeds, and the first cross hand-pollinations were achieved. Biologists, Dobbs and Kölreuter, during the second half of the eighteenth century found that bees foraged flowers in search of nectar and incidentally were carrying the "*Farina*", and thus were pollinating agents for numerous flowering plants (Proctor et al., 1996). Lamarck (1778) discovered the heat produced by spadixs of *Arum italicum*, during this period. On about 500 plant species, the first systematic study of insect pollination described by Sprengel (1793) and he explained many pollination mechanisms, in flowers which temporarily trap insects (e.g., *Aristolochia*). Darwin investigated the plant variation and insect pollination mechanisms, during 1857–1876, e.g., self-pollination avoidance within the frame of his theory of natural selection. Hildebrand (1867) at this time, wrote the first book on floral biology, followed by Delpino (1868) who submitted a flower classification (two volumes) based on different floral types in relation to plant pollination. Delpino investigated several flower pollination systems, including those of *Arum italicum* and in 1870 of *A. Maculatum*. Three books on *plant-pollinator insect relationships, describing many pollination mechanisms,* including *Arum maculatum* written by Muller (1883). Arcangeli in 1883, reported the pollinating fauna of *Arum italicum* and *Arum pictum*. Knuth (1895–1905) compiled a handbook with a description on flower pollination in which *Arum italicum, A. maculatum, A. pictum* and *A. dioscoridis*. Thanks to experimental studies of the pollination of *Arum maculatum* and *A. nigrum* (Schmucker, 1925; Knoll, 1926), the comprehension of the insect trap mechanism was explained 40 years later. In England during the last 60 years, only the pollination of *Arum maculatum* has been reported. Nonetheless, the genus *Arum* is the most documented among Araceae. The great studies on the significance on pollination from the United States was the first finding by Waite (1895) of self-sterility in pears and the need for insect-transfer of pollen between cultivars. Especially in fruit pollination, a new ripple of interest was started, although many investigations on the importance of bee pollination had already appeared (Crane, 1876; Hutchinson, 1886; Muller, 1883). Benton (1896) advocated "... 4 or 5 well-populated hives of honeybees for every hundred large apple trees, the hives to be placed in or near the orchard." The hiring of beehives for orchard pollination service had its beginnings the first decade of this century (Beuhne, 1909; Stricker, 1971). The hard necessity that evolved for legume seed during World War II encouraged the set-up of the USDA Legume Seed Research Laboratory at Logan, Utah. The joint efforts at this research laboratory established the importance of bee pollination in alfalfa seed production (Utah Agr. Expt. Sta., 1950). Currently thousands of beehives are being utilized to pollinate this crop alone. Recently, in the management of pollinating insects largely wild bees are utilized in crop production programs, primarily the gregarious ground-nesting alkali bee (*Nomia melanderi* Cockerell) and the equally gregarious tube-nesting leafcutter bee (*Megachile pacifica* Panzer) (Bohart, 1912; Stephen, 1959). Clements and Long (1923) mentioned about pollination of many plant species, in general. Hooper (1921), Hutson (1926), Kenoyer (1916), and Wellington et al. (1929) reported the pollination of several specific crops, and Farrar (1931) worried about the strength of beehives for crop pollination services. Major research papers published in the United States include those by Bohart (1960), Bohart and Todd (1961), Eckert (1959), Hambleton (1944), Todd and McGregor (1960), and Vansell and Griggs (1952). Many broad subject articles published in other countries include: (Australia) Gale (1897); (England) Butler and Simpson (1953), and Free (1960); (India) Krishnamurthi and Madhava Rao (1963); (Italy) Giordani (1952); (Jamaica) Chapman (1964), and Purseglove (1968); and (Russia) Krishchunas and Gubin (1956), Gubin and Khalifman (1958), Kasiev (1964) and Tanda, 2019a, b, 2020).

None of them excelled the Free (1970) in published work. The author got an opportunity to have discussions on insect pollination of crops with Dr John B Free during his visit to India in 1975-76. We showed him our experiments on Asiatic cotton pollination including my first intervarital cotton trial confining bee colonies at PAU Ludhiana Cotton Farm. Free described in details the insect pollination necessities and the management of pollinating insects for each plant family in crop insect pollination. Gous et al. (2019) reported that small, insect-carried pollen samples from historic bee specimens collected from as early as 1914 can be used to obtain pollen metabarcodes. Also Corneile et al. (2019) described that unlike internally fertilizing animals, the pathway to paternity in insect pollinated plants is a complex barrier along which pollen can be lost at multiple stages before the final stylar race to fertilize ovules. He reported that after considering the potential ramifications of pollen placement and deposition competition, it is perhaps possible to do some comparative studies with animal reproductive strategies. Pollinators can detect the color, shape, scent, and even temperature of the flowers they forage. Khan et al. (2021) reported the previously unappreciated capacity of hover flies

(*Eristalis tenax* and *Cheilosia albipila*) to detect the electric field surrounding flowers. Studies disclosed that rapid variations in a floral electric field were associated to a nectar reward and enhanced the likelihood of the fly's return visits. They observed that thoracic hairs played a role in the polarity of Hover fly charge, revealing their electro-mechanosensory capability, as in bumblebees (*Bombus terrestris*) (Khan et al., 2021). Australian horticulture relies on the honeybee, *Apis mellifera* L. to cross-pollinate major crops. Cook et al. (2020) identified 11 calliphorid, two rhiniid and seven syrphid species as candidates with high potential for use in Australian managed pollination services.

2.7 CONCLUSION

A program for the conservation and the sustainable use of pollinators should be designed globally. This initiative should encourage international cooperation with some important goals:

- Pollinator decline including wild bees, its reasons, and its influence on pollination services should be monitored.
- Information on taxonomic details of potential pollinators be addressed.
- Economic value of insect pollination and the economic impact of the decline of pollinators should be evaluated (Tanda, 2019a, b, 2020).
- Conserve the bee pollinators on your piece of land, garden, and the backyard of your company or your rooftops.
- Towns, schools, corporations, and individuals land pieces can be utilized.
- Stop using any pesticides on your garden flowering plants.
- Grow bee pollinator's preferred flowering plants as they are a big asset for bees and butterflies offering nectar and pollen.
- Plant prairies, keep away from lawns as they desert for bee pollinators.
- Never weed your gardens as dandelions are best source of bee food.
- Install water basins in summer with pebbles or floating corks on water to prevent bees from drowning in every balcony.
- Educate and train the public including your family by showing bee documentaries. Native bees are harmless, and visit flowers.
- Install a beehive and modified handmade devices for wild bees in your backyard or your rooftops.
- Conservation of disappearing species, their restoration and sustainable utilization of pollinator biodiversity in agro and related ecosystems should be enhanced (Tanda, 2020, 2021a-e; Tanda, 2022 a,b).

REFERENCES

- Ames O (1946) The evaluation of the orchid flower. Am Orchid Soc Bull 14: 355–360
- Amici GB (1824) Observations microscopiques sur diverse espèces de plantes. Ann Sci Nat Bot 2(41–70): 211–248

- Amici GB (1830) Note sur le mode d'action du pollen sur le stigmate. Extrait d'une lettre d'Amici à Mirbel. Ann Sci Nat Bot 21: 239–332
- Amici GB (1847) Sur la fécondation des Orchidées. Ann Sci Nat Bot 7(8): 193–205
- Bai, Q et al. (2019) Sex determination during inflorescence bud differentiation in Monoecious *Pistacia chinensis* Bunge. Forests, 10, 202; doi:10.3390/f10030202
- Benton F (1896) The honeybee, a manual of instruction in apiculture. US Dept Agr Div Ent Bul 1 N S Rev Ea, 118 pp
- Beuhne R (1909) Bees and fruit fertilization. Victoria Dept Agr J 7(11): 693–696
- Boavida, Leonor & Maria Vieira, Ana & Becker, Jörg & Feijó, José. (2005) Gametophyte interaction and sexual reproduction: How plants make a zygote. The International Journal of Developmental Biology. 49. 615-32. 10.1387/ijdb.052023lb.
- Bohart GE (1912) Management of wild bees for the pollination of crops. Ann Rev Ent 17: 287–312
- Bohart GE (1960) Insect pollination of forage crops. Adv Agron 12: 72–88
- Brown R (1833) Discovery of cells and the development of cell theory. Trans Linn Soc 16: 685
- Bohart BE, Todd FE (1961) Pollination of seed crops by insectsYearbook of Agriculture, vol. 245. Department of Agriculture, United States
- Butler CG, Simpson J (1953) Bees as pollinators of fruit and seed crops. Report Rothamsted Experimental Station England, UK, pp 167–175
- Cook David F, Sasha C Voss, Jonathan TD, Finch, Romina C Rader, James M Cook, Cameron J Spurr (2020) "The Role of Flies as Pollinators of Horticultural Crops: An Australian Case Study with Worldwide Relevance" Insects 11, no. 6: 341. https://doi.org/10.3390/insects11060341
- Corneile Minnaar, Bruce Anderson, Marinus L de Jager, Jeffrey D Karron, (2019) Plant–pollinator interactions along the pathway to paternity, Annals of Botany, Volume 123, Issue 2, 23 January, Pages 225–245, https://doi.org/10.1093/aob/mcy167
- Clements EE, Long EL (1923) Experimental pollination: an outline of the ecology of flowers and insects, 336th edn. Carnegie Institute of Washington publication, Washington, DC, 274 pp
- Crane, JE. (1876) The relation of bees to fruit culture. In 3rd Biennial report of the Vermont State Board of Agriculture, pp 253–270
- Cresti M, Linskens HF (1999) The discovery of sexual reproduction in higher plants. Acta Biol Crac Ser Bot 41: 19–29
- Darwin C (1862) Fertilization of orchids by insects. Murray, London, p. 15
- Darwin C (1869) Notes on the fertilization of orchids. Annals and magazine of natural history september 1869: 1–19
- Delpino F (1868–1875) Ulteriori osservazioni e considerazioni sulla dicogamia nel regno vegetale. Atti Soc Ital Sci Nat 11,12: 179–233
- Dowden AO (1964) The secret life of flowers. 45 pp. The Odyssey Press, New York

- Eckert JR (1959) The Kangaroo Island Ligurian bees. Gleanings in Bee Culture 86: 660–663, 722–725
- *Encyclopædia Britannica, Inc.* Plant reproductive system. https://www.britannica.com/science/plant-reproductive-system/media/1/498651/48341
- Faegri K, van der Pijl L (1966) The principles of pollination biology. Pergamon Press, New York
- Farrar CL (1931) The evaluation of bees for pollination. Jour Econ Ent 24: 622–627
- Free JB (1960) The pollination of fruit trees. Bee World 41(141–151): 169–186
- Free JB (1970) Insect pollination of crops. Academic, London, New York
- Friedman WE (2001) Developmental and evolutionary hypotheses for the origin of double fertilization and endosperm. C R Acad Sci III 324: 559–567
- Gale A (1897) The influence of bees on crops. Agr Gaz NS Wales 8: 264–276, 337–339, 423–425, 490–493, 581–584, 620–623, 734–736, 831–833, 908–910
- Golicz Agnieszka A. Prem L. Bhalla Mohan B. Singh (2018) lncRNAs in Plant and Animal Sexual Reproduction February 01, https://doi.org/10.1016/j.tplants.2017.12.009
- Gous A, et al., (2019) Plant–pollinator interactions over time: Pollen metabarcoding from bees in a historic collection. Evol Appl. Feb; 12(2): 187–197
- Giordani G (1952) Bees and flower pollination. National Apicultural Institute of Bologna. [in Italian.] Aa-120/54, 59 pp
- Gottlieb Koelreuter J (1733–1806) A German who demonstrated the existence of sexes in plants and produced a plant hybrid by crossing two species of tobacco.
- Goor A (1967) The history of the date through the ages in the holy land. Econ Bot 21: 320–340
- Grant V (1949a) Arthur dobbs (1750) and the discovery of the pollination of flowers by insects. Bull Torr Bot Club 76(3): 217–219
- Grant V (1949b) Pollination systems as isolating mechanisms in angiosperms. Evolution 3: 82–97
- Gubin AF, Khalifman IA (1958) Flowers and bees. Agrobiologiya, Moscow 1–139 USSR, Moscow: Moskovskii rabochii. [in Russian.] Aa-64/59, 194 pp
- Hambleton JI (1944) The role of bees in the production of fruit and seed. Jour Econ Ent 37: 522–525
- Herbert W (1837) Amaryllidaceae with a treatise on crossbred vegetables. In: Muller H (ed) (1883) The fertilization of flowers Macmillan Co., Inc., London, p 371. [Translated and edited by d'arcy thompson.]
- Hildebrand F (1867) Diegeschlechter-vertheilungbei den pflanzen. Wilhelm Engelmann, Leipzig, 92 pp
- Hooper T (ed) (1921) Guide to Bees and Honey: Amazon.Co.UK: Ted Hooper: Books
- Hutchinson WZ (1886) The relation of bees to horticulture, benefits and injuries. Mich Hort Soc Proc 16: 13–15

- Hutson R (1926) The relation of honeybees to fruit pollination in New Jersey. NJ Agr Exp Sta Bull 434: 1–32
- John Ray (1627–1705) was an English naturalist. His classification of plants in his Historia Plantarum, was an important step towards modern taxonomy
- Kasiev TI (1964) Problems of nectar production of cotton and the role of honeybees in increasing its yield. 215 pp. Baku, Azerbaidzhan Goz. lzd. [In Russian.]
- Kenoyer LA (1916) Pollination of economic plants.In Iowa State Apiarist Report 1915: 24–30
- Kerner A (1878) Flowers and their unbidden guests.(trans: Ogle W), C. Kegan Paul & Co, London, 164 pp
- Khan SA, Khan KA, Kubik S et al. (2021) Electric field detection as floral cue in Hover fly pollination. Sci Rep 11, 18781 https://doi.org/10.1038/s41598-021-98371-4
- Knight TA (1799) An account of some experiments on the fecundation of vegetables. Roy Soc London Phil Trans pt I 89: 195–204
- Knoll F (1926) Die Arum-Blütenstände and ihre besucher (Insekten und Blumen IV). Abh. Zool.-Botan. Ges. Wien 12: 381–481.
- Knuth P (1895–1905) Handbuch der blutenbiologie iiii. Engelmann, Leipzig, 2973 pp
- Krishnamurthi S, Madhava Rao VN (1963) Certain problems in pollination of fruit crops. So Indian Hort 11(1, 2): 1–17
- Lamarck [J.P.A.M] (1778) Flore françoise, ou, Description succincte de toutes les plantes qui croissent naturellement en France: disposée selon une nouvelle méthode d'analyse, & à laquelle on a joint la citation de leurs vertus les moins équivoques en médecine, & de leur utilité dans les arts. Tome premier. pp. 1-4, i-cxix, [1]-132, i-xxix, pls I-XVIII. Paris: De l'Imprimerie
- Royale Lutz FE (1918) Fieldbook of insects. G.P. Putnam's sons, New York and London, 509 pp
- Maheshwari P (1950) An introduction to the embryology of angiosperms. McGraw-Hill, New York, 453 pp
- Muller H (1883) The fertilisation of flowers (trans: D'Arcy W. Thompson) Macmillan, London, 669 pp.
- Nehemiah Grew, who, with the Italian botanist Marcello Malpighi, may be considered as co-founder of the science of Plant Anatomy. Grew was drawn to the study of plant structure at the age of twenty-three, and seven years later he produced his earliest work on the subject
- The Anatomy of Vegetables Begun, which was published by the Royal Society in 1672. In 1682 he published his magnum opus, *The Anatomy of Plants*, prefaced by an Epistle Dedicatory
- Proctor M, Yeo P, Lack A (1996) The natural history of pollination. Harper Collins Publishers, London, 479 pp
- Rudolf Jakob Camerarius (1665–1721) German botanist and physician (latinization of "Camerer") (1665–1721) German physician and botanist. Professor at Tubingen (from 1688); demonstrated sexuality in plants (reported 1694)

- Sanders HF (2019) The Origin of Sexual and Asexual Reproduction. April 18 https://answersingenesis.org/biology/origin-sexual-and-asexual-reproduction/
- Schmucker T (1925) Beiträge zur biologie und physiologie von *Arum maculatum*. Flora 118(119): 460–475
- Sinnott EW (1946) Botany – principles and problems, 4th edn. Mcgraw-Hill Book Co., Inc, New York and London, 726 pp
- Sprengel CK (1793) The secret of nature in the form and fertilization of flowers discovered. 4 sects. [in German]
- Stephen WP (1959) Maintaining alkali bee beds for alfalfa seed production. Oreg Agr Expt Sta Bul 568, 23 pp
- Stricker MH (1971) Pollination pioneer: Richard D. Barclay 1888–1938. Gleanings bee cult 99: 358–360
- Station (1950) Growing alfalfa for seed in utah. Utah Agr Expt Sta Cir 125, 72 pp
- Tanda AS (2022a) Chapter 11 Pest management technology and bee pollinators' integration. In: Tanda A S (ed.) Advances in Integrated Pest Management Technology: Innovative and Applied Aspects. Springer Nature Switzerland (in press)
- Tanda AS (2022b) Chapter 1: Mutulistic Plant Associations related to Insect Resistance. In: Tanda A S (eds.) Molecular Advances in Insect Resistance of Field Crops: Modern and Applied Approaches. Springer Nature Switzerland (in press)
- Tanda AS (2021a) Biofloralphenology, foraging behavior and entpollinatological effect of honeybees in pomegranate (*Punica granatum*) fruit quality and yield. Journal of Horticulture 08 2
- Tanda AS (2021b) Bee pollinators matter in crop improvement need immediate conservation strategies. Journal of Insect Science (Accepted)
- Tanda AS (2021c) Comparative pollination efficacies of *Apis mellifera* L. and *Tetragonula carbonaria* (Smith) on peach (*Prunus persica* L.) Indian Journal Entomology (Accepted)
- Tanda AS (2021d) Why insect pollinators important in crop improvement?. Indian Journal Entomology (Accepted)
- Tanda AS (2021e) Insect pollinators matter in sustainable world food production. Indian Journal Entomology (Accepted)
- Tanda AS (2020) Entpollinatology: A strong relationship between plants and insects for crop improvement. *6th Edition of Global Conference on Plant Sciences and Molecular Biology* (*GPMB 2020*)" Paris, France.
- Tanda AS (2019a) Entomofauna enhance the quality and quantity in Okra. Indian Journal Entomology 81(1): 16–17
- Tanda AS (2019b) Entomophilous crops get better fruit quality and yield: An appraisal. Indian Journal Entomology 81(2): 227–234
- Theophrastus (300 BC) Enquiry into plants and minor works on odors and weather signs. Book 2, p. 8, paragraph 3–4, p 155 (trans: sir arthur hort) Harvard University Press, Boston, 475 pp

- Todd FE, McGregor SE (1960) The use of honeybees in the production of crops. Ann Rev Ent 5: 265–278 Utah Agricultural Experiment
- Vansell G, Griggs WH (1952) Honeybees as agents of distribution. US.D.A. Yearbook: 88–107
- Waite MB (1895) The pollination of pear flowers. US Dept Agr Div Veg Path Bul 5, 86 pp
- Watson W (1751) Some observations upon the sex of flowers. Roy Soc London Phil Trans 47: 169–183
- Wellington Ra, Stout AB, Einset O, and others (1929) Pollination of fruit trees. N.Y. (Geneva) Agr Expt Sta Bul 577, 54 pp
- Werkenthin FC (1922) The founders of the art of plant breeding. Iowa Acad Sci 29: 291–310
- Wilhelm Friedrich Benedikt Hofmeister (18 May 1824–12 January 1877)

3

Pollination Concepts and Crop Production

R. Kanagarajan* and K. Nishanthini[1]

ABSTRACT

The extent of our reliance on animal pollination for world crop production for human food has not previously been evaluated and the previous estimates for countries or continents have seldom used primary data. In this review, we expand the previous estimates using novel primary data from 200 countries and found that fruit, vegetable or seed production from 87 of the leading global food crops depend on animal pollination, while 28 crops do not rely upon animal pollination. However, global production volumes give a contrasting perspective, since 60% of global production comes from crops that do not depend on animal pollination, 35% from crops that depend on pollinators, and 5% are unevaluated. Using all crops traded on the world market and setting aside crops that are solely passively self-pollinated, wind-pollinated or parthenocarpic, we then evaluated the level of dependence on animal-mediated pollination for crops that are directly consumed by humans. We found that pollinators are essential for 13 crops, production is highly pollinator dependent for 30, moderately for 27, slightly for 21, unimportant for 7, and is of unknown significance for the remaining 9. We further evaluated whether local and landscape-wide management for natural pollination services could help to sustain crop diversity and production. Case studies for nine crops on four continents revealed that agricultural intensification jeopardizes wild bee communities and their stabilizing effect on pollination services at the landscape scale.

Keywords *Agriculture • Conservation • Pollination • Biodiversity • Pollinators • Crops*

[1]Department of Entomology, Faculty of Agriculture, Annamalai University, Tamil Nadu-608 002, India.

*Corresponding author E-mail address: krkanagarajan@yahoo.com

3.1 INTRODUCTION

Pollination is one of the most fascinating processes in the natural world. Pollination is how flowering plants reproduce. The process involves the transfer of pollen from the male parts to the female parts of the same or another plant. For some plants, this movement of pollen requires the action of another organism, a pollinator. Plants and their pollinators form a mutualistic relationship, a relationship in which each benefits from the other. In the plant-pollinator relationship, the pollinator benefits by feeding on food rewards provided by the flower, primarily nectar and pollen. In return, the plant benefits as the pollinator moves from flower to flower, transferring pollen as it forages for the food rewards. This movement of pollen allows the plant to reproduce and to exchange genetic information with other plants. Most flowering plants require relationships with pollinators to reproduce. Flowers are the tools that plants use to make their seeds. The basic parts of the flower are shown in the diagram below. Seeds can only be produced when pollen is transferred between flowers of the same species. A species is defined a population of individuals capable of interbreeding freely with one another but because of geographic, reproductive, or other barriers, they do not interbreed with members of other species.

3.2. PROCESS OF POLLINATION

Pollen germination has three stages; hydration, activation and pollen tube emergence. The pollen grain is severely dehydrated so that its mass is reduced enabling it to be more easily transported from flower to flower. Germination only takes place after rehydration, ensuring that premature germination does not take place in the anther. Hydration allows the plasma membrane of the pollen grain to reform into its normal bilayer organization providing an effective osmotic membrane. Activation involves the development of actin filaments throughout the cytoplasm of the cell, which eventually become concentrated at the point from which the pollen tube will emerge. Hydration and activation continue as the pollen tube begins to grow (Raghavan, 1997).

In conifers, the reproductive structures are borne on cones. The cones are either pollen cones (male) or ovulate cones (female), but some species are monoecious and others dioecious. A pollen cone contains hundreds of microsporangia carried on (or borne on) reproductive structures called sporophylls. Spore mother cells in the microsporangia divide by meiosis to form haploid microspores that develop further by two mitotic divisions into immature male gametophytes (pollen grains). The four resulting cells consist of a large tube cell that forms the pollen tube, a generative cell that will produce two sperm by mitosis, and two prothallial cells that degenerate. These cells comprise a very reduced microgametophyte, that is contained within the resistant wall of the pollen grain (Campbell and Reece, 2002; Runions and Owens, 1999).

The pollen grains are dispersed by the wind to the female, ovulate cone that is made up of many overlapping scales (sporophylls, and thus megasporophylls), each protecting two ovules, each of which consists of a megasporangium (the nucellus) wrapped in two layers of tissue, the integument and the cupule, that were derived from highly modified branches of ancestral gymnosperms. When a pollen grain lands close enough to the tip of an ovule, it is drawn in through the micropyle (a pore in the integuments covering the tip of the ovule) often by means of a drop of liquid known as a pollination drop. The pollen enters a pollen chamber close to the nucellus, and there it may wait for a year before it germinates and forms a pollen tube that grows through the wall of the megasporangium (nucleus) where fertilisation takes place. During this time, the megaspore mother cell divides by meiosis to form four haploid cells, three of which degenerate. The surviving one develops as a megaspore and divides repeatedly to form an immature female gametophyte (egg sac). Two or three archegonia containing an egg then develop inside the gametophyte. Meanwhile, in the spring of the second year two sperm cells are produced by mitosis of the body cell of the male gametophyte. The pollen tube elongates and pierces and grows through the megasporangium wall and delivers the sperm cells to the female gametophyte inside. Fertilization takes place when the nucleus of one of the sperm cells enters the egg cell in the megagametophyte's archegonium (Runions and Owens, 1999).

In flowering plants, the anthers of the flower produce microspores by meiosis. These undergo mitosis to form male gametophytes, each of which contains two haploid cells. Meanwhile, the ovules produce megaspores by meiosis, further division of these form the female gametophytes, which are very strongly reduced, each consisting only of a few cells, one of which is the egg. When a pollen grain adheres to the stigma of a carpel it germinates, developing a pollen tube that grows through the tissues of the style, entering the ovule through the micropyle. When the tube reaches the egg sac, two sperm cells pass through it into the female gametophyte and fertilization takes place (Runions and Owens, 1999).

3.3. BASIC FLOWER ANATOMY

A knowledge of basic flower anatomy can aid in understanding how plants reproduce (Figure 3.1). Flowers are generally comprised of four whorls [1]of parts: the calyx, corolla, androecium and gynoecium. Each whorl contains a set of organs: sepals, petals, stamens and carpels, respectively.

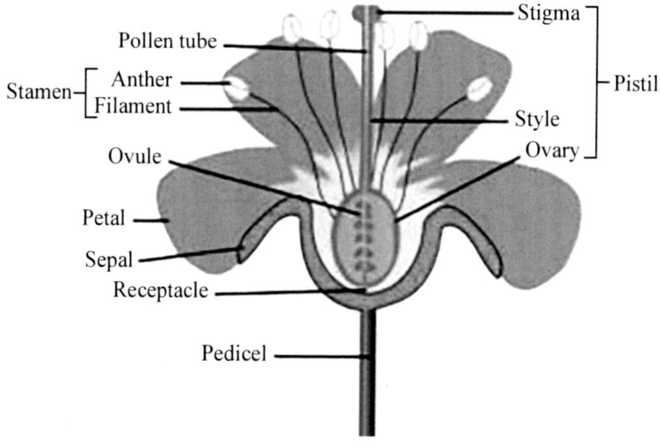

Fig 3.1. Basic flower anatomy: a perfect flower containing male and female organs

1. Calyx

Sepals of the calyx look similar to leaves and surround the unopened bud. The sepals protect the flower before it opens and may remain attached to the base of the flower afterward. Sepals are typically green and smaller than the petals, but they can also be large and brightly colored, resembling petals. When sepals and petals are identical in appearance, they are called tepals.

2. Corolla

The corolla typically encircles the reproductive parts of the flower and may be continuous, lobed, or divided into separate petals. Petals also help protect the reproductive organs of the flower, but primarily serve to attract pollinators. To succeed in attracting pollinators, petals are often brightly colored and fragrant.

3. Androecium

The next whorl in, the androecium, is comprised of stamens, the male reproductive organs of the flowering plant. Stamens are made up of two parts: a thin string like stalk called a filament and the pollen-producing organ at its tip, called the anther.

4. Gynoecium

The innermost whorl of the flower, the gynoecium, contains the female reproductive parts of the flowering plant. The carpels that make up the gynoecium may be separate or fused to form a compound pistil. The pistil is made up of three parts: the ovary, the style and the stigma. The ovary holds the unfertilized seed, or ovules, and typically develops into a fruit once pollination takes place. The stigma is the uppermost section of the pistil at the tip of the style, and receives the pollen. To catch the pollen, the stigma is commonly sticky, or covered with small hairs or grooves.

5. Other Structural Parts

At the base of a flower are two more structural parts, the receptacle and pedicle. The receptacle is simply the base of the flower, and it is supported by the pedicel, or flower stalk.

6. Nectaries and Nectar Secretion

Flowers frequently have one or more nectarines. Nectaries vary in size from microscopic to the 11-in. nectary of the orchid (*Angraecum sesquipedale* Thou.) (Darwin, 1877). The nectary is most often located within the flower, usually at the base of the sexual column inside the circle of petals. In cotton, however, there is a nectariferous ring just outside the base of the petals on the inner base of the calyx. Nectaries are also found outside the flower, on the stem or leaves. Nectar secretion within the flower usually starts about the time the flower opens and ceases soon after fertilization. Secretion of nectar on the stems and leaves is not influenced directly by flowering and may continue for several weeks. The amount of nectar secreted varies from infinitesimal in numerous species to more than an ounce in the orchid *Coryanthes* spp. (Kerner, 1878) and in *Protea mellifera* Thunb, which natives in Africa reportedly remove and drink (Langstroth, 1913; Holmes, 1963); Nichol (1952) reported that the nectar of the *Agave parryi* Engelm. Flower stalk was gathered by Indians in the Southwest and used as a syrup. Numerous bee specialists have calculated the amount of nectar produced in the flowers of various crops (Todd and Mcgregor, 1960).

7. Stigma Receptivity

When the stigma is receptive to pollen, it is coated with a colorless, relatively tasteless stigmatic fluid. If viable, compatible pollen comes in contact with this moist stigma, it adheres, germinates, and sends a pollen tube bearing the tube nucleus and the two sperm nuclei down through the style into the ovary and, finally, into one of the ovules. Fertilization follows this pollination process by the sexual union of one of the two sperm nuclei of the pollen grain and the egg nucleus of the ovule to form the fertilized egg or zygote. Through this process of sexual union, a viable seed is formed that is capable of producing another complete plant. In general, the sooner pollination can occur after a flower opens the greater the likelihood that fertilization of the ovule and seed development will occur. As time elapses, the pollen may be lost to insect foragers, wind, gravity, or damage by heat, moisture, or drying out. Also, processes may set in that result in the shedding of the fruit. Pollination also requires consideration of pollenizers. (The terms "pollinator" and "pollenizer" are often confused: a pollinator is the agent that moves the pollen, whether it be wind, bees, bats, moths, or birds; a pollenizer is the plant that provides the pollen.) Some plants are self-fertile or self-compatible and can pollinate themselves. Other plants have chemical or physical barriers to self-pollination and need to be cross-pollinated: with these self-infertile plants, not only pollinators must be considered but pollenizers as well. In pollination management, a good pollenizer is a plant that provides compatible, viable and plentiful pollen and blooms at the same time as the plant that is to be pollinated.

3.4 POLLINATION AND POLLENIZER

Pollination can be cross-pollination with a pollinator and an external pollenizer, self-pollenization with a pollinator, or self-pollination without any pollinator.

1. Cross-pollination (syngamy)

Pollen is delivered to a flower of a different plant. Plants adapted to outcross or cross-pollinize have taller stamens than carpels to better spread pollen to other flowers.

2. Self-pollination (autogamy)

Pollen moves to the female part of the same flower, or to another flower on the same individual plant. This is sometimes referred to as self-pollination, but this is not synonymous with autogamy. Clarity requires that the term "self-pollination" be restricted to those plants that accomplish pollination without an external pollinator (example: the stamens actually grow into contact with the pistil to transfer the pollen). Most peach varieties are autogamous, but not truly self-pollinated, as it is generally an insect pollinator that moves the pollen from anther to stigma. Plants adapted to self-fertilize have similar stamen and carpel length.

3. Cleistogamy

Pollination that occurs before the flower opens is always self-pollination. Some cleistogamous flowers never open, in contrast to chasmogamous flowers that open and are then pollinated.

Cleistogamous flowers must of necessity be self-compatible or self-fertile plants. Other plants are self-incompatible. These are end points on a continuum, not absolute points. Hybridization is effective pollination between flowers of different species of the same genus, or even between flowers of different genera (as in the case of several orchids).

3.5 POLLINATION AND FERTILIZATION

Pollination is usually considered in the combined sense of transfer of pollen and set of fruit or seed. In fact, it is not so, for this purpose two sets of factors are involved: (1) transfer of viable pollen from the anther to a receptive stigma, and (2) sprouting of the pollen grain and growth of the pollen tube down the style into the ovary, and ultimately the union of male nuclei of the pollen grain with female germ cells in the ovule that results in seed development. Pollination is of no value without fertilization. A plant may be spoken of as self-fertile or self-compatible if it can produce fruit without the need for the transfer of pollen to it from another cultivar. If the plant is not receptive to its own pollen, it is self-sterile. Even self-pollinating plants are frequently benefited by cross-pollination, the transfer of pollen from one flower to another. They may also benefit from having the pollen more thoroughly transferred and distributed over the stigma at the most receptive period. A plant is cross-compatible if it can normally be pollinated with pollen of another cultivar, but it is cross-incompatible if it is not receptive to pollen of certain cultivars.

3.6 TYPES OF POLLINATION

Two types of pollination may occur: self- and cross-pollination.

1. Self-Pollination

It is the transfer of pollen from anther to the stigma of same flower or to the flowers of the same or other plants of the identical genetic material such as apple flowers.

2. Auto-Pollination

In some plants pollen is automatically deposited on the stigma when the flower opens and this type of self-pollination may be called auto-pollination.

3. Cross-Pollination

It is the transfer of pollen between plants which are not of identical genetic material, i.e., pollen gametes are not genetically identical to the ovule gametes, e.g., cross variety 39 pollination of apples, cross inbred lines pollination for hybrid seed production. The most important and efficient carrier of pollen from anther to stigma is the honeybee. An understanding of the activities of honeybees is essential if the best return is to be obtained in fruit production.

4. Self Fruitful

Self fruitful are capable of setting fruit with its own pollen. The pollen in such cases is heavy, sticky and for commercial crop production, it must be moved from flower to flower by bees.

5. Self Unfruitful or Self in Compatible

Varieties which do not set commercial fruit/seed when self-pollinated are called self-unfruitful or self-incompatible. Most apple varieties are in this category. There is a need to plant other varieties with compatible pollen so that cross variety pollination takes place.

6. Cross Fruitful Varieties

Varieties which set a commercial crop when properly interplanted for cross-pollination. They have compatible pollen.

7. Cross Unfruitful or Incompatible

Varieties this condition usually refers to a physiological incompatibility. The pollen may not send down a pollen tube to deliver the male gamete to the ovule or the embryo may abort after union of gametes. Some varieties have sterile (non-viable) pollen. As a rule physiological incompatibilities exist between varieties with a close genetic relationships. There are varying degrees of incompatibilities which may be altered by different environmental conditions.

8. Monoecious Plants

Monoecious plants are those plants which staminate (male) flowers and the pistillate (female) flowers are separate but on the same plant, e.g., maize, cucumbers, etc.

9. Dioecious Plants

Dioecious plants are those plants which staminate and the pistillate flowers are on different plants, e.g., willow, poplars.

10. Gynoecious Plants

The term gynoecious F1 hybrids has been applied to certain cultivars of picking cucumbers and they are often referred to as "gynoecious" varieties. The cultivars have predominantly pistillate flowers. Many of the fruit crops, legumes, vegetables, and oilseed crops depend upon pollinating insects for set of fruit or seeds.

In spite of the best agronomic practices proper seedbed preparation, fertilization, soil moisture, cultivation, pest control, and harvesting methods-bountiful harvests are not obtained if pollination is neglected.

3.7 POLLINATING AGENTS

The process of pollination requires pollinators as agents that carry or move the pollen grains from the anther to the receptive part of the carpel. Faegri and van der Pijl (1979), in their classic book on pollination ecology, defined two major types of pollen dispersal: Biotic pollination in which the pollen dispersal agent is an animal (i.e., either an invertebrate or a vertebrate); and abiotic pollination where pollen is dispersed by an inanimate physical agent, such as wind or water. The various flower traits that attract different pollinators are known as pollination syndromes. Methods of pollination, with common pollinators or plants, are:

3.7.1. Biotic Pollination

Invertebrates beetles	:	Cantharophily
Flies	:	Myophily
Hymenopterans wasps	:	Sphecophily
Bees	:	Melittophily
Ants	:	Myrmecophily
Lepidopterans butterflies	:	Psychophily
Moths	:	Phalaenophily
Vertebrates birds	:	Ornithophily
Mammals	:	Therophily
Marsupials	:	Metatherophily
Placentals	:	Eutherophily
Bats	:	Chiropterophily
Rodents	:	Sminthophily

3.7.2. Abiotic Pollination

Self-pollination	:	Autogamy + Cleistogamy
Wind-pollination	:	Anemophily
Water-pollination	:	Hyphydrophily
Gravity pollination		

3.8. POLLINATORS IN AGRICULTURE

An evaluation of experimental evidence for pollination requirements of 107 globally traded fruit and vegetable crops (representing 40 percent of global plant-based food production) by Klein et al. (2007) shows that animal pollination improves production in 75 percent of the crops studied. Most cultivars of another 10 percent of the crops require animal pollination. Another 8.5 percent of the crops do not benefit from animal pollination and its role in production of the remaining 6.5 percent crops is not known. Many crops, however—notably the staple grains that form the foundation of most human diets (rice, wheat, maize, sorghums, millets, rye, barley)—are self-pollinating or pollinated by the wind. Together, species that do not rely on pollinators account for most of the world's food supply by weight.

Pollinator declines, therefore, do not fundamentally threaten the world's caloric supplies. However, fruits and vegetables, which add diversity to the human diet and provide essential nutrients, tend to depend heavily on pollinators (Prescott-Allen and Prescott-Allen, 1990; Roubik, 1995). Seven of the nine crops that provide at least 50 percent of the vitamin C available to the human diet globally depend partially or entirely on animal pollination for the production of fruits or seeds (oranges, cabbages, green peppers, tomatoes, melons, tangerines, watermelon) (Free, 1993; McGregor, 1976).

Animal-pollinated crops also tend to have greater economic value than do crops that are not animal-pollinated, and they provide relatively more income to farmers per unit of land. Moreover, several oilseed crops depend on pollinators, and bee pollination is required to produce the seeds of major forage and hay crops, such as alfalfa and clover, that feed the animals that supply meat and dairy products. Thus, despite the fact that they do not provide the bulk of the human diet, animal-pollinated plants contribute importantly to the quality of the human diet. Although estimates of the proportion of the human diet that is attributable to animal pollination are occasionally attempted and frequently cited (for example, McGregor's 1976 estimate that one-third of the human diet can be traced directly or indirectly to animal pollination), the proportion likely varies among countries and regions and depends on dietary preferences, seasonal availability, cultural practices, and economic status of consumers.

3.8.1. Major Crops

Plant species grown as agricultural commodities display a wide variety of breeding systems. Some readily self-pollinate; others are dioecious, monoecious, or self-incompatible and require animal pollination (Free, 1993; McGregor, 1976; Table 3.1). Dependence on pollinators also varies greatly among crops. In the United States, about 130 agricultural crops benefit from insect pollination (McGregor, 1976; see examples in Table 3.1). Most North American crops can produce some fruits and seeds without animal pollination, but pollination often increases their value through higher yields or improved quality (Klein et al., 2007). Coffee self-pollinates, but yields increase as a result of pollinator visits (Klein et al., 2003a,b; Ricketts et al., 2004; Roubik, 2002). A few U.S. crops—melons, cucurbit squashes, almond, and most pome fruits—are completely dependent on animal pollination to set fruit. Crops with many-seeded fruits, such as watermelon, pumpkin, and kiwi, have hundreds of ovules so they require many pollen grains per stigma. Thus, they are more susceptible to pollinator shortages than are those with few seeds per fruit (Free, 1993).

Table 3.1 Value of common crops that require or benefit from insect pollination

Commodity	Bearing acres	Production unit	Utilized production (thousand units)	Value of utilized production ($ thousand U.S.)
Apple	395,000	Ton	4,261.6	1,783,952
Blueberry, cultivated	41,720	Ton	94.4	221,610
Blueberry, wild	NA	Ton	40.2	28,540
Cranberry	39,600	Ton	309.2	208,025
Cherry, sweet	73,940	Ton	246.8	344,471

Cucumber, fresh	58,600	Hundred weight	55,000.0	187,391
Melon, cantaloupe	86,000	Hundredweight	22,107.0	371,721
Melon, watermelon	149,600	Hundredweight	38,208.0	342,918
Nut, almond	550,000	Ton	833.3	1,600,144
Peach	145,530	Ton	1,205.1	454,532
Pear	64,150	Ton	922.5	270,385
Squash, all	50,700	Hundredweight	7,685.0	197,020

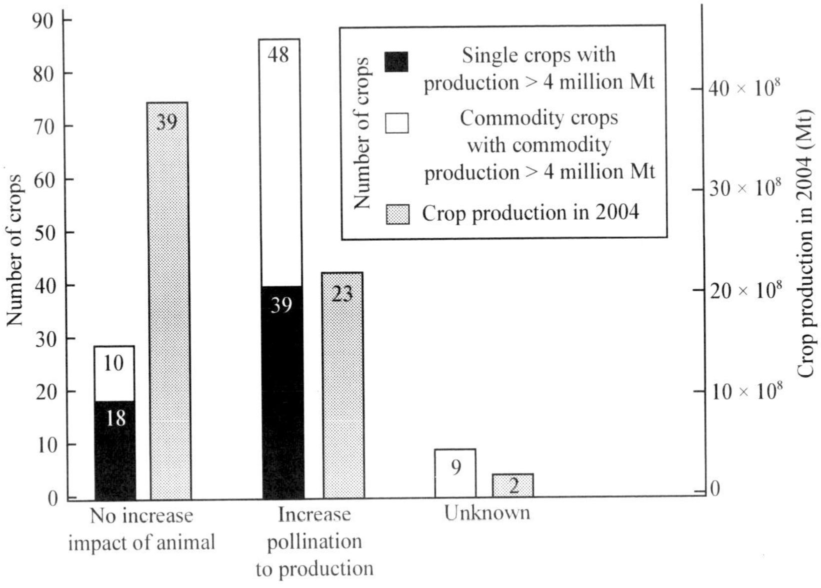

Figure 3.2 Relative importance of animal pollination for the leading global crops and commodities used for human food and selected by their annual production in 2004.

We considered crops and commodities with an annual production greater than 4000000 metric tons (Mt) as these comprise 99% of the 2004 total crop production listed for human food. The number of crops and the production are listed according to their production increase with pollinators (see electronic supplementary material 1 for details). Single crops and commodity crops in NES* commodities are separated. The category 'unknown' includes only commodity crops for the number of crops while the 'unknown' production is the production of the leading commodities, as the production value of each commodity crop is not known. Crops in the 'increase' category could be classified into three sub-categories with the following number of species and total production figure for the individual crops: production increase with pollinators for plant parts that we consume (fruits and/or seeds: 26 crops with 12108Mt = 55%); increase in seed production with pollinators to produce the vegetative parts that we consume (six crops with 2108Mt = 9%); and increase in seed production with animals for breeding alone, as the plants reproduce vegetatively and we consume the vegetative parts (seven crops with 8108 Mt = 36%). NES* is an abbreviation for not elsewhere specified; leading commodities are fresh vegetables NES, fresh fruits NES, fresh tropical fruits NES, roots and tubers NES and pulses NES. Commodity crops are included based on a questionnaire that countries fill out to include important crops for the world market which are not listed as single crops.

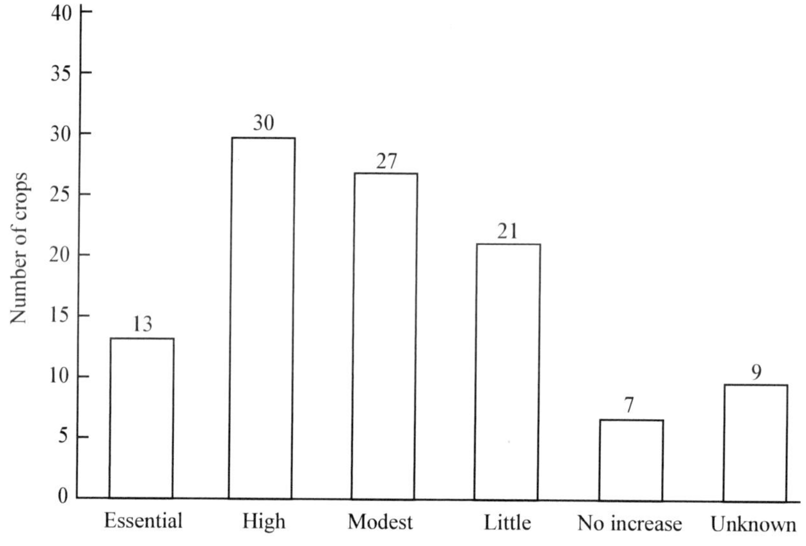

Figure 3.3 Level of dependence on animal-mediated pollination. The selected crops are those included directly in the production list published by the FAO for 2004 (FAOSTAT 2005).

We further included commodity crops for which the production was pooled in commodities with an annual 2004 commodity production greater than 4000000 metric tons (Mt). Only crops that produce fruits or seeds for direct human use as food were considered. We did not include: (i) crops for which seeds are only used for breeding or to grow vegetable parts for direct human use or for forage, and (ii) crops known to be only wind-pollinated, passively self-pollinated or reproduced vegetatively. Essential, pollinators essential for most varieties (production reduction by 90% more, comparing experiments with and without animal pollinators); high, animal pollinators are extreme (40 to less than 90% reduction); modest, animal pollinators are clearly beneficial (10 to less than 40% reduction); little, some evidence suggests that animal pollinators are beneficial (greater than 0 to less than 10% reduction); no increase, no production increase with animal-mediated pollination; unknown, empirical studies are missing.

The pollination of plants, essential in the perpetuation of most species, is so unobtrusively accomplished that it is often overlooked." Although pollinators are critical to many Agroecosystems, their importance varies not only among crop species but even among varieties of the same crop. In addition, wild-pollinator abundances can fluctuate dramatically among locations and over time (Kremen et al., 2002). Reliable data on the specific contributions of pollinators to crop yields are often lacking, and obtaining this information requires large-scale,

Table 3.2 Pollinator and pollination limitations in crop plants in response to land-use and landscape changes. (Significance $*p < 0.05$; $**p < 0.01$; $***p < 0.001$.)

Species name (common crop name)	Land-use and landscape variable	Pollination variable and significance level of reduction	Reference
Annona squamosa × *A. cherimola* (sugar apple)	Comparison of sites near and far from forest fragments	Pollinator diversity*** (fruit set reduction with Pollinator exclusion***)	Blanche & Cunningham (2005)
Brassica napus and *B. rapa* (turnip rape, canola and oilseed rape)	Comparison of organic, conventional and genetically modified (GM) fields	Number of seeds per silique from a flower sample***	Morandin & Winston (2005)
	Proportional area of uncultivated land around fields within a 750 m radius	Number of seeds per silique from a flower sample*	Morandin & Winston (2006)
Citrullus lanatus (watermelon)	Comparison of organic versus conventional fields	Number of pollen grains/stigma, n.s.	Kremen et al. (2002, 2004)
	Proportional area of oak woodland and chaparral habitat	Number of pollen grains/ stigma***	
Citrus paradisi (grapefruit)	Distance from forest	Number of pollen grains/stigma* number of pollen tubes/stigma*	Chacoff & Aizen (2006)
Coffea arabica (coffee)	Coffee plants near, intermediate and far from forest fragments	Number of pollen grains/ stigma***, fruit set*, seed mass**	Ricketts (2004) and Ricketts et al. (2004)
	Distance from forest	Fruit set**	Klein et al. (2003a)
	Plant diversity	Fruit set**	
	Coffee monocultures versus agroforestry	Fruit set*	De Marco & Coelho (2004)
	Comparison sites near and far from forest fragments	Fruit set*	
Coffea canephora (coffee)	Distance from forest	Fruit set**	Klein et al. (2003b)
Dimocarpus longan (longan fruit)	Comparison sites near and far from forest fragments	Number of fruits per centimetre panicle*	
Helianthus annuus (sunflower)	Proportional area of natural habitat	Wild bee diversity and abundance*** (estimated increase in seed set via single visit studies)	Greenleaf & Kremen (2006)
	Organic versus conventional farm management	Wild bee diversity and abundance, n.s.	

(Cont.)

Species name (common crop name)	Land-use and landscape variable	Pollination variable and significance level of reduction	Reference
Lycopersicon esculentum (tomato)	Distance to natural habitat	*Bombus vosnesenskii* abundance***; *Anthophora urbana* abundance, n.s. (fruit set and fruit weight reduction with pollinator exclusion for variety with exerted stigma)	Greenleaf & Kremen (2006)
Macadamia integrifolia (macadamia nut)	Percentage of eucalyptus forest surrounding orchards	*Trigona* abundance (seed set reduction with pollinator exclusion* and only *Trigona* pollinated*)	Heard & Exley (1994)
	Comparison of sites near and far from forest fragments	number of fruits/raceme*	

long-term studies. Listed in Table 3.2 are the relative area and value of several major crops known to benefit from pollinators. Because managed honeybees are often used to pollinate them, the extent to which managed and unmanaged pollinators provide adequate pollination service for optimal yields remains for the most part undetermined.

The adequacy of pollination services provided by managed and wild pollinators, including feral honeybees, can vary widely and is poorly documented for many crops (Free, 1993). Greenleaf and Kremen (2006) studied field-grown tomatoes for which growers did not import honeybee colonies. They found that native bees (such as *Anthophora* spp. and *Bombus* spp.) were buzz-pollinating the crops. Although wild species are either known or likely to be important pollinators in many crop systems, the relative contributions of different groups of pollinators to crop yields are rarely defined. With such a paucity of relevant research, it is useful to review evidence from a sample of well-studied crops to evaluate the availability and importance of wild pollinators. For these exemplars, pollinator shortages often can be overcome by providing sufficient numbers of managed pollinators, especially honeybees, although in a few cases the crop production depends completely on specific pollinators (for example, midges or bats).

3.8.2 Watermelon

Watermelon (*Citrullus lanatus*) is grown in several parts of the United States and Mexico. U.S. production in 2005 was worth $410 million. Because watermelon has separate male and female flowers, animal pollination is critical. Watermelon pollination requirements are high: 500–1,000 pollen grains must be deposited on stigmas to produce marketable fruits (McGregor, 1976), so the plant relies on multiple insect visits to each female flower for successful pollination (Stanghellini et al., 1997). In most areas of cultivation within North America, pollination is provided by managed honeybees and a diverse set of native bee species (Kremen et al., 2002).

In the central valley of California, Kremen and colleagues (2002) measured pollination at 14 watermelon farms that differed in degree of isolation from natural habitat. At farms near natural habitat, the native bee community was sufficient to provide adequate pollination. At farms far from natural habitat, however, average pollen deposition by native bees alone was insufficient. The inadequacy was the result of a reduction in overall pollinator abundance and of local extinction of the most efficient pollinators (Kremen et al., 2004; Larsen et al., 2005).

3.8.3 Blueberry

Blueberries (*Vaccinium* spp.) grow in several Canadian provinces and in several U.S. states. High-bush blueberries are actively cultivated; low-bush blueberries are harvested from semicultivated wild plants. In the United States, about 44,000 acres (17,800 hectares) of high-bush blueberries are planted to produce a crop worth more than $275 million annually. Like many other species in their family, blueberries are buzz-pollinated, primarily by Bumblebees (Buchmann, 1983; Free, 1993). Introduced honeybees do not buzz-pollinate, so high-bush blueberries depend largely on native wild bees for pollination (Buchmann, 1983; Free, 1993).

Several research groups have investigated the dependence of blueberry production on wild bees. Cane and Payne (1988), for example, listed four species of wild bumblebee (*Bombus* spp.) and the anthophorine *Habropoda laboriosa* that buzz-pollinate high-bush blueberry fields in Alabama and Georgia. Honeybees from colonies at field borders were found to carry little or no blueberry pollen. In New Brunswick, Canada, pesticide applications in nearby forests were followed by devastating reductions in native populations of bumblebees, andrenid bees, and halictid bees, which might have contributed to a severe drop in low-bush blueberry production (Kevan, 1975b). When a less harmful insecticide was introduced, native bees appeared to recover (Kevan et al., 1997), and production returned to normal (Kevan and Plowright, 1995). These findings highlight both the dependence of the crop on wild pollinators and the interconnectedness of contiguous natural and cultivated habitats.

3.8.4 Sunflower

Sunflowers are grown throughout the world for oil and confection seed, and in 2005, 2.7 million acres were under production in North America, mostly in the United States. Sunflower, *Helianthus annuus*, is a widely distributed native plant in North America, visited by many of its native pollinator species, including some bee species that specialize on sunflower pollen (for example, *Svastra obliqua*). Unlike the sunflower, many crops are grown outside their native ranges and therefore are visited by novel pollinators. The sunflower head consists of many individual florets, each of which passes first through a male stage and then a female stage (Free, 1993). Although the plant is self-compatible, bees often increase sunflower seed set by moving pollen from male-stage florets to female-stage florets within the same flower head and through cross-pollination between individual plants (Greenleaf, 2005; Greenleaf and Kremen, 2006). Most sunflower varieties are sold as hybrid seeds, which require cross-pollination between pollen-producing lines and those that lack pollen. The hybrid sunflower seed production industry is worth $26 million annually in the United States (Greenleaf, 2005), and more than 90 percent of seed production occurs in northern California.

Several studies have documented improvements in sunflower seed production as a result of insect pollination, although the degree of improvement varies among cultivars. For example, seed set on flower heads increased from 1 to 35 percent when bees provided self-pollination, transferring pollen among florets; when bees moved pollen between florets on different sunflower heads, seed set increased to 63 percent (Free, 1993). Although managed honeybee hives are stocked in hybrid sunflower fields in the United States, non-Apis bees also pollinate hybrid sunflowers and might do so far more efficiently than honeybees (Greenleaf, 2005; Parker, 1981). Non-Apis bee interactions with honeybees increase the pollination efficiency on hybrid sunflower by as much as five times, by disturbing them as they forage on florets and causing them to move among plants more often and cross-pollinate plants more effectively (Greenleaf, 2005). From those findings, Greenleaf (2005) calculated that the direct pollination provided by wild bees nationally to the hybrid sunflower seed industry was worth $1.9 million, the interspecific interaction between wild and honeybees was worth $10.4 million, and the direct contribution of honeybees was worth $13.8 million. Both proximity to natural habitat and crop rotation practices affect the amount of sunflower pollination provided by wild bees (Greenleaf, 2005; Greenleaf and Kremen, 2006).

3.8.5 Canola

Canola is Canada's fourth largest crop in terms of area under cultivation, and annual exports are currently worth more than $2 billion Canadian (Morandin and Winston, 2005). The two species in the mustard family grown to produce canola oil, *Brassica napus* and *B. rapa*, differ in dependence on pollinators. *B. napus* is self-compatible, but some cultivars produce better yield with cross-pollination (Kevan and Eisikowitch, 1990). In contrast, B. rapa is self-incompatible and therefore requires pollinators for seed set. Although managed honeybees are used, many native species also pollinate canola crops (Morandin and Winston, 2005).

Morandin and Winston (2005) studied pollination of self-incompatible canola in Alberta, Canada, and compared results from fields with different management regimes. They reported that farms with more wild bees also showed smaller pollination deficits, which were defined as the extent to which yields were depressed because of insufficient pollination. The farms with the fewest bees were those planted in genetically modified, herbicide resistant plants. Because those fields were treated heavily with herbicides, they had a lower diversity and abundance of flowering weeds, which could account for their reduced capacity to support wild bees. Using those results, Morandin and Winston (2006) calculated that profits lost by retiring up to 30 percent of the field area from production would be more than offset by increased yield that would result from better pollination by the more diverse and abundant pollinators found in fields within 750 meters of uncultivated areas.

3.8.6 Coffee

Coffee (*Coffea arabica* and *C. robusta*) is grown extensively in Mexico and is one of the developing world's most valuable export commodities. Worldwide, 11 million hectares are planted in coffee, and the industry employs 25 million people (O'Brien and Kinnaird, 2003). *C. arabica*, the species thought to produce the highest-quality coffee, self-pollinates, but yields increase 15 to 50 percent when flowers are visited by bees (Free, 1993; Klein et al., 2003b; Roubik, 2002). Despite that, few coffee farmers recognize pollination as an issue for their crops.

Several recent studies have investigated the importance of wild bees to coffee production. Ricketts and colleagues (Ricketts, 2004; Ricketts et al., 2004) reported that in Costa Rica a diverse community of wild bees pollinates coffee. The pollinators were mainly wild Africanized honeybees and 10 species of social stingless bee (tribe Meliponini). The species depend on forest habitat for nest sites and food resources but flew into nearby coffee fields when they were in flower. Coffee fields near forest patches received more visits by a more diverse set of bees and had 20 percent higher yields than did fields more than 1 km away from a forest. Klein and colleagues (Klein et al., 2002) also noted a diverse community of wild coffee pollinators in Indonesia. They reported that coffee yields increased with increased diversity of pollinators and that pollinator diversity increased near forest remnants and in less intensive farms managed with a more diverse canopy of shade trees. Although these studies were performed outside North America, their findings are informative for coffee production in Mexico.

3.8.7 Cacao

Chocolate is made from the seeds of the cacao tree (*Theobroma cacao*), which is native to the western Amazon basin, but now grown in many humid tropical regions, including Mexico (Clay, 2004). Cacao trees can be either self-compatible or self-incompatible, but even self-compatible trees require insect visits to move pollen from anthers to stigmas and pollinate the flowers. Honeybees do not appear to be effective pollenators (Free, 1993). Although the identity of the principal pollinators of cacao eluded growers and scientists for many years, midges in the family Ceratopogonidae have been identified as the most important group (Free, 1993; Young, 1985). Many other wild insects also have been observed visiting cacao (bees in the subfamilies Halictinae, Meliponinae), but their pollinating efficiency is either low or undetermined (Young, 1985).

There is substantial evidence that farm management practices influence the amount and quality of habitat for midges within and around cacao farms and therefore can affect yields (Free, 1993). Midges prefer shady, moist conditions; their eggs are laid and larvae develop in rotting wood and leaves on the ground (Free, 1993). Young (1982, 1986) showed that adding pieces of rotting wood to the leaf litter within cacao plantations increased the abundance of several midge species and that cacao plants near shade trees or rotting material often had more fruits than did trees that were farther away.

3.8.8 Agave

Tequila—one of several alcoholic beverages derived from agave is produced exclusively from *Agave tequilana* var. *azul* in select regions of Mexico. After 1989, and at least until 2002, the tequila industry faced major agave shortages caused by several diseases that affected about 30 percent of the plants at different stages of development (Jiménez-Hidalgo et al., 2004).

Although many species of agave, including *A. tequilana*, are naturally pollinated by bats and other animals (Rocha et al., 2005), the industry uses only asexually reproduced plants to replenish the producing population. This practice has prevented outcrossing of individuals and resulted in loss of genetic diversity compared to other agaves (Dalton, 2005).

3.8.9 Other Crops

Ecologists have investigated a variety of crops, both in North America and elsewhere, to determine the degree of pollination services provided by wild bees. Some studies demonstrate extensive pollination by wild pollinators (Roubik, 1995), whereas others find little support for this assertion (reviewed in Free, 1993). Additional studies on pollination of muskmelon (cantaloupe), almond, and tomato in California, New Jersey, and elsewhere are under way (Greenleaf and Kremen, 2006). The growing body of research will contribute to elucidating the degree to which wild pollinators benefit agricultural production and, more important, the circumstances and management regimes that affect pollination of crops by unmanaged wild pollinators.

3.9 CONCLUSION

A high diversity of wild pollinators contributes to increased stability in pollination, even when managed bees are present in high numbers. Crop yields depend on both wild and managed species. The western honeybee is the most widespread managed pollinator in the world, producing an estimated 1.6 million tonnes of honey annually. Decreasing exposure of pollinators to pesticides by reducing their usage, seeking alternative forms of pest control, and adopting a range of specific application practices, including technologies to reduce pesticide drift; Improving managed bee husbandry for pathogen control, coupled with better regulation of trade and use of commercial pollinators. We also emphasize the collection of data for understanding the effects of spatial and temporal pollinator resource availability and for interaction effects between honeybees and other bee species for crop pollination to recommend future management applications. Therefore, we urgently need more research in crop pollination along with better coordination of the research efforts at the community level in different producing areas to help sustain production of the diverse crops that nourish humanity.

REFERENCES

- Adamson, NL (2011) An Assessment of Non-Apis Bees as Fruit and Vegetable Crop Pollinators in Southwest Virginia (PDF) (Ph.D. thesis)
- Blanche, R and Cunningham, SA (2005) Rainforest provides pollinating beetles for atemoya crops. J. Econ. Entomol., 98: 1193–1201
- Buchmann SL (1983) Buzz pollination in angiosperms. Pp. 73–114 in Handbook of Experimental Pollination Biology, C.E. Jones and R.J. Little, (eds.) New York: Scientific and Academic Editions
- Campbell NA and JB Reece (2002) Biology (6th ed.). Pearson Education. pp. 600–612
- Cane JH and Payne JA (1993) Regional, annual and seasonal- variation in pollinator guilds — intrinsic traits of bees (Hymenoptera, Apoidea) underlie their patterns of abudance at *Vaccinium ashei* (Ericaceae). Ann. Entomol. Soc. Am., 86: 577–588
- Chacoff NP and Aizen MA (2006) Edge effects on flower visiting insects in grapefruit plantations bordering premontane subtropical forest. J. Appl. Ecol., 43: 18–27

- Clay J (2004) World agriculture and the environment. Washington: Island Press
- Dalton R (2005) Alcohol and science: saving the agave. Nat., 438: 1070–1071
- Darwin C (1877) The different forms of flower on plants of the same species. Murray, London
- De Marco P and Coelho FM (2004) Services performed by the ecosystem: forest remnants influence agricultural cultures' pollination and production. Biodivers. Conserv., 13: 1245–1255
- Faegri K and L van der Pijl (1979) The principles of pollination ecology, 3rd edn. Pergamon, Oxford
- Free JB (1993) Insect pollination of crops. London, UK: Academic Press
- Greenleaf SA and Kremen C (2006) Wild bees enhance honeybees' pollination of hybrid sunflower. Proc. Natl Acad. Sci., 13: 890–895
- Greenleaf SS (2005) Local-scale and foraging-scale affect bee community abundances, species richness, and pollination services in Northern California. Ph.D. thesis. Princeton University.
- Heard TA and Exley E (1994) Diversity, abundance and distribution of insect visitors to macadamia flowers. Environ. Entomol., 23: 91–100
- Holmes EO (1963) *Protea mellifera*, queen of nectar producing plants. Am. Bee J., 103: 216–217
- Jimenez-Hidalgo I, Virgen G, Martinez D, Vandemark G, Alego J and V Olalde (2004) Identification and characterization of soft rot bacteria of *Agave tequilana* var *Azul*. J. Plant Pathol., 110: 317–331
- Kerner A (1878) Flowers and their unbidden guests. The translation revised, and edited by W. Ogle. Kegan, London. pp 164
- Kevan PG (1975b) Forest application of the insecticide fenitrothion and its effects on wild bee pollinators (Hymenoptera: Apoidea) of lowbush blueberries (*Vaccinium* spp.) in southern New Brunswick, Canada. Biolog. Conserv., 7: 301–309
- Kevan PG and RC Plowright (1995) Impact of pesticides on forest pollination. Pp. 607–618 in Forest Insect Pests in Canada, J.A. Armstrong and W.G.H. Ives, (eds.) Ottawa: Natural Resources Canada
- Kevan PG, Greco CF and S Belaoussoff (1997) Log-normality of biodiversity and abundance in diagnosis and measuring of ecosystemic health: pesticide stress on pollinators on blueberry heaths. J. Appl. Ecol., 34: 1122–1136
- Kevan PG and D Eisikowitch (1990) Self- and cross-pollination in canola (*Brassica napus* L. var. O.A.C. Triton) and its implication on seed germination. Euphytica, 45: 39–41
- Klein AM, Steffan-Dewenter I and Tscharntke T. (2003a) Fruit set of highland coffee increases with the diversity of pollinating bees. Proc. R. Soc. B., 270: 955–961
- Klein AM, Steffan-Dewenter I and Tscharntke T (2003b) Pollination of *Coffea cenephora* in relation to local and regional agroforestry management. J. Appl. Ecol., 40: 837–845

- Klein AM, Vaissière BE, Cane, JH., Steffan-Dewenter I, Cunningham SA, Kremen C, Tscharntke T (2007) Importance of pollinators in changing landscapes for world crops. Proc. R. Soc. B Biol. Sci., 274: 303–313
- Kremen C, Williams NM and Thorp RW (2002) Crop pollination from native bees at risk from agricultural intensification. Proc. Natl Acad. Sci., 16: 812– 816
- Kremen C, Williams NM, Bugg RL, Fay JP and Thorp RW (2004) The area requirements on an ecosystem service: crop pollination by native bee communities in California. Ecol. Lett., 7: 1109–1119
- Kuehn F (2015) Farming for native bees. Sustainable Agriculture Research & Education (SARE). Archived from the original on 30 September 2015
- Langstroth LL (1913) The hive and the honeybee. Rev. By C.P. Dadant, Dadant and Sons, Hamilton, III, pp 575
- Larsen TH, Williams N and Kremen C (2005) Extinction order and altered community structure rapidly disrupt ecosystem functioning. Ecol. Lett., 8: 538–547
- Losey, J. E. and M. Vaughan. (2006) The economic value of ecological services provided by insects. BioScience, 56 (4): 311–23
- McGregor, S E (1976) Insect pollination of cultivated crop plants. U.S.D.A. Agriculture Handbook No. 496: 93–98
- Morandin, LA and ML Winston (2005) Wild bee abundance and seed production in conventional, organic, and genetically modified canola. Ecolog. Appl., 15: 871–881
- Morandin, LA and ML Winston (2006) Pollinators provide economic incentive to preserve natural land in Agroecosystems. Agric. Ecosys. Environ., 116: 289–292
- Nichol AA (1952) The natural vegetation of Arizona. Ariz. Agr. Expt. Sta. Tech. Bul., 127: 188–230
- O'Brien TG and MF Kinnaird (2003) Caffeine and conservation. Sci., 300: 587
- Prescott-Allen R and Prescott-Allen C (1990) How many plants feed the world? Conserv. Biol., 4: 366–374
- Raghavan V (1997) Molecular embryology of flowering plants. Cambridge University Press. pp. 210–216
- Ricketts T (2004) Tropical forest fragments enhance pollinator activity in nearby coffee crops. Conserv. Biol., 18: 1262–1271
- Ricketts T, Daily GC, Ehrlich PR and Michener CD (2004) Economic value of tropical forest to coffee production. Proc. Natl Acad. Sci., 12: 579–582
- Rocha M, Valera A and LE Eguiarte (2005) Reproductive ecology of five sympatric *Agave littaea* (Agavaceae) species in Central Mexico. Am. J. Bot., 92: 330–1341
- Roubik DW (1995) Pollination of cultivated plants in the tropics. Food and agriculture organization of the United Nations, Rome, Italy. Bull. 118
- Roubik DW (2002) The value of bees to the coffee harvest. Nat., 417:708
- Runions CJ and JN Owens (1999) Sexual reproduction of interior spruce (Pinaceae). I. Pollen germination to archegonial maturation. Int. J. Plant Sci., 160(4): 631–640

- Stanghellini MS, Ambrose JT and JR Schultheis (1997) The effects of honeybee and bumblebee pollination on fruit set and abortion of cucumber and watermelon. Am. Bee J., 137: 386–391
- Todd FE and SE Mcgregor (1960) The use of honeybees in the production of crops. Ann. Rev. Ent., 5: 265–278
- Young AM (1982) Effects of shade cover and availability of midge breeding sites on pollinating midge populations and fruit-set in 2 cocoa farms. J. Appl. Ecol., 19(1): 47–63
- Young AM (1985) Pollen-collecting by stingless bees on cacao flowers. Experientia., 41(6): 760–762
- Young AM (1986) Distribution and abundance of Diptera in flypaper traps at *Theobroma cacao* L (Sterculiaceae) flowers in Costa Rican cacao plantations. J. Kansas Entomol. Soc., 59(4): 580–587

4

Supplementary Pollinator Services in Yield Optimization of Crop Plants through Management of Bees

Ujjwal Layek[1], Sourabh Bisui[2], Arijit Kundu[2], and Prakash Karmakar[2*]

ABSTRACT

Pollination (i.e., deposition of an adequate amount of viable pollen on the receptive stigma for fertilization) is essential for most agricultural plants to achieve desired yield and quality. Crop plants mainly depend on diverse insect pollinators like bees, flies, butterflies, wasps, moths, and beetles for successful pollination. Honeybees and stingless bees (Tetragonula iridipennis Smith) are potent pollinators of several crop plants due to their high floral constancy, large colony size, high flower visitation rate, and possess of recruitment behavior. Unfortunately, the abundance of these pollinators is gradually decreasing day by day in nature. Therefore, many agricultural plants are facing pollination limitations. The management of honeybees and stingless bees is one of the effective strategies to increase pollinator services. The fruit set of the selected crops increased significantly with the utilization of managed honeybee and stingless bee colonies compared to the open-pollination system. The quality of fruits (in terms of length, girth, and weight) also increased in some crops with supplementary bee pollination.

Keywords • *Crop quality* • *Crop yield* • *Honeybee* • *Pollinator* • *Stingless bee*

4.1 INTRODUCTION

Pollination is the act of transferring pollen grains from the male organ (anther) of a flower to the female organ (stigma), enabling fertilization and the production of seeds. It is helpful to generate offspring for the next generation. Pollination is an important input of crop production, comparable to any other input such as fertilizer, farm labour or pesticides (McGregor, 1976). Though, some crop plants which supply staple food to humankind (rice, wheat, maize, potato) are wind-pollinated, self-pollinated or propagated by vegetative reproduction. Still, the majority of them depend on biotic pollinators. There are around 150,000 pollinators, mainly consisting

[1]Department of Botany, Rampurhat College, Birbhum-731224, India

[2]Department of Botany & Forestry, Vidyasagar University, Midnapore- 721102, India

[*]Corresponding Author: Prakash Karmakar (E-mail address: prakashbot1973@gmail.com)

of bees, flies, butterflies, wasps, moths, and beetles (Hein, 2009). Insects are considered to be responsible for 80–85% of all pollination (Williams, 1996; Johannsmeier and Mostert, 2001), including fruits, vegetables, oilseeds, legumes and fodder (Richards, 2001). Honeybees and stingless bees visit various plant species and serve as generalist pollinators. When an individual worker bee forages, it visits the flowers of one plant species typically at a time and ignores other flower types, even if they are much closer to the target flower type. Their visitation rate (number of flowers visited per unit time) is relatively higher than other insect groups like beetles, butterflies, flies, and moths. When they visit male or hermaphrodite flowers to collect rewards, some pollen grains adhere to their hairy body parts like the thorax, abdomen, or sometimes on the head. Pollen grain holding areas may touch the receptive stigma during the same bout or subsequent bout over female or hermaphrodite flowers. Some pollen grains shed off on the stigma, and the plant achieves pollination.

We have unexpectedly little information on the actual degree of pollinator dependence for some major crops. Few crop plants depend entirely on insect pollinators for fruit and seed sets. Other crops are partially dependent on insects for their pollination and can produce more than 90% of the maximum yield without pollinators (Klein et al., 2007). Fruit numbers can be a good proxy for measuring the yield of a crop plant. In addition to the number, weight is also counted as a significant parameter to measure yields because plants allocate resources for producing fruits of variable sizes based on the number of fruits per plant and the level of pollination received (Gonzalez et al., 1998; Bashir et al., 2018; Abrol et al., 2019). Furthermore, similar levels of fruit set can also differ in total crop yield due to different fruit sizes (Bos et al., 2007). In some fruit plants, quality can be negatively correlated with quantity when the fruit load on a tree or a vine is too high (Ferguson and Watkins, 1992; Bussi et al., 2005; Serra et al., 2016). However, it may not be applicable to crops with indeterminate flowering such as oilseed rape (Bommarco et al., 2012) and soybean (Kato et al., 2015).

Both wild and managed pollinators have a significant role in crop pollination, although their relative contributions differ according to the crops' nature and their cultivation location. Management of several bee species for commercial pollination purposes began in the 19th century (Parker et al., 1987; Batra, 1994; Stubbs et al., 1994). In developing countries, farmers can rent managed bees for crop pollination purposes (Van Heemert et al., 1990). The honeybees regard as the most important commercial pollinators for monofloral crops, some seeded crops, fruits and nuts worldwide (Klein et al., 2007). Other bee species (viz., alkali bees, mason bees, leafcutter bees and Bumblebees) are also used for more specific pollination services. In some cases, non-Apis bees are more efficient pollinators than honeybees. For example, tomatoes are commercially pollinated by managed bumblebees (Velthuis and Van Doorn, 2006; Graystock et al., 2016) because they can effectively vibrate the anthers of a flower (i.e., buzz pollination) to transfer the pollen grains over stigma. The alkali bees (*Nomia melanderi* Cockerell) and leafcutter bees (*Megachile rotundata* Fabricius) more effectively pollinate alfalfa than honeybees. Recently, stingless bees have been used to pollinate greenhouse and field crops (Bisui et al., 2020; Layek et al., 2021b, 2022a). In this chapter, we have focused on the result of greater pollination success in the yield enhancement of some crop plants.

4.2 CROP DIVERSITY IN INDIA

Crop diversity is the variability in genetic and phenotypic traits of plants used in agriculture. Even within a single crop species, many varieties vary in height, branching pattern, flower characteristics, fruiting time, fruit, and seed size. Corp plants may also vary in their response to abiotic factors such as response to heat, cold, drought or their ability to resist specific diseases and pests. India shows various physiographic features concerning climate, rainfall, landform and soil types. Consequently, wide varieties of crops are grown in different parts of the country. Northern states of montane and humid subtropical climate zone are suitable for cultivating fruit trees like apple, cherry and pear. In addition, several food crops (viz., rice, maize, wheat, millets and pulses) are also cultivated. Southern states revealed tropical wet and semi-arid climates and cultivated the crops like rice, cotton, groundnut, sunflower and chickpeas. Uttar Pradesh and certain eastern states like West Bengal, Bihar, and Assam experience 1000–2000 mm rainfall annually and grow crops such as rice, sugarcane, jute and tea. Many regions of the western side of India (Gujarat, Rajasthan, South Punjab) experience less than 500 mm of rainfall annually and farming systems are restricted to cultivating crops (viz., jowar, bajra and peas) that can withstand drought conditions (Aggarwal, 2008). Around 1500 wild edible plant species are widely exploited by native tribes. These include 145 species of roots and tubers, 521 leafy vegetables, 101 buds and flowers, 647 fruits and 118 seeds and nuts (NAAS, 1998). In addition, Indian agriculture has been enriched by a continuous stream of the introduction of new crop plants. Among the introduced varieties, some good examples are cereals (wheat, barley, oats, maize), pulses (chickpea, French bean, peas), vegetables (potato, onion, cauliflower, cabbage, carrot, tomato), fruits (apple, pear, cherry, peach, apricot), oilseeds (soybean, sunflower) and fiber plants (cotton). Thus, both indigenous and better adapted exotic sets of materials constitute a well-balanced matrix of crop diversity in India.

Table 4.1 Season-wise crop types in India

Parameter	Crop types		
	Kharif	**Rabi**	**Zaid**
Sowing time	June-July	October-November	March-June
Harvesting time	September-October	April-May	June-July
Climate	Hot, rainy weather	Warm climate for seed germination; cold climate for growth	Warm, dry weather
Example	Bajra, cotton, jute, maize, rice	Gram, mustard, onion, pea, potato, sunflower, wheat	Bitter gourd, cucumber, moong dal, pumpkin, watermelon

Crop diversity depends not only on different geographic locations but also on seasonal variations in weather conditions. Several crops were cultivated on a single agricultural land during different periods. According to seasons, the cultivated crops can be divided into three types, viz., rabi, kharif and zaid (Table 4.1). The crop plants' reproductive ecology (flowering phenology, pollination ecology and seed ecology) varied in each of the above classes. There is no unique floral character by which we can demark the seasonal crop categories.

4.3 REPRODUCTIVE STRATEGIES OF CROP PLANTS

The diversity of plant breeding systems is well represented among the cultivated crop plants. Natural parthenocarpic fruit sets are less common and occur only in a few crops like banana, figs and pineapple. However, it can be induced with the application of hormones to young inflorescences (Nitsch, 1952; Schwabe and Mills, 1981; Chowdhury et al., 2007) or even by genetically modification of several crops like cucumber (Yin et al., 2006), eggplant (Acciarri et al., 2002; Liming et al., 2016) and tomato (Carmi et al., 2003; Hazra et al., 2010). From the adaptive point of view, the parthenocarpic fruit development is an intriguing phenomenon because seedless fruits are costly to the mother plant and do not contribute to the production of offspring. Concerning fruit and seed sets of different crops, they are dependent on successful pollination. For pollination purposes, a few crop plants (viz., black cumin, common bean and peanut) were independent of abiotic and biotic vectors. Most of the crop plants exhibit an allogamous mode of pollination. Some of them achieve such a kind of breeding system by avoiding selfing through spatial (herkogamy, monoecy or dioecy) or temporal (dichogamy: protandry, protogyny) separation between male and female reproductive structures or self-incompatibility. Though, maximum crop plants retained their self-fertility. A few crop plants (e.g., cashew, eggplant) exhibit andromonoecy (Table 4.2). It has been suggested that andromonoecy evolved from hermaphroditism by losing the female reproductive structures, which is the first step in the evolution of a plant breeding system towards monoecy, androdioecy and dioecy (Bertin, 1982; Zhang and Tan, 2009). In these cases, the male flowers increase their floral display and attractiveness and provide more rewards to pollinators (Charlesworth and Morgan, 1991; Barrett, 2002). Therefore, the presence of staminate flowers can increase the outcrossing rate and female fitness of individuals.

For successful pollination, crop plants depend on the wind (i.e., anemophilous plants), insects (i.e., entomophilous plants) or both wind and insects (i.e., amphiphilous plants). Cereals (barley, corn, oats, rice and wheat) are anemophilous. Their flowers are small, drab, unscented and inconspicuous. Anthers and stigma often project from the flower to be exposed to the wind. Wind pollination is very wasteful concerning the loss of pollen grains. Major crop plants are entomophilous. Several floral attributes like flower architecture, flower colour, and floral scent act as attractants and floral resources perform as rewards for the incoming visitors. Only a few crop plants, especially members of Arecaceae, are amphiphilous. Their flowers are small-sized, compactly arranged and produce enormous pollen grains to ensure wind pollination. They also have nectar, which is attractive to diverse insect groups, thereby perceiving additional pollinator services. However, the degree of pollination achievement is debatable because male flowers of these plants remain more attractive than female flowers.

4.4 FLORAL VISITORS OF CROP PLANTS

Floral visitors of crop plants are diverse, ranging from small-sized insects to large birds, bats, and squirrels. Among some selected crop plants, the most diverse are the members of Hymenoptera, especially bees (Table 4.3). The most common visitors are honeybees, solitary

Table 4.2 Reproductive traits of some crop plants

Crop species	Floral biology				Resources to visitor	Pollination status
	Reproductive system	Size	Shape	Color		
Anacardium occidentale	andromonoecy	medium	funnel	cream	mainly nectar	entomophilous
Brassica juncea	hermaphrodite	medium	cruciform	yellow	nectar, pollen	entomophilous
Capsicum annuum	hermaphrodite (a certain degree of dichogamy)	small	dish	white	nectar, pollen	entomophilous
Citrullus lanatus	monoecy	medium	saucer	yellow	nectar, pollen	entomophilous
Citrus × aurantiifolia	hermaphrodite	medium	stellate	yellow	nectar, pollen	entomophilous
Coccinia grandis	dioecious	large	funnel	white	nectar, pollen	entomophilous
Cocos nucifera	andromonoecy	small	stellate	cream	nectar, pollen	amphiphilous
Coriandrum sativum	hermaphrodite	small	–	pinkish white	nectar, pollen	entomophilous
Foeniculum vulgare	hermaphrodite	small	funnel	yellow	nectar, pollen	entomophilous
Mangifera indica	hermaphrodite	small	stellate	whitish yellow	nectar	entomophilous
Momordica charantia	monoecy	medium	saucer	yellow	nectar, pollen	entomophilous
Oryza sativa	hermaphrodite	small	–	greenish white	–	anemophilous
Phaseolus vulgare	hermaphrodite	medium	pea-shaped	cream	nectar	entomophilous
Sesamum indicum	hermaphrodite	large	gullet	white	nectar, pollen	entomophilous
Solanum lycopersicum	hermaphrodite	medium	dish	yellow	nectar, pollen	entomophilous
Solanum melongena	andromonoecy	medium	dish	purple	pollen	entomophilous

Table 4.3 Floral visitors of some crop plants

Floral visitors	Crop plants										
	Brassica juncea	*Citrus × aurantiifolia*	*Citrus × limon*	*Coccinia grandis*	*Coriandrum sativum*	*Foeniculum vulgare*	*Mangifera indica*	*Momordica charantia*	*Phaseolus vulgare*	*Sesamum indicum*	*Solanum melongena*
Coleoptera											
Coccinella septumpunctata	+	–	–	–	+	+	–	–	–	–	–
Raphidopalpa foveicollis	–	–	–	–	+	+	–	+	–	–	–
Diptera											
Chrysomya bezziana	+	–	–	–	+	+	–	+	–	–	–
Episyrphus balteatus	+	+	+	–	+	+	+	+	–	–	–
Eristalinus tabanoides	+	–	–	–	–	–	–	+	–	–	–
Eristalis tenax	–	–	–	–	+	+	+	–	–	–	–
Helophilus sp.	–	+	–	–	–	–	–	–	–	–	–
Ischiodon scuttellaris	–	–	–	–	–	–	–	+	–	–	–
Metasyrphus corollae	+	–	–	–	–	–	–	–	–	–	–
Sarcophaga sp.	+	–	–	–	–	–	–	–	–	–	–
Stomorhina discolor	+	+	+	–	–	–	–	–	–	–	–
Stomorhina obsoleta	+	+	+	–	–	–	–	–	–	–	–

Floral visitors	Crop plants										
	Brassica juncea	*Citrus × aurantiifolia*	*Citrus × limon*	*Coccinia grandis*	*Coriandrum sativum*	*Foeniculum vulgare*	*Mangifera indica*	*Momordica charantia*	*Phaseolus vulgare*	*Sesamum indicum*	*Solanum melongena*
Heteroptera											
Leptocoris augar	–	–	–	–	+	–	–	–	–	–	–
Hymenoptera											
Amegilla zonata	–	–	–	–	–	–	–	–	–	+	+
Andrena leaena	+	–	–	–	–	–	–	–	–	–	–
Andrena ilerda	+	–	–	–	–	–	–	–	–	–	–
Apis cerana	+	+	+	+	+	+	+	+	+	+	+
Apis dorsata	+	+	+	+	+	+	+	+	+	+	+
Apis florea	+	+	+	+	+	+	+	+	+	+	+
Apis mellifera	+	+	+	+	+	+	+	+	+	+	+
Camponotus compressus	–	–	–	–	–	–	–	–	–	+	–
Camponotus sericeus	–	–	–	–	+	+	–	+	+	–	+
Ceratina sexmaculata	+	–	–	–	+	+	–	–	–	–	–
Delta sp.	+	–	–	–	–	–	–	–	–	–	–
Halictus sp.	+	+	+	+	–	–	–	+	–	+	+
Lasioglossum sp.	–	–	–	–	–	–	–	+	–	+	–
Nomia curvipes	–	–	+	–	–	–	–	–	–	+	–
Polistes hebraceus	+	–	–	–	+	+	–	–	–	–	–
Sphex sp.	+	–	–	–	–	–	–	–	–	–	–

Floral visitors	Crop plants										
	Brassica juncea	*Citrus × aurantiifolia*	*Citrus × limon*	*Coccinia grandis*	*Coriandrum sativum*	*Foeniculum vulgare*	*Mangifera indica*	*Momordica charantia*	*Phaseolus vulgare*	*Sesamum indicum*	*Solanum melongena*
Tetragonula iridipennis	+	+	+	+	+	+	+	+	+	+	+
Vespa orientalis	–	–	–	–	–	+	–	+	–	–	–
Xylocopa fenestrata	+	–	+	–	–	–	–	–	+	+	+
Lepidoptera											
Amata bicincta	+	–	–	–	–	–	–	–	–	–	–
Borbo cinnara	–	–	–	–	–	–	–	+	–	–	–
Danaus chrysippus	+	–	–	–	+	–	–	+	–	–	–
Eurema hecabe	–	–	–	–	–	–	–	+	–	–	–
Hypolimnas sp.			+								
Pelopidas mathias	–	–	–	–	–	–	–	+	–	–	–
Pieris rapae	+	–	–	–	+	–	–	–	–	–	–
Precis iphita	–	–	–	–	–	–	–	+	–	–	–

bees and stingless bees (*Tetragonula iridipennis* Smith). Behaviors concerning floral fidelity, the most important pollinators are honeybees (Free, 1963; Layek et al., 2015; Martin and Farina, 2016) and stingless bees (Layek and Karmakar, 2018). These bees legitimately foraged on several crop plants (Fig. 4.1). Their flower visitation rates also remain higher than other visitors like butterflies, flies and moths. Occasionally a few honeybee foragers are also illegitimately visiting some crop plants like *Sesamum indicum* L., acting as nectar robbers (Layek et al., 2022b). Their robbing behavior depends on floral size, architecture (especially for tube, gullet and campanulate flower), resource availability of the flower, visiting time and colony demands.

Fig. 4.1 Some crop plants visited by managed bees: (a) *Brassica juncea* (by *Tetragonula iridipennis*); (b) *Citrullus lanatus* (by *Apis cerana*); (c) *Coccinia grandis* (by *Apis cerana*); (d) *Coriandrum sativum* (by *Tetragonula iridipennis*); (e) *Cucumis sativus* (by *Tetragonula iridipennis*); (f) *Foeniculum vulgare* (by *Tetragonula iridipennis*); (g) *Helianthus annuus* (by *Apis mellifera*); (h) *Luffa aegyptica* (by *Tetragonula iridipennis*); (i) *Momordica charantia* (by *Tetragonula iridipennis*); (j) *Papaver somniferum* (by *Apis mellifera*); (k) *Punica granatum* (by *Tetragonula iridipennis*); and (l) *Sesamum indicum* (by *Apis mellifera*). Scale bar = 10 mm

4.5 POLLINATION DEFICIENCY

Habitat fragmentation and adverse climatic conditions lead to a decline in the wild pollinator populations. In addition, the number of wild migratory honeybee (*Apis dorsata* Fabricius and

Apis florea Fabricius) colonies significantly reduced during monsoon and autumn in some parts of India, including West Bengal (Layek and Karmakar, 2020). As a result, most entomophilous crop plants show significant pollination limitations in open conditions (Bisui et al., 2020; Layek et al., 2020, 2021b, 2022a). Pollinator density and crop yield are much correlated in the case of entomophilous and amphiphilous crop plants. The decline of wild pollinator populations and diversity has resulted in declining crop yields in many parts of the world (Garibaldi et al., 2011; Partap and Ya, 2012). To overcome the problem of pollination deficit, different alternatives have been developed. Some farmers, such as passion fruit growers in Brazil and apple growers in Sichuan-China, pollinate their crops by hand (Klein et al., 2007; Partap and Ya, 2012). Tomato growers use hand-held electrical shakers in greenhouses (Bispo dos Santos et al., 2009). Some farmers use pollen sprays (Bahadur et al., 2015). All the methods mentioned above are much more difficult and expensive. Therefore, in modern agricultural practices, farmers are intended to use managed bees to ensure crop pollination.

4.6 YIELD ENHANCEMENT OF CROP PLANTS BY INCREASING POLLINATOR SERVICES

Pollinator's abundance can increase by installing managed bee colonies near the crop fields or by utilising co-blooming bee-friendly plants (Layek et al., 2021a). Higher pollination services result in the enhancement of crop yield. Layek et al. (2021a) selected three types of cashew vegetations (i.e., sparsely distributed cashew individuals, densely distributed individuals, and cashew individuals associated with co-blooming bee-friendly plants) for their study. The abundance of wild pollinators on cashew flowers significantly varied among these cashew vegetations. The abundance of the primary pollinator (i.e., *Apis dorsata*) was comparatively lower (1.45 ± 1.05 visitors/panicle/30 min) in sparsely distributed individuals and higher (2.80 ± 1.68 visitors/panicle/30 min) in densely distributed individuals. The higher pollinators' density within the densely distributed cashew individuals associated with co-blooming bee-friendly plants resulted in a higher fruit set and a greater yield of cashew nuts than the open-pollination treatment (Table 4.4). The spices coriander (*Coriendrum sativum* L.) and fennel (*Foeniculum vulgare* Mill.) are partially dependent on insect pollinators for their fruit set. In the absence of a biotic pollinator, the yield of coriander and fennel was very low (coriander: 9.83 ± 0.52 gm/m^2 area; fennel: 379.47 ± 13.54 kg/ha). The yield increased up to 72.15% in coriander and 19.31% in fennel (Layek et al., 2022a), employing managed stingless bee colonies. Fruit quality (in terms of fruit weight) also significantly differed among different pollination treatments. The weight of fruits was lower in the pollinator exclusion treatment than in the open-pollination and open-pollination with added managed bee colonies. The members of Cucurbitaceae are obligatory dependent on biotic pollinators as most of them are monoecious. In bitter gourd (*Momordica charantia* L.), both quantity and quality of fruits are dependent on pollinator services. The pollination success rate increased significantly with the installation of the managed stingless bee colonies. Accordingly, the percentage of fruit set increased (Bisui et al., 2020). The quality of fruits (length, girth, and weight) remained better in hand-pollination and managed bee pollination treatments where more pollen deposition on stigma occurs. The fruit set of watermelon differed significantly among the five pollination systems, viz., manual cross-pollination, manual selfing, open-pollination, open-pollination

Table 4.4 Pollinator services affect the yield of crop plants

Crop plant	Measure	Pollinator exclusion	Hand–pollination	Open–pollination	Open–pollination with stingless bee colonies	Open–pollination with western honeybee colonies
Anacardium occidentale	Fruit set (%)	0	$72^{a} \pm 13.04$	$28^{c} \pm 8.37$	–	$58^{b} \pm 10.33$
	Yield (matured fruits/ panicle)	0	$8.10^{a} \pm 2.47$	$4.15^{c} \pm 2.68$	–	$6.20^{b} \pm 2.44$
	Weight of fleshy thalamus (g)	–	39.45 ± 11.22	38.08 ± 10.04	–	38.64 ± 10.39
	Weight of matured nut (g)	–	4.38 ± 0.70	4.29 ± 0.62	–	4.28 ± 0.64
Citrullus lanatus	Fruit set (%)	0	$88^{a} \pm 4.47$	$64^{c} \pm 6.52$	$78^{bc} \pm 10.37$	$81^{ab} \pm 8.22$
	Fruit quality (after 40 days of flower open)					
	Length (mm)	–	$28.07^{a} \pm 1.99$	$24.48^{c} \pm 3.26$	$26.04^{b} \pm 2.24$	$26.44^{b} \pm 2.04$
	Girth (mm)	–	$70.10^{a} \pm 3.55$	$63.20^{c} \pm 5.81$	$66.37^{b} \pm 4.19$	$67.26^{b} \pm 3.73$
	Weight (g)	–	$3.72^{a} \pm 0.26$	$3.01^{c} \pm 0.40$	$3.36^{b} \pm 0.40$	$3.48^{b} \pm 0.38$
Coriandrum sativum	Yield (gm/ m^2 area)	$9.84^{c} \pm 0.52$	–	$15.01^{b} \pm 2.21$	$16.94^{a} \pm 1.94$	–
	Weight of fruit (mg/ 10 fruits)	$92.2^{b} \pm 6.09$	–	$108.5^{a} \pm 7.63$	$111.5^{a} \pm 7.63$	–
Momordica charantia	Fruit set (%)	0	$94^{a} \pm 8.43$	$75^{b} \pm 12.69$	$86^{a} \pm 11.74$	–
	Fruit quality (after 10 days of flower open)					
	Length (mm)	–	$122.9^{a} \pm 8.31$	$108.4^{b} \pm 7.17$	$115.3^{b} \pm 8.84$	–
	Girth (mm)	–	$96.7^{a} \pm 5.81$	$80.3^{b} \pm 2.41$	$92.6^{a} \pm 5.32$	–
	Weight (g)	–	$35.45^{a} \pm 3.84$	$25.90^{c} \pm 2.83$	$30.91^{b} \pm 5.39$	–

Values given as mean ± standard deviation. Means in the row followed by same letters do not differ significantly by DMRT at 5%.

with added stingless bee colonies, and open-pollination with added western honeybee colonies. The percentage of fruit set was the highest in the manual cross-pollination system and lowest in the open-pollination system (Table 4.4). For the open-pollination system, the fruit set increased by approximately 14% and 17% after introducing stingless bees and western honeybees (Layek et al., 2021b). In terms of growth parameters (length, girth, and weight), fruit quality also differed significantly among the different pollination systems. Regarding the open-pollination system, fruit quality was significantly enhanced by adding stingless bees and western honeybees. Crop yield was increased approximately by 36.05% and 46.32% after introducing the managed stingless bees and western honeybees, respectively.

4.7 CONCLUSION

Diverse landscape and weather conditions support a rich diversity of agri-horticultural plants. Maximum crop plants depend on insect pollinators for their successful fruit and seed sets. Honeybees, solitary bees (*Halictus acrocephalus*) and stingless bees (*Tetragonula iridipennis*) were prevalent visitors of cultivated crop plants. The abundance of these bee species is gradually decreasing day by day. To overcome the pollinator crisis, the use of managed bees will be very effective in fulfilling the crop plants' pollinator demand. The introduction of managed honeybees, *Apis cerana* and *Apis mellifera* has been used from ancient times throughout the world. Recently, colonies of stingless bees are demanding as managed pollinators due to their floral constancy, polylectic foraging behavior, and easy domestication. As pollinator service increases, the crop yield (both quality and quantity) will be significantly increased.

Acknowledgements

The authors are indeed thankful to the authorities of Vidyasagar University for providing the necessary library and laboratory facilities. WBDSTBT partially offered financial support for the study by sponsoring a research project to the corresponding author (Memo No. 417(Sanc.)/ ST/P/S&T/1G-15/2018) has gratefully acknowledged. Thanks are also due to UGC for partial funding through DRS-SAP-II [F.5-2/2018/DRS-II]. The authors are very much thankful to Dr. Amarjit S. Tanda (Member of ASPS, AAPA, ANBA; Ex-Pest control consultant, Flick Pest Control/Rentokil Sydney, Australia) for his invitation as well as cooperation as an editor.

REFERENCES

- Abrol DP, Gorka AK, Ansari MJ, Al-Ghamdi A, Al-Kahtani S (2019) Impact of insect pollinators on yield and fruit quality of strawberry. Saudi J Biol Sci 26: 524–530. https://doi.org/10.1016/j.sjbs.2017.08.003
- Acciarri N, Restaino F, Vitelli G, Perrone D, Zottini M, Pandolfini T, Spena A, Rotino GL (2002) Genetically modified parthenocarpic eggplants: improved fruit productivity under both greenhouse and open field cultivation. BMC Biotechnol 2: 1–7. https://doi.org/10.1186/1472-6750-2-4
- Aggarwal PK (2008) Global climate change and Indian agriculture: impacts, adaptation and mitigation. Indian J Agri Sci 10: 911–919

- Bahadur B, Rajam MV, Sahijram L, Krishnamurthy KV (2015) Plant biology and biotechnology: volume I: plant diversity, organization, function and improvement. Springer, India, p 827
- Barrett SCH (2002) The evolution of plant sexual diversity. Nat Rev Genet 3: 274–284. https://doi.org/10.1038/nrg776
- Bashir MA, Alvi AM, Khan KA, Rehmani MIA, Ansari MJ, Atta S, Ghramh HA, Batool T, Tariq M (2018) Role of pollination in yield and physicochemical properties of tomatoes (*Lycopersicon esculentum*). Saudi J Biol Sci 25: 1291–1297. https://doi.org/10.1016/j.sjbs.2017.10.006
- Batra SWT (1994) *Anthophora pilipes villosula* Sm (Hymenoptera: Anthophoridae), a manageable Japanese bee that visits blueberries and apples during cool, rainy, spring weather. Proc Entomol Soc Was 96: 98–119
- Bertin RI (1982) The evolution and maintenance of andromonoecy. Evol Theory 6: 25–32
- Bispo dos Santos SA, Roselino AC, Hrncir M, Bego LR (2009) Pollination of tomatoes by the stingless bee *Melipona quadrifasciata* and the honeybee *Apis mellifera* (Hymenoptera, Apidae). Genet Mol Res 8: 751–757
  https://doi.org/10.4238/vol8-2kerr015
- Bisui S, Layek U, Karmakar P (2020) Utilization of Indian dammar bee (*Tetragonula iridipennis* Smith) as a pollinator of bitter gourd. Acta Agrobot. 73: 7316. https://doi.org/10.5586/aa.7316
- Bommarco R, Marini L, Vaissière BE (2012) Insect pollination enhances seed yield, quality, and market value in oilseed rape. Oecologia 169: 1025–1032. https://doi.org/10.1007/s00442-012-2271-6
- Bos MM, Veddeler D, Bogdanski AK, Klein AM, Tscharntke T, Steffan-Dewenter I, Tylianakis JM (2007) Caveats to quantifying ecosystem services: fruit abortion blurs benefits from crop pollination. Ecol Appl 17: 1841–1849. https://doi.org/10.1890/06-1763.1
- Bussi C, Lescourret F, Genard M, Habib R (2005) Pruning intensity and fruit load influence vegetative and fruit growth in an early maturing peach tree (cv. Alexandra). Fruits 60: 133–142. https://doi.org/10.1051/fruits:2005017
- Carmi N, Salts Y, Dedicova B, Shabtai S, Barg R (2003) Induction of parthenocarpy in tomato via specific expression of the rolB gene in the ovary. Planta 217: 726–735. https://doi.org/10.1007/s00425-003-1052-1
- Charlesworth D, Morgan MT (1991) Allocation of resources to sex functions in flowering plants. Philos Trans R Soc Lond B: Biol Sci 332: 91–102. https://doi.org/10.1098/rstb.1991.0036
- Chowdhury RN, Rasul MG, Islam AKMA, Mian MAK, Ahmed JU (2007) Effect of plant growth regulators for induction of parthenocarpic fruit in kakrol (*Momordica dioica* Roxb.). Bangladesh J Pl Breed Genet 20: 17–22. https://doi.org/10.3329/bjpbg.v20i2.17030

- Ferguson LB, Watkins CB (1992) Crop load affects mineral concentrations and incidence of bitter pit in 'Cox's Orange Pippin' apple fruit. J Am Soc Hortic Sci 117: 373–376. https://doi.org/10.21273/JASHS.117.3.373
- Free JB (1963) The flower constancy of Honeybees. J Anim Ecol 32: 119–131
- Garibaldi LA, Aizen MA, Klein AM, Cunningham SA, Harder LD (2011) Global growth and stability of agricultural yield decrease with pollinator dependence. Proc Natl Acad Sci U S A 108: 5909–5914. https://doi.org/10.1073/pnas.1012431108
- Gonzalez MV, Coque M, Herrero M (1998) Influence of pollination systems on fruit set and fruit quality in kiwifruit (*Actinidia deliciosa*). Ann Appl Biol 132: 349–355. https://doi.org/10.1111/j.1744-7348.1998.tb05210.x
- Graystock P, Blane EJ, McFrederick QS, Goulson D, Hughes WHO (2016) Do managed bees drive parasite spread and emergence in wild bees? Int J Parasitol Parasites Wild 15: 64–75. https://doi.org/10.1016/j.ijppaw.2015.10.001
- Hazra P, Dutta AK, Chatterjee P (2010) Altered gibberellins and auxin levels in the ovaries in the manifestation of genetic parthenocarpy in tomato (*Solanum lycopersicum*). Curr Sci 99: 1439–1443
- Hein L (2009) The economic value of the pollination service, a review across scales. Open Ecol J 2: 74–82. https://doi.org/10.2174/1874213000902010074
- Johannsmeier MF, Mostert AJN (2001) Crop pollination. In: Johannsmeier MF (ed.) Beekeeping in South Africa, 3rd edition. Agricultural Research Council of South Africa, Plant Protection Research Institute Pretoria, pp 235–240
- Karmakar P (2013) Pollination biology of *Citrus aurantiifolia* (Christm.) Swingle: a m-edicinally important fruit plant. Int J Innovat Res Dev 2: 138–142
- Kato S, Fujii K, Yumoto S, Ishimoto M, Shiraiwa T, Sayama T, Kikuchi A, Nishio T (2015) Seed yield and its components of indeterminate and determinate lines in recombinant inbred lines of soybean. Breed Sci 65: 154–160. https://doi.org/10.1270/jsbbs.65.154
- Klein AM, Vaissiere BE, Cane JH, Steffan-Dewenter I, Cunningham SA, Kremen C, Tscharntke TD (2007) Importance of pollinators in changing landscapes for world crops. Proc Royal Soc B 274: 303–313. https://doi.org/10.1098/rspb.2006.3721
- Layek U, Bera K, Bera B, Bisui S, Pattanayek SK (2021a) Assessment of yield enhancement in cashew (*Anacardium occidentale* L.) by the pollinator sharing effect of magnetic bee-friendly plants in India. Acta Ecol Sin 41: 243–252. https://doi.org/10.1016/j.chnaes.2021.05.003
- Layek U, Bhakat RK, Karmakar P (2015) Foraging behavior of *Apis florea* Fabricius during winter and spring-summer in Bankura and Paschim Medinipur districts, West Bengal. G J B B 4: 255–263
- Layek U, Das A, Karmakar P (2022a) Supplemental stingless bee pollination in fennel (*Foeniculum vulgare* Mill.): an assessment of impacts on native pollinators and crop yield. Front Sustain Food Syst 6: 820264. https://doi.org/10.3389/fsufs.2022.820264

- Layek U, Das U, Karmakar P (2022b) The pollination efficiency of a pollinator depends on its foraging strategy, flowering phenology, and the flower characteristics of a plant species. J Asia-Pac Entomol 25: 101882. https://doi.org/10.1016/j.aspen.2022.101882
- Layek U, Karmakar P (2018) Nesting characteristics, floral resources, and foraging activity of *Trigona iridipennis* Smith in Bankura district of West Bengal, India. Insect Soc 65: 117–132. https://doi.org/10.1007/s00040-017-0593-4
- Layek U, Karmakar P (2020) Distribution, nesting biology and floral resources of red dwarf honeybee (*Apis florea* Fabricius) in West Bengal, India. In: Abrol DP (ed.) The future role of dwarf honeybees in natural and agricultural systems. CRC Press (Taylor & Francis Group), pp 301–309. https://doi.org/10.1201/9781003033936-23
- Layek U, Kundu A, Bisui S, Karmakar P (2021b) Impact of managed stingless bee and western honeybee colonies on native pollinators and yield of watermelon: A comparative study. Ann Agric Sci 66: 38–45. https://doi.org/10.1016/j.aoas.2021.02.004
- Layek U, Kundu A, Karmakar P (2020) Floral ecology, floral visitors and breeding system of Gandharaj lemon (*Citrus* × *limon* L. Osbeck). Bot Pac 9: 113–119. https://doi.org/10.17581/bp.2020.09208
- Liming D, Chonglai B, Tianhua H, Qinmei Z, Haijiao H, Qunyan H, Weihai M (2016) SmARF8, a transcription factor involved in parthenocarpy in egg plant. Mol Genet Genomics 291: 93–105. https://doi.org/10.1007/s00438-015-1088-5
- Martin CS, Farina WM (2016) Honeybee floral constancy and pollination efficiency in sunflower (*Helianthus annuus*) crops for hybrid seed production. Apidologie 47: 161–170. https://doi.org/10.1007/s13592-015-0384-8
- McGregor SE (1976) Insect pollination of cultivated crop plants. Agricultural Research Service, US Dept Agriculture
- NAAS (1998) Conservation, management and use of agrobiodiversity. Policy paper 4. National Academy of Agriculture Sciences, New Delhi, p 7
- Nitsch JP (1952) Plant hormones in the development of fruits. Q Rev Biol 27: 33–57
- Parker FD, Batra SWT, Tepedino VJ (1987) New pollinators for our crops. Agric Zool Rev 2: 279–304
- Partap U, Ya T (2012) The human pollinators of fruit crops in Maoxian Country, Sichuan, China, a case study of the failure of pollination services and farmers' adaptation strategies. Mt Res Dev 32: 176–186. https://doi.org/10.1659/MRD-JOURNAL-D-11-00108.1
- Richards AJ (2001) Does low biodiversity resulting from modern agricultural practice affect crop pollination and yield? Annals Bot 88: 165–172. https://doi.org/10.1006/anbo.2001.1463
- Schwabe WW, Mills JJ (1981) Hormones and parthenocarpic fruit set: a literature survey. Hort Abstr 51: 661–698
- Serra S, Leisso R, Giordani L (2016) Crop load influences fruit quality, nutritional balance, and return bloom in 'honeycrip' apple. HortScience 51: 236–244. https://doi.org/10.21273/HORTSCI.51.3.236

- Stubbs CS, Drummond FA, Osgood EA. (1994). *Osmia ribifloris biedermannii* and *Megachile rotundata* (Hymenoptera: Megachilidae) introduced into the low bush blueberry agroecosystem in Maine. J Kans Entomol Soc 67: 173–185
- Van Heemert C, De Ruijter A, Van Den Eijnde J, Van Der Steen J (1990) Year-round production of bumblebee colonies for crop pollination. Bee World 71: 54–56. https://doi.org/10.1080/0005772X.1990.11099036
- Velthuis HHW, Van Doorn A (2006) A century of advances in bumblebee domestication and the economic and environmental aspects of its commercialization for pollination. Apidologie 37: 421–451. https://doi.org/10.1051/apido:2006019
- Williams IH (1996) Aspects of bee diversity and crop pollination in the European Union. In: Matheson A, Buchmann SL, O'Toole C, Westrich P, Williams IH (eds.) The conservation of bees. New York, Academic Press, pp 63–68
- Yin Z, Malinowski R, Ziokowska A, Sommer H, Plder W, Malepszy S (2006) The DefH9-iaaM containing construct efficiently induces parthenocarpy in cucumber. Cell Mol Bio Lett 11: 279–290. https://doi.org/10.2478/s11658-006-0024-4
- Zhang T, Tan DY (2009) An examination of the function of male flowers in an andromonoecious shrub *Capparis spinosa*. J Integr Plant Biol 51: 316–324. https://doi.org/10.1111/j.1744-7909.2008.00800.x

5

Climate Change and Pollinators and Conservational Aspects of Pollinators

Gouri Shankar Giri*, Chenesh Patel[1], Deeksha M.G.[2], and Kaushal Kishor[3]

ABSTRACT

Insect pollinators are censorious critters for the survival of plant diversity and also whole organisms across this globe. As they provide major ecological service by pollinating the crops, thus increasing the crop yields. But the survival of this pollinators has become enigmatic due to various circumstances, among which climate change is also an important factor. Change in climate has a significant impact on the decline of pollinators' diversity. Climate change affects pollinators in various ways which can be stratified under direct effects and indirect effects. Direct effects of climate change are mainly associated with the pollinators, where factors like effect on diversity, distribution, morphology, physiology, biology, disease and pest incidence, at last behavioral activity are dealt in detail. Whereas under indirect effects are mainly related with the host plants interacting with the pollinators' diversity, where effects on diversity of bee flora, their distribution, morphology, floral reward production, nutritional value, bee attractants or repellents produced by host plants and pesticide mobility on foraging plants are discussed. As impacts of climate change on pollinators are very intricate issue, it requires righteous contemplation to figure out this matter. Accordingly conservative plans should be framed, where a thorough knowledge on decline rate and policies applicable on pollinator conservation is to be known. The blueprint for conservation is prepared according the ecosystem where pollinators exist, like urban area, agricultural land and forest ecosystem. Thus, the

* Assistant Professor, Department of Entomology, Tirhut College of Agriculture, Dholi, RPCAU, Pusa, Bihar-843121

E-mail: gsgiri@rpcau.ac.in

[1] Ph.D. Scholar, GB Pant University of Agriculture and Technology, Pantnagar, Uttrakhand-263145

E-mail: cheneshpatel28@gmail.com

[2] Research Scholar, DST Inspire Fellow, Indian Agricultural Research Institute, New Delhi-110012

E-mail: deekshamudagadde@gmail.com

[3] Assistant Professor, Department of Agronomy, PG College of Agriculture, RPCAU, Pusa, Bihar-848125

E-mail: kaushal@rpcau.ac.in

complication of climate change on pollinators can be foreshortened, and existence of pollinators can be eased.

Keywords *Climate change • Pollinators • Distribution • Diversity • Phenology • Physiology • Biology • Foraging • Plant pollinator interaction • Conservation*

5.1 INTRODUCTION

The past few decades had witnessed the drastic changes in global climate due to adverse effects of human activities and unsustainable use of natural resources of the ecosystem (Nova et al., 2018). The adverse effects of climate change can be easily seen as changes in precipitation level, CO_2 level and temperature regimes of the world. A report on climate change by IPPC, 2014 (Intergovernmental Panel on Climate Change) suggested that since the year 1750, the concentration of greenhouses gases such as methane (CH_4), carbon dioxide (CO_2), and nitrous oxide (N_2O) have been increased by 150%, 40% and 20%, respectively. The share of greenhouse gases in atmosphere reached about 36.14 billion metric tons in 2014 as compared to 22.15 billion metric tons (1990) (Abeydeera et al., 2019). The average global temperature has increased at an average rate of 0.15-0.20°C per decade since 1975, and is expected to increase by 1.4-5.8°C by 2021 (Arora et al., 2005). These data are suggesting that the climate change is severely affecting the natural ecosystems which ultimately results in a negative impact on human health and food security (Arunanondchai et al., 2018).

Insects are the one of the beautiful creature of nature that represents major part of the kingdom Animalia. Some of them are beneficial while some of them are harmful to human beings. Beneficial insects are may be a natural enemy, or may be a productive one, or may be a pollinators whereas the insects that cause nuisance to the human beings are designated as pests which serve as a prime constraint in successful agricultural cultivation and forestry production (Oerke, 2006; Pureswaran et al., 2018). With an increase in global human demand for food and nutritional security, the impact of climate change on insect pests assumes to have greater importance (Godfray et al., 2010) as they cause an estimated loss of approximately 18% of total global annual crop production (Oerke, 2006). On the other hand, beneficial insects which include different kinds of pollinators and natural enemies increase food production and ensure nutritional security by pollination service and manipulating the pest density respectively.

The present chapter addresses the information on four important issues:

- Importance of pollinators for leading global crops production and their level of dependence
- The effect of climate changes on pollinators
- The effect of climate change on plant-pollinators interaction
- Conservation of pollinator for sustainable crop production

5.2 IMPORTANCE OF INSECT POLLINATORS FOR LEADING GLOBAL CROPS PRODUCTION

In the present scenario, fulfilling the nutritional requirements of growing global population is a major concern for all of us. There are probably three known methods for increasing the global food production and thus ensuring nutritional security. These are use of good agronomic practices, use of biotechnological approaches to overcome biotic and abiotic stress, use of pollinators for enhancing yield and yield attributes. Among them, the first two were already exploited by us, so the only and one option left for us is to use these pollinators, which perform a major role in important ecosystem services which are essential for ecosystem functioning and rise in food production, thus ensure food and nutritional security. Pollinator may be biotic or abiotic, of which biotic pollinators have a major importance as agriculture and food production is concern. Among biotic pollinators, insect pollinators' accounts for 80% of the pollination services recorded worldwide. The pollination in many agriculturally important crops (McGregor, 1976; Crane and Walker, 1984; Free, 1993; Williams, 1994; Nabhan and Buchmann, 1997; Westerkamp and Gottsberger, 2000) as well as in wild plants (Burd, 1994; Kearns et al., 1998; Larson and Barrett, 2000; Ashman et al., 2004) depends on the activity of insect pollinator. Insect pollinators are responsible for pollination of about 35 percent of cultivated crops (Allen-Wardell et al., 1998; Richards, 2001; Klein et al., 2007 and Gallai et al., 2009) and the total economic value of crop pollination worldwide has been estimated at €156 billion annually (Gallai et al., 2009). During the past 50 years, the area covered by pollinator-dependent crops has increased by more than 300 percent (Aizen and Harder, 2009).

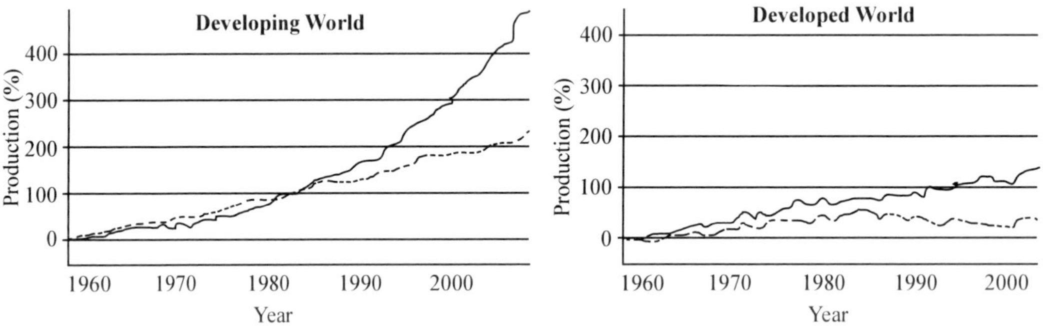

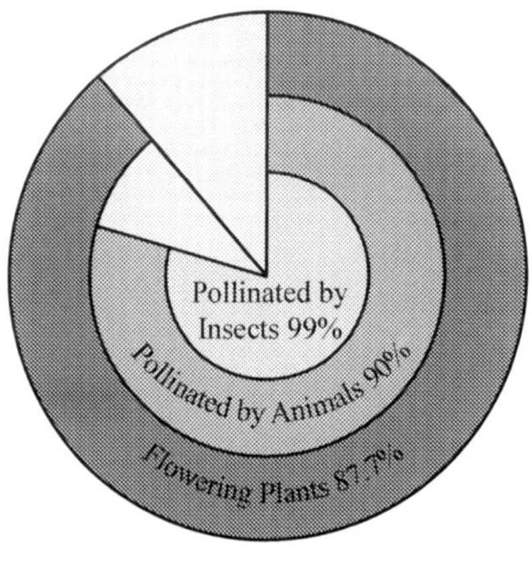

Percentage of plants pollinated by insects. Pollinating animals other than insects are bird and bats, and the majority of insect pollinators are bees represented by honeybees

(*Source:* Hoshiba and Sasaki, 2008)

5.3. THE EFFECT OF CLIMATE CHANGES ON POLLINATORS AND THEIR SERVICES

The climate change is drastically affecting the population of insect pollinators both directly and indirectly. Directly it affects by modulation of their diversity, distribution, phenology, physiology, biology, flight and foraging activity. Indirectly, it affects by altering the diversity, distribution, morphology and chemical properties of plant species (Bartomeus et al., 2011; Burkle et al., 2013; Scaven et al., 2013; Kerr et al., 2015; Miller-Struttmann et al., 2015) or by rendering them more susceptible to attacked by parasite and pathogens. These direct and indirect effect of climate change ultimately altered the plant-pollinators interaction by mismatch that may lead to absence of mutualistic partners (Stenscth and Mysterud 2002; Visser and Both, 2005).

5.3.1 Direct Effects

5.3.1.1 Effect on diversity of pollinators

Insect pollinators are playing a key role both in natural as well as agricultural ecosystems and are known to provide a valuable ecological service which we called pollination. There are about 25000-30000 species of bees which are known to be effective pollinators and together with the moth, butterfly, beetle, wasp and flies, make up the major pollinating agents (Neelam, 2017). Climate change, macroclimate in general and microclimate in particular may have significant effects on the relative abundance and diversity of pollinators as they are very sensitive and unable to acclimatize (Thomas, 2014). This leads to extinction of some species from particular areas that are unable to acclimatize those frequent changes (Martins, 2015). Because of climate change, most of the bee species lost their occurrence area.

Studies reveal that climate change, an increase in temperature in particular, becoming the one of the potential cause of declining the population of *Bombus bellicosus* in South America (Martins, 2015). Change in precipitation gradient in one of the factor associated with the spatial turnover of major pollinators from the temperate forest of South America (Devote, 2009). Frequent change in temperature and rainfall patterns leads to a greater decline in species occupancy and richness as well as increase local extinction rates, reducing colonisation and site occupancy and decreasing species richness within a region (Soroye et al., 2020). Because of climate change, there was decline in diversity of bee in Caatinga as compared to other areas of South America (187 species and 77 genera) (Zanella, 2000). Change in climate along with other anthropogenic activities may result in decline in species richness of bumblebee in North America by 2070 (Delisle and Kerr, 2018).

Giannini et al. (2020) predicted that there will be a severe reduction in a number of bee species in general and pollinator species having more restricted distribution in particular from the Carajas region. Decline in diversity of wild bees and hover fly having narrow habitat requirements and dietary specialisation is more prominent in some part of Europe. Climate change along with anthropogenic activity is one of the prime causes responsible for extinction of some species of hover fly from the alpine region of southern Europe (Biesmeijer, 2006). Decline of species richness as xwell as land occupancy of wild pollinators and hover fly has been reported (Powney et al., 2019). In contrast to this, though there was decline in the butterfly and wild bee species from Britain as a result of climate warming, no decline in hover fly diversity over a past decade from the Netherlands and Beligium (Carvalheiro et al., 2013). As a result of global warming, there was about 20% decline in abundance and 80% decline in the species richness of hover fly in different parts of Germany (Hallmann, 2019).

5.3.1.2 Effect on distribution of pollinators

Insects are known to occur in this earth prior to the inhabitation of human beings and exists a co-evolution of plants and pollinators. So insect pollinators are well distributed all over the world starting from tropical to temperate climate, from high altitude to lower altitude as well as from agricultural ecosystems to forest ecosystems. Wild species of bees as well as other pollinators are well distributed in the forest ecosystem. Their distribution is greatly influenced by the availability of pollinating plants as well as suitability of climatic conditions. Because of climate change, areas that are known to be suitable for a particular species of pollinators became vulnerable to inhabitation by other species of organisms leading to shifting of native place by the insect pollinators.

As insect pollinators and distribution is concern, bees are most prominent and distributed over a large geographical area. The distribution is mainly depends on their thermal tolerance and plasticity to temperature changes. For example, European dark bee is restricted to an area where the average temperature during the month of July ranges from 15-200C. Similarly, the Eastern honeybee (*Apis cerana*), popularly known as Indian bee is distributed in some part of Asia, whereas the giant honeybee (*Apis dorsata*) is restricted to the tropical climate and higher altitudes (Frank et al., 2000; Le Conte and Navjas, 2008).

Studies reveal that climate change may be one of the major causes for the potential reduction of population and shifting the habitat by an organism. The stingless bee (*Mellipona subnitida*) is widely distributed in the semi-arid region of Brazil and known to be effective local pollinators of wild and crop plants. Because of climate change, the species may find their climatically suitable habitats towards the external boarder of the current pollinator's distribution (towards the higher altitude) with a continuous central disconnection, which may hamper species dispersal and gene flow (Giannini et al., 2017). Native pollinator of acerola and cashew (*Centris hyptidis* and *Centries trigonoides*) as well as of avocado and guava (*Melipona scutellaris*) in Northeastern Brazil represents a 9-35% reduction in their occurrence area (Giannini et al., 2012).

Bumblebees, important pollinators of native and agricultural crops in the temperate and high altitude area, shifting their habitats towards the northern and higher elevation over the last 100 years while losses their range and abundance in southern region (Bedford et. al., 2012; Kerr et. al., 2015). Local extinction of these species from southern range may be associated with their vulnerability to frequent extreme temperature events as a result of climate change (Chen et al., 2011; Herreara et al., 2015). It has been predicted that over half of the species of bumblebees will face significant losses of ranges by 2050 (Delisle and Kerr, 2018).

Carpenter bee belonging to genus *Xylocopa*, an efficient local pollinator of passion fruits, showed up to 90% reduction in their native distribution area in the Brazilian savana (Giannini et al., 2013). The distribution of tomato pollinators such as *Bombus morio* showed the greatest reduction up to 71% in Brazil (Elias et al., 2017). Rise in temperature results expansion of geographical distribution of many species of butterflies in cooler margins and an eventual contraction of distributional ranges in warmer margins (Parmesan et al., 1999; Warren et al., 2001; Parmesan and Yohe, 2003; Wilson et al., 2007).

5.3.1.3 Effect on morphology of pollinators

Morphological traits such as body size and proboscis length are directly connected with the foraging efficiency of pollinators. Pollen load carried out by pollinators is directly proportional to the body size whereas proboscis length is a important aspect for sucking the nectar from the nectar gland of a flower. Length of proboscis also directly related to the resource acquisition and pollination effectiveness.

In general, rise in temperature is known to decrease the size and body weight of an insect probably because of increase in metabolic rate and accelerated development (Kingslover et al., 2008; Oliveira et al., 2016). This decrease in size negatively influenced the thermoregulatory ability of the insect in general. Though thickness and color of body hair plays an important role in thermoregulation (Heinrich, 1947), it has been reported that bees with larger body size are able to thermoregulate more efficiently than the bees with smaller body size (Bishop and Armbruster, 1999). Decrease in size also reduced the pollination efficiency of pollinators because of reduced pollen delivery on the stigma of the female flower (Howlett et al., 2011). At the same time, pollinators with large body size probably have more contact with anther and stigma of the flower, thus having greater pollination efficiency (Adler and Irwin, 2006; Solis-Montero and Vallejo-Marin, 2017). Exposing the larvae and pupa of solitary bees (*Osmia bicornis*) to elevated temperature, whether constant or fluctuating, results in to the development of adults with smaller body size partially due to the incomplete consumption of provision (Radmacher and Strohm, 2010) thus adversely influencing the activity. Rise in temperature leads to a decline in the body mass of the adult drone of *Apis mellifera* (Czekonska et al., 2020). It has been reported that, during the period of development, a minor deviation from optimum temperature can reduce the fitness of adult worker bees (Jones et al., 2005; Tautz et al., 2003). Reduction in proboscis length in two bumblebee species as a result of climate warming resulted into decoupling of mutualism between bumblebees and pollinating plants (Miller-Struttmann et al., 2015)

5.3.1.4 Effect on physiology of pollinators

Climate change is also known to adversely affect the physiology of insect pollinators. Various physiological activities of insect such as food consumption, digestion, utilisation, diapause, hibernation, etc. are affected by change in temperature as well as concentration of CO_2. Rise in temperature causes rise in the rate of metabolism in *Apis melleifera ligustica*, which is a crucial physiological trait, adapted in the evolutionary process to environment and climatic conditions (Kovac et al., 2014). Being social insects, bees live in the colony and they are able to maintain their hive temperature in an optimum range by a physiological process called thermoregulation.

Free and Spencer-Booth (1958) reported that a rise in temperature in winter as a result of climate change up to 400 C may causes death of the bees that are live in the colony by affecting the thermoregulation. They also found that, there was a decline in food consumption rate (sugar syrup) and increase in water up take with respect to rise in temperature. The timing of the brood onset of hibernating bee colonies is an important physiological event that determined the colony fitness, growth and subsequent reproduction and is highly influenced by climatic condition (Seeley and Visscher, 1985). Even a slight change in climatic condition in general, temperature and photoperiod in particular, adversely affect the timing of the brood onset of honeybee followed by the emergence of adult bees. For example, premature brood onset may result in death of bees due to starvation or attacked by *Varroa destructor* while late brood onset leads to death of bees by producing asynchrony between emerged bee and available bee flora (Numberger et al., 2018).

5.3.1.5 Effect on biology of pollinators

Most of the insect pollinators belong to holo-metabolous group, i.e., they have four stages in their life cycle. Any change in temperature, rainfall or concentration of CO_2 is known to affect the growth, development, survival as well as reproduction of the insect pollinators. The immature stages are more sensitive to change in weather parameters as a result of climate change as compared to adult stages. In general for all insect, rise in temperature leads to increase in the fecundity and decreased in the incubation period. An increase in temperature reduces the percent hatchability because of destruction of the embryo present inside the egg.

Rise in temperature leads to loss of fertility in queen bees of *Apis mellifera*. It is well known that the queen bee mates only once in her life cycle and store sufficient amount of spermatozoa in her spermatotheca that is sufficient for egg laying for her rest of life. It has been reported that temperature of 15-38°C is safe for queen with tolerance threshold of 11.5% loss of sperm viability (McAfee et al., 2020). Since, bees maintain their hive temperature as well as humidity through thermoregulation and fanning behavior, the impact of temperature on immature stages is not well documented. However, a study reported that, lowering the brood rearing temperature had a negative impact on the emergence of adult bee which is more susceptible to biotic and abiotic stress (Medrzycki et al., 2010).

Butterflies are one of the most important pollinators in our ecosystems. The developmental period of larvae of a butterfly is also affected by the rise in temperature. In general, rise in temperature results in quick growth because of the increase rate of food consumption and metabolism. The pupal period is also decline as a result of rise in temperature. Therefore, elevated temperature results in the shorter developmental period, which in turn, increases the number of generation per year or season. However, rise in temperature results in reducing the lifespan of most of the insect pollinators. For example, the lifespan of Hover fly is about 21.2 days at a temperature of 26.6°C whereas 19.6 days when temperature rise to 29.9°C (Hong and Hung, 2010).

Rise in temperature and fall in humidity is also known reduced the survival of honeybee (Abou-Shaar et al., 2012). King-solver and Huey (2008) reported that rise in the temperature lead to quick development and the emergence of adults that are smaller in size whereas Radmacher and Strohm (2010) reported that the larvae and pupa of bees are of lesser in weight because of increase in temperature. Because of smaller body size and weight they cannot go to a longer distance for foraging as the effectiveness of foraging distance is directly related to body size of the forager bees (Greenleaf et al., 2007). It was reported that the solitary bee, *Osmia lignaria* exposed to simulated longer summers, leads to accumulations of greater degree-day resulted in shorter lifespans that had overwintered (Sgolastra et al., 2011) while exposed to constant elevated pre-wintering temperature results in the reduction of adult lifespan up to several days in this species (Bosch et al., 2000). It has been reported that the fig wasp which has developed a symbiotic and obligatory relationship with fig is more susceptible to the changing climate because of its small size and short lifespan. An increase in temperature of 3°C or more above is known to decrease the lifespan adult fig wasp. Increased in concentration of carbon dioxide causes the bee to enter in to anaesthesia and reduced their survival (Czekonska, 2009).

5.3.1.6 Effect on flight and foraging activity of pollinators

Climate change is known to affect the behavior of inset pollinators by a variety of ways such as by affecting foraging and flight activity or by affecting communication. During the extreme of climatic event, whether during the excess hot or cold, or rain, bees do not go for collecting food and waters. If such conditions continued for the prolonged period, then bees automatically die. For each and every pollinator, there is an optimum temperature for carrying out foraging activity. Even a small change in temperature may adversely affect the flight activity and foraging behaviour of insect pollinators (Heard and Hendrikz, 1993; Stone, 1994). For example, the optimum temperature for foraging activity of bumblebee (*Bumbus terrestres*) is 25.7°C. Rise in temperature up to 32.7°C results in 69.7% reduction in number forager bees visited to a plant species.

Bumblebees prefer to collect pollen rather than nectar during the warmer and sunny climate at the foraging site otherwise prefers to collect nectar. Since the numbers of bees go out from hive for foraging strongly depends on temperature and solar radiation, rise in temperature may negatively influence the foraging activity, which in turn, leads to less pollination and reduced crop productivity. It has been reported that insect pollinators with large body size avoids foraging during the hottest part of the days and prefer to forage either in the morning or afternoon part of the day (Willmer, 1983).

Visit by pollinators to flowers earlier in the day lead to receive fewer insect visitors by plants with flowers that open later in the day, which would result in poor pollination and ultimately reduced the fruit and seed set (Wilcock and Neiland, 2002). Insect visitors of flowers may limit their foraging trips to a shorter distance in order to reduce the excess heating during flight on very warm days, which results in the alternation in patterns of pollen flow. As they restrict their foraging distance, the plants in the nearby area received less pollen from conspecific located in a larger distance (Herrera, 1987). This serves as a major constraint in the setting of superior quality of seed and better seeding survivorship (Price and Waser, 1979). Pollinators are known to change their behaviour and activity pattern as a result of rise in temperature, which in turn, reduces the efficiency of pollen removal and deposition, a part of foraging activity. Adults of *Apis mellifera* experienced high temperature during their pupal period showed a loss of short-term learning abilities and memory, thus affecting the foraging activity (Jones et al., 2005).

Though temperature is positively correlated with activity of pollinators, pollination services of most of the pollinators was reduced after experiencing the threshold temperature as they spend less time in foraging and more time in coolonig. Rise in ambient temperature may alter the frequency of the bee visited by altering the floral temperature in case of Australian native stingless bee (*Trigona carbonaria*). Plant species like *Hibertia* influence the visit by responding to change in ambient temperature in a simpler way by increasing the petal temperature whereas plant species like *Anigozanthos flavidus* and *Esallonia macrantha* do not respond significantly to rise in temperature, thus have no effect on the number of visits. Some plant species like *Erysimum linifolium* and *Polygala grandiflora* modulate their temperature with respect to change in weather, thus are able to adapt better and maintained interaction with pollinators in the changing climatic scenario (Norgate et al., 2010). Interaction among butterflies pollinators and plant they pollinate was potencially influenced by rise in temperature due to prolonged flight duration (Roy and Sparks, 2000).

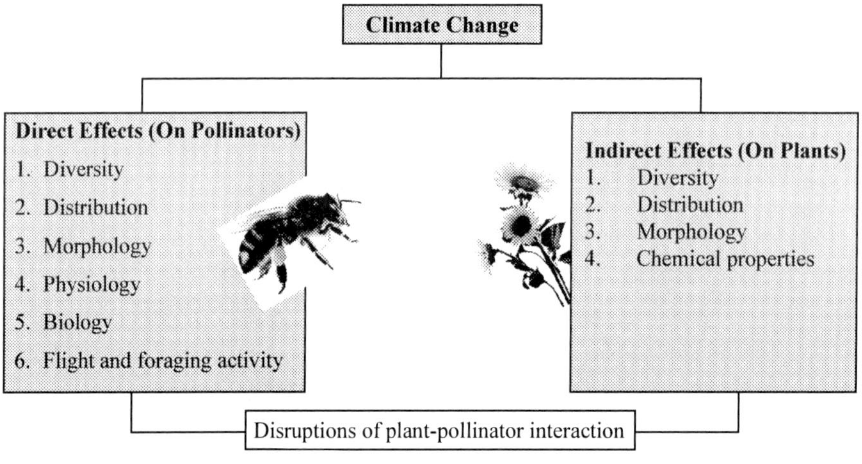

5.3.2 Indirect Effects

5.3.2.1 By affecting the flowering plants

Climate change is known to affect the insect pollinators indirectly in various ways such as change in diversity, distribution, morphology of pollinating plants or by altering the production of pollen and nectar or by altering the concentration of nutritional value of the foraging plants. The response of plant species to changing climatic scenario is complex phenomena and is species specific. Plants pollinated by insect react strongly to climate warming as compared to those of wind pollinated plants. Interaction among plant and pollinators mainly depends on the warmness of the flower as well as nectar. Some plants respond positively whereas some plants respond in a negative manner and known to influence the plant-pollinators interaction. Even, some plant species are able to modify their flower temperature, thus provide reward to the pollinators. Most of the pollinating species prefer warmer flower and nectar when the ambient temperature is cool (below 30°C), but alter their preference with respect to rise in temperature of above 30°C (Nogate et al., 2010). As the body temperature is directly related to body size, smaller bee prefers cooler flower when there was rise in ambient temperature (Corbet et al., 2016). Rise in ambient temperature may alter the frequency of the bee visited by altering the floral temperature in the Australian native stingless bee (*Trigona carbonaria*). The absence of cooling mechanism in plant species makes them less attractive to potential plant pollinators in the scenario of climate change where ambient temperature rises above 30°C.

Because of climate change, temperature, relative humidity, atmospheric carbon dioxide concentration increases which lead to the occurrence of the extremity of climatic events such as drought, flood and several other climatic phenomena. These extreme climatic phenomena are known to alter in the flowering period of the several plant species, which, in turn, the availability of pollen and nectar to the pollinators. Pollen and nectar are the main food for most of the pollinators as they derive protein from pollen and vitamin, energy from nectar. Alternation in the amount of nectar produced and concentration of sugar in nectar in response to change in climate is a major concern for survival of insect pollinators (Corbet, 1990). A study shows that in a Mediterranean plant (*Thymus capitatus*), the volume of nectar and sugar concentration increased with an increase in temperature up to a point (38°C) (Petanidou and Smets, 1996), whereas temperature (23°C vs. 19°C) affected the ratio of glucose to fructose in the nectar of pumpkin plants in a negative manner (Hoover et al., 2012). Besides this, the size and number of flower is also affected by this extreme phenomenon.

Temperature along with rainfall and photoperiod are the predominant factor responsible for initiation of flowering in many plant species (Barrett and Brown, 2021). Reductions in the flower production lead to the reduction in food availability for insect visitors which ultimately led to a reduction in the reproductive output of plant species (Boggs and Ross, 1993; Minckley et al., 1994) and population densities of insect pollinators (Westphal et al., 2003). Because of these extreme phenomena, the flower produced by plant species differ in some morphological traits such as colour, size, length, etc. that could reflect attractiveness, accessibility and profitability of their rewards for the insect visitors. For example, under warm (23°C) temperature pumpkin plant produced small sized flower (Hoover et al., 2012). Similarly, the length of corolla tube in flowers of coastal morning glory (*Ipomoea richocarpa*) during the period of low temperature were several mm shorter than the flowers that produced during the period of warm temperature. Climate change also known to affect the time of anthesis as some flowers open 2-3 hour earlier on warmer mornings (Murcia, 1990).

Change in climate may also affect the floral fitness. Exposure of plants to elevated temperature results in damage to floral parts like petal, pollen, gynaecium (Patirio et al., 2002) as well as decreased the reproductive output also (Ladinig et al., 2015). It may have negative impact on plant-pollinator interaction by affecting the future plant community as survival, growth and reproduction of most plant species is strongly influenced by the rise in temperature (Ensslin et al., 2017). Rise in temperature may lead to the reduction in richness of the bee flora or extinction of bee flora, as some of the plant species wilted when ambient temperature exceeds 36°C indicating that temperature plays a key role in plant survival and reproduction (Chehetka, et al., 2017). Global warming may affect the plant-pollinator's interaction by producing asynchrony between the availability of bloom and the emergence of insect pollinators. Warmer climate leads to the early flowering in most of the plant species which results in unavailability of pollen and nectar at the time of the emergence of brood after winter.

5.3.2.2 By increasing the disease and pest incidence

Nature plays an important in maintaining the population density of each and every organism within certain upper and lower limits. For every organism there exist a variety of natural enemies that kept the population within control. Insect pollinators, being living organisms, also has a number of natural enemies. Like other organisms, climate change is known to change the geographical distribution of the disease causing organism. The expression of these organisms as well as their pathogenecity is totally depend on climatic factors. The important disease and pest that are known to suppress the population of bees are mites, wax moth, nosema and brood diseases. Some of them are known to distribute worldwide such as *Varroa destructor*, *Nosema apis* and bacteria that causing brood diseases as well as some of the viruses.

Climate change is known to change the virulence of this pathogen and encourages its spread within honeybee population. Some pathogen like tracheal mite (*Tropolailis clarae*) have limited distribution and known to infest primarily a wild species of bee (*Apis dorsata*) only in tropical Asia (Sammataro et al., 2000). Change in geographical distribution of bees as a result of climate change brings the other species of bees in contact with that species of bee, which in turn, leads to transfer of pathogen among them. During the last few decades, this pathogen had known to invade all over the distributional area of *Apis mellifera* (Solignac et al., 2005). Invading of chalk brood diseases caused by *Ascospharea apis* to other area, which is mainly occurred in the humid environment, as a result of climate change has also been reported (Le Conte and Navajas, 2008). Climate change may facilitate the invasion of bee parasite and predator into an area where they do not our previously. For example, the small hive beetle (*Aethina tumida*), which is parasite of social bee (Cervancia, 2016), bumblebee (Spiewok and Neumann, 2006), stingless bee (Greco et al., 2010) as well as solitary bee (Gonthier et al., 2019) and native to South Africa, had known to expand its geographical distribution to all continents except Antarctica and presently became a problem to all over world (Cornelissen et al., 2019). Expansion of geographical distribution of a predator, i.e., bee eater from Mediterranean area to other parts of the world has also been documented.

Recently the colony collapse disorder emerges as a major threat to honeybee causes rapid decline in population particularly in western countries. Though lots of factors were associated with it, prevalence of the disease causing organism along with climate change is one of the prime and important one. Change in climate weakened the bee species by affecting the nectar and pollen production (Thuiller et al., 2005) and causing them more susceptible to attack of diseases and pest. When these sick bees go out for foraging, they were eventually killed and unable to return to the hives. We can say that climate change is known to modify the interaction among the pathogens as well as between the pathogen and bee species.

5.3.3 Effects on Plant-Pollinators Interaction

The direct and indirect effects of warming climate lead to the severe disruption of the plant pollinator's interaction. Loss of diversity of any one of these two results in loss of the diversity of another one. There exists a co-evolutionary mutualistic relation between plant and pollinator species and many plant species solely depend on their pollinator species for reproduction, seed sets and subsequent expansion of range. Shift in the geographical distribution of such pollinator species results in the loss of diversity of dependent plant species or vice versa. Body size and proboscis length of pollinators also determined the plant species to which they visit. They mainly visit to those flowers where they can sit comfortable and able to collect pollen and nectar. Thus, any change in morphology, either in proboscis length or in body size, negatively influenced the plant-pollinators interaction. Likewise, change in many plant characters such as the number of flower produced by plant, amount of pollen and nectar per flower, concentration of the sucrose in nectar, length of the corolla tube, area of petal surface and many other are known to affect the plant pollinators interaction. The mismatch between timing of the flower onset and the first brood emergence as a result of warming climate is prime and most important one affecting plant-pollinators interaction in a negative way. Even a slight disturbance in plant pollinator interaction is known to affect the abundance of both plants by reducing the visit followed by less pollen deposition and subsequent failure of reproduction as well as pollinator species by altering the availability of pollen, nectar followed by reproduction and survival (Hegland et al., 2009).

5.4 CONSERVATION OF POLLINATOR FOR SUSTAINABLE CROP PRODUCTION

5.4.1. Monitoring the Pollinators Decline Rate

The comprehensive surveillance on the decline rate of pollinators due to climate change is lacking, but the decline of pollinators due to various factors has been tracked from several decades. This insight is sufficient enough to perceive that pollinators' diversity is facing pitfall, for which climate change is also a superintended influencer. In recent years, the proportion of the pollinators dependent of the agricultural crop is rising expeditiously > 300%, depicting in the paucity of pollinators to fulfil the demand (Aizen and Harder, 2009). Evaluation of biodiversity risk due to environment changes is carried out based on species modelling for current and future circumstances (Segurado and Araujo, 2004; Thuiller et al., 2003), this may help in devising conservative strategies for biodiversity(Araujo, 2002; Araujo and Williams, 2001; Araujo and Williams, 2000).

Currently, carried out systematic reviews and meta-analysis across the globe reveals that climate change has its impact on the species decline (Wilson et al., 2005; Parmesan and Yohe, 2003; Walther et al., 2001; Parmesan et al., 1999). Knowledge on the status of pollinator's diversity in the empirical form is must to decide the future manoeuvre (National Research Council, 2007; Ghazoul 2005; Steffan-Dewenter 2005; Cane and Tepedino, 2001; Kremen, and Ricketts 2000; Kearns et al. 1998). The report on the regional decline of the honeybee in the USA depicted that around 59% of the colonies were lost between 1947-2005 (van Engelsdorp et al., 2008; National research council 2007) and in Europe nearly 25% decline was observed form 1985 to 2005 (Potts et al., 2010). In Belgium and UK, the bumblebees (*Bombus* sp.) diversity has been reported declining in progress (Goulson et al., 2008; Rasmont and Mersch, 1988). The above intuition is more than enough to take right away decisions on the conservative plan to action, in order to avoid the upcoming complication on pollinators' existence and food security of this world.

5.4.2 Policies Regarding Need for Conservation of Pollinators

Currently the decline in pollinators' species has engaged scientific community to plan effective ways to conserve pollinators. As several techniques are developed they are constructively enacted by policies, as they provide strong pillar to be supported nationwide. In this concern several policies are framed regarding need for pollinators' conservation. As strong political commitment to support and conserve pollinators by formulating policies in favor is must (Dicks et al., 2016).

The establishment of International Pollinators Initiative by the Convention on Biological Diversity along with the coordination of Food and Agriculture Organization of the United Nations (FAO) framed initial guidance on policy development (Rose et al., 2015). The Millennium Ecosystem Assessment (MEA) by the United Nations Secretary-General Kofi Annan in 2000 used the term "Ecosystem services" in their currency in order enlighten people on multiple benefits provided by the nature. Another initiative is "The Intergovernmental Science-Policy Platform on Biodiversity and Ecosystem Services (IPBES)" is an intergovernmental body came into force on 2012, it first subject matter accessed was on pollinators, pollination and food production, provided compendium of policymakers, this was approved by all member states of IPBES in 2016 (Potts et al., 2016).

The UK government has developed Wales Pollinator Action Plan (WG, 2013) and UK National Pollinator Strategy (DEFRA, 2014), whereas England has National Pollinator Strategy (DEFRA, 2014), Ireland developed All-Ireland Pollinator Plan (All-Ireland Pollinator Plan 2015-2020) and US has approved Pollinator Partnership Action Plan (Pollinator Health Task Force, 2016). Also Royal Horticultural Society's annual 'Britain in Bloom' competition in UK has created interest in people for gardening and in US 'Million Pollinator Garden Challenge' has engaged millions of people to establish network of garden to support pollinators (Saunders et al., 2018; Roy et al., 2016; Deguines et al., 2012). In US, state level legislature has endorsed 110 new laws regarding conservation of pollinators from 2000 to 2017 (Hall and Steiner, 2019). Recently bio culture approaches is gaining importance in pollinators conservation indigenous peoples and local communities (IPLCs) across the globe (Hill et al., 2019). These polices played a prime role in operating the strategies on conservation programs of pollinators in many countries.

5.4.3 Blueprint for Conservative Activity of Pollinator

Pollinators are predominantly widespread species found across the globe, playing an important role in continuance of several floras associated with them by providing their ecological services. As pollinators are mostly terrestrial organism, so the planning of conservative strategy should be according to the ecosystem where pollinators are occupied. Pollinators are important influence in the agroecosystem and the forest ecosystem, beside this they are also found in urban ecosystem also. These three ecosystems are highly vulnerable to impact of the climate change, which in turn, has the effect on pollinators' diversity. So care is to be taken in ecosystem-wise to conserve the pollinators. Following ideas are framed according to ecosystem requirement in turn to preserve the pollinators.

5.4.4 Conservation in Urban Ecosystem

Urban areas constitute a complex variety of ecological habitats and different land uses. So, the planning for conservation need to be variegated according to the multiple stakeholders residing there. Present days Citizen Science programs and conservation projects have enlightened society regarding the service of pollinators and need of their conservation, this can be achieved in urban area by (Baldock, 2020). Establishment of community garden, which can support bees and Hover fly (Baldock et al., 2019; Leve et al., 2019; Kaluza et al., 2016). In public green space flowering plants should be planted and care need to be taken during mowing the lawn (Aguilera, 2019; Baldock et al., 2019; Wastian et al., 2016). Area with limited green space can adopt green roof method, it also helps in sustaining the pollinators' diversity (Hofmann and Renner, 2018). Construction of artificial trap nest (bee hotels) on terrace is also an efficient method to conserve bees (Fortel et. al., 2016; MacIvor and packer, 2015).

5.4.5 Conservation in Agroecosystem

In agroecosystem pollinators provide a prime ecosystem services by pollinating crops and thus increasing food production (Ollerton et al., 2011). So the conservation of pollinators in the agroecosystem is a paramount necessity. The preliminary effort for this is to educate farmers regarding pollinators' services, threats on pollinators and their conservation via ecological intensification (Waddington, et al., 2014).Training farmers to adopt diversified farming system in order accomplish the hostile environment for pollinators safe guarding (Pywell et al., 2015). Sustaining the patches of untilled, barren land in order to provide, dry branches or logs, or sandy or earth bank for nesting of soil dwelling bees (Filipiak, 2018; Vaudo et al., 2015; Willmer, 2011; Dicks et al., 2010). Planting copiously blooming ornamental or wild flower plants on the bunds of the field act as a flower bank, which result in habitat enhancement and restoration for pollinators' upkeep (Pontin et al., 2006; Balzan, 2017).

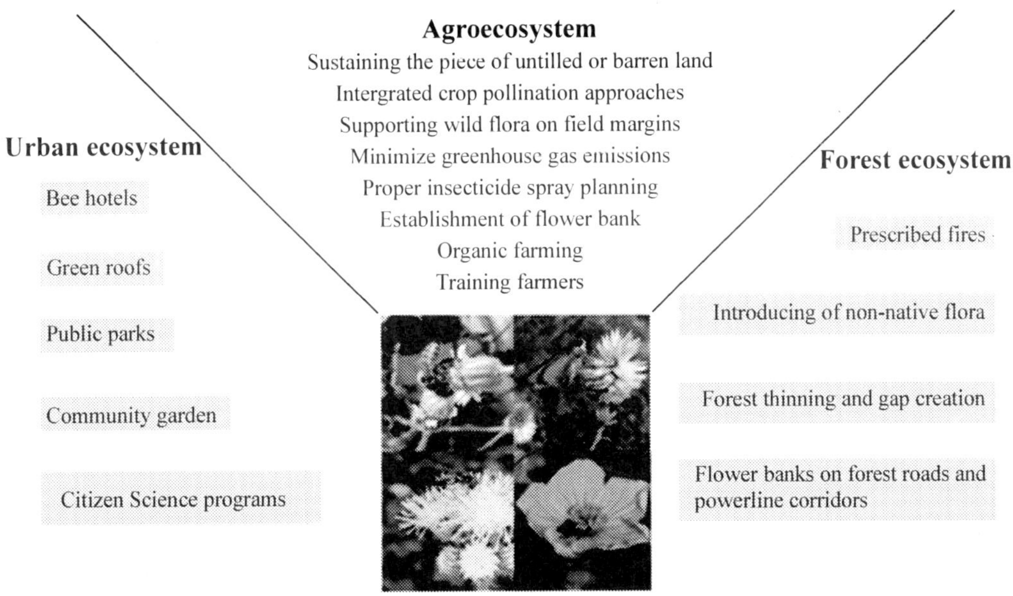

Adopting organic farming in agriculture land has proved safer for pollinators' survival (Tuck et al., 2014). Growing crop varieties which are rich in floral nutrition and has good interaction with pollinators' diversity (Ziska et al., 2016) may aids in conservation of important native pollinators. Use of managed bees for pollination of crops has shown negative impact on wild bees, so integrated crop pollination approaches, and by bee-friendly farming practices to achieve pollination effectiveness in modern intensive agroecosystem (Decourtye, 2019; Isaacs et al., 2017). Avoid the use of insecticides harmful to pollinators and circumvent the pesticide spray during floweringtime of the crop and weeds existing in field to keep away the pollinators exposure (Vighi et al., 2010). Supporting the tolerate level of weed flora on the field margins and adjacent wasteland to harbor pollinators diversity (Requier et al., 2015; Morandin and Kremen, 2013; Nicholls and Altieri, 2013; Hannon and Sisk, 2009). Minimize emissions of greenhouse gasses in agricultural land to take care of pollinators (Settele et al., 2008).

5.4.6. Conservation in Forest Ecosystem

Forest ecosystem act as a well-connected resource donor for pollinators' diversity. The distribution of pollinators varies across the forest area. In forest ecosystem pollinators are diversified and found abundantly in open forest than compare to closed forest (Hanula et al., 2016; Winfree, 2010). So, forest thinning and gap creation was found to be advantageous for pollinators' diversity and abundance (Taki et al., 2010; Romey et al., 2007; Kleintjes et al., 2004; Waltz and Covington, 2004). Supervised prescribed fires (controlled fire in order to influence herbaceous vegetation and reducing risk of wild fire due to fuel load) in forest ecosystem has ensued better forest habitat for pollinator persistence (Strahan et al., 2015; Nyoka, 2010; Wagner et al., 2003). Maintenance of flowering plants on forest roads and powerline corridors has proved efficacious way of conserving pollinators in forest area (Jackson et al., 2014; Skórka et al., 2013; Haddad et al., 2011). Introduction of non-native flora also enhance the pollinator survivability specially butterfly by supporting their larval stage (Schweitzer et al., 2011).

By following these above conservative strategies pollinators can be protected from the impact of climate change and other attributes responsible for their decline.

5.5. CONCLUSION

It is very difficult to predict what is and what will be happening with the species living on this earth. Again it is very difficult to control the anger of the mother earth. The only and one we can control is the activities that we performed for improving the living standards. Activity like deforestation aids to change in climate to a great extent which affects each and every living organism to some extent sooner or later. Pollinators in particular insect pollinators are unable to acclimatize themselves into a changing climatic condition because of their high sensitivity to climatic factors. Loss of diversity of pollinator may leads to loss of biodiversity as many of the wild as well as cultivated plant depends on pollinators for fertilisation and fruit set. So there is need to develop strategies for protecting the pollinators from the harmful effects of warmer climate. Emphasis should be put on the conservation of pollinators as well as pollinating plants. Otherwise, there may be severe disturbance of plant-pollinators interaction which may have negative impact on global food production and nutritional security.

REFERENCES

- Abeydeera LHUW, Mesthrige JW, Samarasinghalage TI (2019) Global research on carbon emissions: A scientometric review. Sustainability 11: 3972
- Abou-Shaara HF, Al-Ghamdi AA, Mohamed AA (2012) Tolerance of two honeybee races to various temperature and relative humidity gradients. Environ Exp Biol 10: 133-138
- Adler LS, Irwin RE (2006) Comparison of pollen transfer dynamics by multiple floral visitors: experiments with pollen and fluorescent dye. Ann Bot 97: 141-150
- Aguilera G, Ekroos J, Persson AS, Pettersson LB, Ockinger E (2019) Intensive management reduces butterfly diversity over time in urban green spaces. Urban Ecosyst 22: 335-344
- Aizen MA, Harder LD (2009) The global stock of domesticated honeybees is growing slower than agricultural demand for pollination. Curr Biol 19: 915-918
- Allen-Wardell G. et al. (1998) The potential consequences of pollinator declines on the conservation of biodiversity and stability of crop yields. Conserv Biol 12: 8-17
- All-Ireland Pollinator Plan 2015-2020. National Biodiversity Data Centre Series No. 3, Waterford.
- Araujo MB (2002) Biodiversity hotspots and zones of ecological transition. Conserv Biol 6: 1662-1663
- Araujo MB, Williams PH (2000) Selecting areas for species persistence using occurrence data. Biol Conserv 96: 331-451
- Araujo MB, Williams PH (2001) The bias of complementary hotspots toward marginal populations. Conserv Biol 15: 1710-1720
- Arora M, Goel NK, Singh P (2005) Evaluation of temperature trends over India/ Evaluation de tendances de temperature en Inde. Hydrol Sci J 50: 81-93
- Arunanondchai P, Fei C, Fisher A, McCarl BA, Wang W, Yang Y (2018) How does climate change affect agriculture. Handbook of Agricultural Economics. Abingdon-on-Thames, UK
- Ashman TL, Knight TM, Steets JA, Amarasekare P et al. (2004). Pollen limitation of plant reproduction: Ecological and evolutionary causes and consequences. Ecol 85(9): 2408-2421
- Baldock KC (2020). Opportunities and threats for pollinator conservation in global towns and cities. Curr Opin in Insect Sci 38: 63-71
- Baldock KC, Goddard MA, Hicks DM, Kunin WE, Mitschunas N, Morse H, Osgathorpe LM, Potts SG, Robertson KM, Scott AV, Staniczenko PP (2019) A systems approach reveals urban pollinator hotspots and conservation opportunities. Nature Ecol Evol 3(3): 363-373
- Balzan MV, Bocci G, Moonen AC (2014) Augmenting flower trait diversity in wildflower strips to optimise the conservation of arthropod functional groups for multiple agroecosystem services. J. Insect Conserv, 18: 713-728

- Barrett A and Brown, L (2021) .Effects or rainfall, temperature and photoperiod on the phenology of ephemeral resources for selected bushveld woody plant species in Southeren Africa. PLoS One, 16 (5): e0251421
- Bartomeus I, Ascher JS, Wagner D, Danforth BN, Colla S, Kornbluth S, Winfree R (2011) Climate-associated phenological advances in bee pollinators and bee-pollinated plants. Proc Natl Acad Sci USA 108: 20645-20649
- Bedford FE, Whittaker RJ, Kerr JT (2012) Systemic range shift lags among a pollinator species assemblage following rapid climate change. Botany 90: 587-597
- Biesmeijer JC (2006) Parallel declines in pollinators and insect-pollinated plants in Britain and The Netherlands. Sci 313: 351-354
- Bishop JA, Armbruster WS (1999) Thermoregulatory abilities of Alaskan bees: effects of size, phylogeny and ecology. Funct Ecol 13: 711-724
- Bosch J, Kemp WP, Peterson SS (2000) Management of *Osmia lignaria* (Hymenoptera: Megachilidae) populations for almond pollination: methods to advance bee emergence. Environ Entomol 29: 874-883
- Burd M (1994) Bateman's principle and reproduction: the role of pollinator limitation in fruit and seed set. Bot Rev 60: 83-139
- Burkle LA, Martin JC, Knight TM (2013) Plant-pollinator interactions over 120 years: loss of species, co-occurrence, and function. Sci 339: 1611-1615
- Cane JH, Tepedino VJ (2001) Causes and extent of declines among native North American invertebrate pollinators: detection, evidence, and consequences. Conserv Ecol 5: [online] URL: http://www.ecology and society.org/vol5/iss1/com
- Carvalheiro LG (2013) Species richness declines and biotic homogenisation have slowed down for NW-European pollinators and plants. Ecol Lett 16: 870-878
- Cervancia CR, de Guzman LI, Polintan EA, Dupo ALB, Locsin AA (2016) Current status of small hive beetle infestation in the Philippines. J Apic Res 55(1): 74-77
- Chechetka SA, Yu Y, Tange M, Miyako E (2017) Materially engineered artificial pollinators. Chem. 2: 224-239
- Chen I, Hill JK, Ohlemuller R, Roy DB, Tomas CD (2011) Rapid range shifts of species of climate warming. Sci 333: 1024-1026
- Corbet SA (1990) pollination and the weather. Israel J Bot 39: 13-30
- Corbet SA, Huang S (2016) Small bees overheat in sunlit flowers: do they make cooling flights? Ecol Entomol 41: 344-350
- Cornelissen B, Neumann P, Schweiger O (2019) Global warming promotes biological invasion of a honeybee pest. Glob Change Biol 25: 3642–3655
- Crane E, Walker P (1984) Pollination directory for world crops. International Bee Research Association. Bucks, UK
- Czekońska K (2009) The effect of different concentrations of carbon dioxide (CO_2) in a mixture with air or nitrogen upon the survival of the honeybee. Apic Res Bee World 48(1): 67-71

- Czekonska K, Tofilski A (2020) Body mass of honeybee drones developing in constant and in changing temperatures. Apidologie 51: 510-518
- Decourtye A, Alaux C, Le Conte Y, Henry M (2019) Toward the protection of bees and pollination under global change: present and future perspectives in a challenging applied science. Curr Opin in Insect Sci 35: 123-131
- Defra (2014) The National Pollinator Strategy: For Bees and Other Pollinators in England. Department for Environment, Food and Rural Affairs.
- Deguines N, Julliard R, de Flores M, Fontaine C (2012) The whereabouts of flower visitors: contrasting land-use preferences revealed by a country-wide survey based on citizen science. PLoS One 7: e45822
- Delisle CS, Kerr JT (2018) Climate change driven range losses among bumblebee species are poised to accelerate. Scientific Reports 8: 14464
- Devoto M, Medan D, Roig-Alsina A, Montaldo NH (2009) Patterns of species torn over in plant pollinator communities along a precipitation gradient in Patagonia (Argentina). Aust Ecol 34: 848-857
- Dicks LV, Showler DA, Sutherland WJ (2010) Bee conservation: evidence for the effects of interventions. Pelagic Publication
- Dicks LV, Viana B, Bommarco R, Brosi B, del Coro Arizmendi M, Cunningham SA, Galetto L, Hill R, Lopes AV, Pires C, Taki H. (2016). Ten policies for pollinators. Sci 354 (6315): 975-976
- Elias MAS, Borges FJA, Bergamini LL, Francceschineli EV, Sujii ER (2017) Climate change threatens pollination services in tomato crops in Brazil. Agric Ecosyst Environ 239: 257-264
- Elias MAS, Borges FJA, Bergamini LL, Franceschinelli EV, Sujii ER (2017) Climate hange threatens pollination services in tomato crops in Brazil. Agric Ecosyst Environ 239: 257-264
- Ensslin A, Mollel NP, Hemp A, Fischer M (2017) Elevational transplantation suggests different responses of African submontane and savanna plants to climate warming. J Ecol 45: 1-10
- Filipiak M (2018) A better understanding of bee nutritional ecology is needed to optimize conservation strategies for wild bees the application of ecological stoichiometry. Insects 9(3): 85
- Fortel L, Henry M, Guilbaud L, Mouret H, Vaissie`re BE (2016) Use of human-made nesting structures by wild bees in an urban environment. J Insect Conserv 20: 239-253
- Free JB (1993) Insect pollination of crops. London, UK: Academic Press
- Free JB, Spencer-Booth Y (1958) Observations on the temperature regulation and food consumption of honeybee (*Apis mellifica*). J Exp Biol 35(4): 930-970
- Gallai N, Settele JM, Vaissiere BE (2009) Economic valuation of the vulnerability of world agriculture confronted with pollination decline. Ecol Econ 68: 810-821
- Ghazoul J (2005) Response to Steffan-Dewenter et al.: Questioning the global pollination crisis. Trends Ecol. Evol. 20: 652-653

- Giannini TC, Acosta AL, Garofalo CA, Saraiva AM, Alves-dos-Santos I, Imperatriz-Fonseca VL (2012) Pollination services at risk: Bee habitats will decrease owing to climate change in Brazil. Ecol Model 244: 127-131
- Giannini TC, Acosta AL, Silva CI, Oliveira PEAM, Imperatriz-Fonseca VL, Saraiva AM (2013) Identifying the areas to preserve passion fruit pollination service in Brazilian Tropical Savannas under climate change. Agric Ecosyst Environ 171: 39-46
- Giannini TC, Acosta WF, Borges RC (2020) Climate change in the Eastern Amazon: Crop pollinator and occurrence-restricted bees are potentially more affected. Reg Environ Change 20: 9
- Giannini TC, Silva CI, Acosta AL, Jaffe R, Carvalho AT, Martins CF, Zanella FCV, Carvalho CAL, Hrncir M, Saraiva AM, Siqueira JO, Fonseca I (2017) Protecting a managed bee pollinators against climate change: Strategies for an area with extreme climatic conditions and socioeconomic vulnerability. Apidologie 48: 784-794
- Gonthier J, Papach A, Straub L, Campbell JW, Williams GR, Neumann P (2019) Bees and flowers: How to feed an invasive beetle species. Ecol Evol 9: 6422–6432
- Goulson D, Lye GC, Darvill B (2008) Decline and conservation of bumblebees. Annu. Rev. Entomol. 53: 191-208
- Greco MK, Hoffmann D, Dollin A, Duncan M, Spooner-Hart R, Neumann P (2010) The alternative Pharaoh approach: Stingless bees mummify beetle parasites alive. Naturwissenschaften, 97(3): 319–323
- Greenleaf SS, Williams NM, Winfree R, Kremen C (2007) Bee foraging ranges and their relationship to body size. *Oecol* 153: 589–596
- Haddad NM, Hudgens B, Damschen EI, Levey DJ, Orrock JL, Tewksbury JJ, Weldon AJ (2011) Assessing positive and negative ecological effects of corridors. Sources, sinks and sustainability, 475–503
- Hall DM, Steiner R (2019) Insect pollinator conservation policy innovations at subnational levels: lessons for lawmakers. Environ Sci Policy 93: 118–128
- Hallmann CA (2019) A tragedy of the common wild bird and insect decline in the current era. Nijmegen, The Netherlands: Radboud University
- Hannon LE, Sisk TD (2009) Hedgerows in an agri-natural landscape: potential habitat value for native bees. Biol Conserv 142(10): 2140–2154
- Hanula JL, Ulyshen MD, Horn S (2016) Conserving pollinators in North American forests: a review. Nat Areas J 36(4): 427–439
- Heard TA, Hendrikz JK (1993) Factors influencing flight activity of colonies of the stingless bee *Trigona carbonaria* (Hymenoptera, Apidae). Aust J Zool 41: 343-353
- Hegland SJ, Nielsen A, La´zaro1 A, Bjerknes, AL, Totland O (2009) How does climate warming affect plant-pollinator interactions. Ecol Lett 12: 184–195
- Heinrich B (1974) Thermoregulation in endothermic insects. Sci 185: 747-756
- Herrera CM (1987) Components of pollinator "quality": Comparative analysis of a diverse insect assemblage. Oikos 50:79-90

- Herrera JM, Ploquin EF, Rodriguez-Perez J, Obeso, JR (2014) Determining habitat suitability for bumblebees in a mountain system: a baseline approach for testing the impact of climate change on the occurrence and abundance of species. J. Biogeogr 41: 700-712
- Hill R, Nates-Parra G, Quezada-Euán JJG, Buchori D, LeBuhn G, Maués MM, Pert PL, Kwapong PK, Saeed S, Breslow SJ, da Cunha MC (2019) Biocultural approaches to pollinator conservation. Nat Sustain 2(3): 214-222
- Hofmann MM, Renner SS (2018) Bee species recorded between 1992 and 2017 from green roofs in Asia, Europe, and North America, with key characteristics and open research questions. Apidologie 49:307-313
- Hong BM, Hung HQ (2010) Effect of temperature and diet on the life cycle and predatory capacity of *Episyrphus balteatus* (Syrphidae: Diptera) cultured on *Aphis gossypii* (Glover). Int Soc Southeast Asian Agric 16: 98-103
- Howlett BG, Walker MK, Rader R, Butler RC, Newstrom-Lloyd LE, Teulon DAJ (2011) Can insect body pollen counts be used to estimate pollen deposition on pakchoi stigmas? N Z Plant Prot 64: 25-31
- IPCC. Climate Change 2014: Synthesis Report; Pachauri RK, Meyer, LA, (eds.); Contribution of Working Groups I, II and III to the Fifth Assessment Report of the Intergovernmental Panel on Climate Change; IPCC: Geneva, Switzerland, 2014; pp.151
- Isaacs R, Williams N, Ellis J, Pitts-Singer TL, Bommarco R, Vaughan M (2017) Integrated crop pollination: combining strategies to ensure stable and sustainable yields of pollination dependent crops. Basic Appl Ecol 22: 44-60
- Jackson MM, Turner MG, Pearson SM (2014) Logging legacies affect insect pollinator communities in Southern Appalachian forests. South Eastern Nat 12: 317-336
- Jones JC, Helliwell P, Beekman M, Maleszka R, Oldroyd BP (2003) The effects of rearing temperature on developmental stability and learning and memory in the honeybee, *Apis mellifera*. J Comp Physiol 191(12): 1121-1129.
- JuJones JC, Helliwell P, Beekman M, Maleszka R, Oldroyd BP (2005) The effects of rearing temperature on developmental stability and learning and memory in the honeybee, *Apis mellifera*. J Comp Physiol 191: 1121-1129
- Kaluza BF, Wallace H, Heard TA, Klein AM, Leonhardt SD (2016) Urban gardens promote bee foraging over natural habitats and plantations. Ecol Evol 6:1304-1316
- Kearns CA, Inouye DW, Waser N (1998) Endangered mutualisms: the conservation of plant–pollinator interactions. Annu Rev Ecol Syst 29:83-112
- Kerr JT, Pindar A, Galpern P, Packer L, Potts SG, Colla S (2015) Climate change impacts on bumblebees converge across continents. Sci 349:177-180
- Kingsolver JG, Huey RB (2008). Soil, temperature and fitness: three rule. Evol Ecol Res 10(2):251-268
- Klein AM, Vaissière BE, Cane JH, Steffan-Dewenter I, Cunningham SA (2007) Importance of pollinators in changing landscapes for world crops. Proc R Soc Lond B Biol Sci 274: 303–313

- Kleintjes PK, Jacobs BF, Fettig SM (2004) Initial response of butterflies to an overstory reduction and slash mulching treatment of a degraded piñon-juniper woodland. Restor Ecol 12: 231–238
- Kovac H, Käfer H, Stabentheiner A Costa MC (2014) Metabolism and upper thermal limits of *Apis mellifera carnica* and *Apis melliferra* L. Apidologie 45(6): 664–677
- Kremen C, Ricketts T (2000) Global perspectives on pollination disruptions. Conserv Biol 14: 1226–1228
- Ladinig U, Pramsohler M, Bauer I, Zimmermann S, Neuner G, Wagner J. Is sexual reproduction of high-mountain plants endangered by heat? Oecologia177: 1195–1210
- Larsen TH, Williams N, Kremen C (2005) Extinction order and altered community structure rapidly disrupt ecosystem functioning. Ecol Lett 8: 538–547
- Le Conte Y, Navajas M (2008) Climate change: impact on honeybee populations and diseases. Rev Sci Tech Off Int Epiz 27 (2): 499–510
- Leve M, Baudry E, Bessa-Gomes C (2019) Domestic gardens as favorable pollinator habitats in impervious landscapes. Sci Total Environ 647: 420–430
- MacIvor JS, Packer L (2015) 'Bee hotels' as tools for native pollinator conservation: a premature verdict? PLoS One 10: e0122126
- Martins AC, Silva DP, Marco PJ, Melo GAR (2015) Species conservation under future climate change: the case of *Bombus bellicosus*, a potentially threatened South American bumblebee species. J Insect Conserve 19: 33–43
- Martins KT, Gonzalez A, Lechowicz MJ (2015) Pollination services are mediated by bee functional diversity and landscape context. Agric Ecosyst Environ 200: 12–20
- McAfee A, Chapman A, Higo H (2020) Vulnerability of honeybee queens to heat-induced loss of fertility. *Nat Sustain* 3: 367–376
- McGregor SE (1976) Insect pollination of cultivated crop plants. U.S.D.A. Agriculture Handbook 496: 93–98
- Medrzycki P, Sgolastra F, Bortolotti L, Bogo G, Tosi S, Padovani E, Porrini C, Sabatini AG (2010). Influence of brood rearing temperature on honeybee development and susceptibility to poisoning by pesticides. J Apic Res 49(1): 52–59
- Miller-Struttmann NE, Geib JC, Franklin JD, Kevan PG, Holdo RM, Ebert-May D, Lynn AM, Kettenbach JA, Hedrick E, Galen C (2015) Functional mismatch in a bumblebee pollination mutualism under climate change. Sci 349: 1541–1544
- Morandin LA, Kremen C (2013) Hedgerow restoration promotes pollinator populations and exports native bees to adjacent fields. Ecol Appl 23(4): 829–39
- Nabhan GP, Buchmann S (1997) Services provided by pollinators. In: Daily GG (ed) Nature's services: societal dependence on natural ecosystems. Island Press:Washington, DC
- National Research Council (2007) Status of pollinators in North America. The National Academies Press. Washington, DC 303
- Neelam M (2017) Impact of global climate change on pollinator diversity and conservation: Current issues and future challenges. J Res 8(1): 76–88

- Nicholls CI, Altieri MA (2013) Plant biodiversity enhances bees and other insect pollinators in agroecosystems. A review. Agron Sustain Dev 33(2): 257–74
- Norgate M, Boyd-Gerny S, Simonov V, Rosa MGP, Heard TA, Dyer AG (2010) Ambient temperature influences Australian native stingless bee (*Trigona carbonaria*) preference for warm nectar. PLoS One 5: e12000
- Noya I, González-García S, Bacenetti J, Fiala M, Moreira MT (2018) Environmental impacts of the cultivation-phase associated with agricultural crops for feed production. J Clean Prod 172: 3721–3733
- Nurnberger F, Hartel F, Dewenter IS (2018) The influence of temperature and photoperiod on the timing of brood onset in hibernating honeybee colonies. Peer J 6: e4801
- Nyoka S (2010) Effects of fuel-reduction treatments on pollinators in a Pinyon-Juniper woodland (Arizona). Ecol Restor 31: 119–121
- Oliveira MO, Freitas BM, Scheper J, Kleijn D (2016) Size and sex-dependent shrinkage of Dutch bees during one-and-a-half centuries of land-use change. PLoS One 11: e0148983
- Ollerton J Winfree R, Tarrant S (2011) How many flowering plants are pollinated by animals? Oikos 120(3): 321–326
- Parmesan C, Ryrholm N, Stefanescu C, Hill JK, Thomas CD, Descimon H, Huntley B, Kaila L, Kullberg J, Tammaru T, Tennent WJ, Thomas JA, Warren MS (1999) Poleward shifts in geographical ranges of butterfly species associated with regional warming. Nature 399: 579–583
- Parmesan C, Yohe G (2003) A globally coherent fingerprint of climate change impacts across natural systems. Nature 421: 37–42
- Patino S, Grace J (2002) The cooling of convolvulaceous flowers in a tropical environment. Plant Cell Environ 25: 41–51
- Pollinator Health Task Force (2016): Pollinator Partnership Action Plan
- Pontin DR, Wade MR, Kehrli P, Wratten SD (2006) Attractiveness of single and multiple species flower patches to beneficial insects in agroecosystems. Ann App Biol 148: 39–47
- Potts SG, Ngo HT, Biesmeijer JC, Breeze TD, Dicks LV, Garibaldi LA, Hill R, Settele J, Vanbergen A (2016) The assessment report of the Intergovernmental Science-Policy Platform on Biodiversity and Ecosystem Services on pollinators, pollination and food production, Secretariat of the Intergovernmental Science-Policy Platform on Biodiversity and Ecosystem Services, Bonn, Germany
- Potts SG, Roberts SP, Dean R, Marris G, Brown MA, Jones R, Neumann P, Settele J (2010) Declines of managed honeybees and beekeepers in Europe. J Apic Res 49(1): 15–22
- Powney GD, Carvell C, Edwards M, Morris RKA, Roy HE, Woodcock BA, Isaac NJB (2019) Widespread losses of pollinating insects in Britain. Nat Commun 10: 1–6
- Price MV, Waser NM (1979) Pollen dispersal and optimal out-crossing in *Delphinium nelson*. Nature. 277:294-297
- Pywell RF, Heard MS, Woodcock BA, Hinsley S, Ridding L, Nowakowski M, Bullock JM (2015) Wildlife-friendly farming increases crop yield: evidence for ecological intensification. Proc R Soc 282:20151740

- Radmacher S, Strohm E (2010) Factors affecting offspring body size in the solitary bee *Osmia bicornis* (Hymenoptera, Megachilidae). Apidologie 41:169-177
- Radmacher S, Strohm E. (2011) Effects of constant and fluctuating temperatures on the the development of the solitary bee *Osmia bicornis* (Hymenoptera: Megachilidae) Apidologie. 42: 711-720
- Rasmont P, Mersch P (1988) Premie`re estimation de la derive faunique chez les bourdons de la Belgique (Hymenoptera, Apidae). Ann Soc R Zool Belg 118:141-147
- Reed M (2008) Stakeholder participation for environmental management: a literature review. Biol Conserv 141: 2417-2431
- Requier F, Odoux JF, Tamic T, Moreau N, Henry M, Decourtye A, Bretagnolle V (2015) Honeybee diet in intensive farmland habitats reveals an unexpectedly high flower richness and a major role of weeds. Ecol App 4: 881-90
- Richards AJ (2001) Does low biodiversity resulting from modern agricultural practice affect crop pollination and yield? Ann. Bot 88:165-172
- Romey WL, Ascher JS, Powell DA, Yanck M (2007) Impacts of logging on midsummer diversity of native bees (Apoidea) in a northern hardwood forest. J Kans Entomol Soc 80: 327-338
- Rose T, Kremen C, Thrupp A, Gemmill-Herren B, Graub B, Azzu N (2016) Mainstreaming of Biodiversity and Ecosystems Services with a Focus on Pollination. Food and Agriculture Organization of the United Nations
- Roy HE, Baxter E, Saunders A, Pocock MJO (2016) Focal plant observations as a standardised method for pollinator monitoring: opportunities and limitations for mass participation citizen science. PLoS One 11: e0150794
- Sammataro D, Gerson U, Needham G (2000) Parasitic mites of honeybees: life history, implications, and impact. Annu Rev Entomol 45: 519-548
- Sathaye J, Shukla PR, Ravindranath NH (2006) Climate change, sustainable development and India: Global and national concerns. Curr Sci 90: 314-325
- Saunders ME, Roger E, Geary WL, Floret M, Welbourne DL, Bako A, Canavan E, Herro F, Herron C, Hung O, Kunstler M (2018) Citizen science in schools: engaging students in research on urban habitat for pollinators. Austral Ecol 43: 635-642
- Scaven VL, Rafferty NE (2013) Physiological effects of climate warming on flowering plants and insect pollinators and potential consequences for their interactions. Curr Zool 59: 418-426
- Schweitzer DF, Minno MC, Wagner DL (2011) Rare, Declining, and Poorly Known Butterflies and Moths (Lepidoptera) of Forests and Woodlands in the Eastern United States. Forest Health Technology Enterprise Team, Washington, DC.
- Seeley TD, Visscher PK (1985) Survival of honeybees in cold climates: the critical timing of colony growth and reproduction. Ecol Entomol 10(1): 81–88
- Segurado P, Araujo MB (2004) An evaluation of methods for modelling species distributions. J Biogeogr 31: 1555–1568

- Settele J, Kudrna O, Harpke A, Kühn I, Van Swaay C, Verovnik R, Warren MS, Wiemers M, Hanspach J, Hickler T, Kühn, E. (2008) Climatic risk atlas of European butterflies, Sofia: Pensoft. pp.710
- Skórka P, Lenda M, Möron D, Kalarus K, Tryjanowski P (2013) Factors affecting road mortality and the suitability of road verges. Biol Conserv 159: 148–157
- Solignac M, Cornuet JM, Vautrin D, Le Conte Y, Anderson D, Evans J, Cros-Arteil S, Navajas M (2005) The invasive Korea and Japan types of *Varroa destructor*, ectoparasitic mites of the Western honeybee (*Apis mellifera*), are two partly isolated clones. Proc roy Soc Lond B biol Sci 272 (1561): 411–419
- Solis-Montero L, Vallejo-Marin M (2017) Does the morphological fit between flowers and pollinators affect pollen deposition? An experimental test in a buzz-pollinated species with anther dimorphism. Ecol Evol 7: 2706–2715
- Soroye P, Newbold T, Kerr J (2020) Climate change contributes to widespread declines among bumblebees across continents. Sci 367(6478): 685–688
- Spiewok S, Neumann P (2006) Infestation of commercial bumblebee (*Bombus impatiens*) field colonies by small hive beetles (*Aethina tumida*). Ecol Entomol 31(6): 623–628
- Steffan-Dewenter I, Potts SG, Packer L (2005) Pollinator diversity and crop pollination services are at risk. Trends Ecol Evol 20: 651–652
- Stenseth NC, Mysterud A (2002) Climate, changing phenology, and other life history and traits: nonlinearity and match-mismatch to the environment. Proc Natl Acad Sci 99: 13379–13381
- Stone GN (1994) Activity patterns of females of the solitary bee Anthophora plumipes in relation to temperature, nectar supplies and body size. Ecol Entomol 19: 177–189
- Strahan RT, Stoddard MT, Springer JD, Huffman DW (2015) Increasing weight of evidence that thinning and burning treatments help restore understory plant communities in ponderosa pine forests. For Ecol Manag 353: 208–220
- Taki H, Inoue T, Tanaka H, Makihara H, Sueyoshi M, Isono M, Okabe K (2010) Response of community structure, diversity, and abundance of plants and insect assemblages to thinning in plantations. For Ecol Manag 259: 607–613
- Tautz J, Maier S, Groh C, Rossler W, Brockmann A (2003) Behavioural performance in adult honeybees is influenced by the temperature experienced during their pupal development. Proc Natl Acad Sci USA 100: 7343–7347
- Thomas C (2014) Extinction risk from climate change Extinction risk from climate change. Nat Clim Change 4: 17–27
- Thuiller W, Araujo MB, Lavorel S (2003) Generalized models versus classification tree analysis: a comparative study for predicting spatial distributions of plant species at different scales. J Veg Sci 14: 669–680
- Thuiller W, Lavorel S, Araujo MB, Sykes MT, Prentice C (2005) Climate change threatens to plant diversity in Europe. PNAS 102(23): 8245–8250

- Tuck SL, Winqvist C, Mota F, Ahnströ M J, Turnbull LA, Bengtsson J (2014) Land-use intensity and the effects of organic farming on biodiversity: a hierarchical meta-analysis. J Appl Ecol 51: 746–755
- van Lexmond MB, Bonmatin JM, Goulson D et al. (2015) Worldwide integrated assessment on systemic pesticides. Environ Sci Pollut Res 22, 1–4
- vanEngelsdorp D, Hayes Jr J, Underwood RM, Pettis JS (2010) A survey of honeybee colony losses in the United States, fall 2008 to spring 2009. J Apic Res 49(1): 7–14
- Vaudo AD, Tooker JF, Grozinger CM, Patch HM (2015) Bee nutrition and floral resource restoration. Curr Opin Insect Sci 10: 133–141
- Vighi M, Settele J, Potts SG (2010) The impact of an insecticide on insect flower visitation and pollination in agricultural landscape. Agric For Entomol 12: 259–266
- Visser ME, Both C (2005) Shifts in phenology due to global climate change: the need for a yardstick. Proc R Soc B 272: 2561–2569
- Waddington H, White H (2014) Farmer field schools: From agricultural extension to adult education. 3ie systematic review summary 1. London: International Initiative for Impact Evaluation
- Wagner DL, Nelson MW, Schweitzer DF (2003) Shrubland Lepidoptera of southern New England and southeastern New York: Ecology, conservation, and management. For Ecol Manag 185: 95–112
- Walther GR, Burga CA, Edwards PJ (eds.) (2001) "Fingerprints" of climate change – Adapted behavior and shifting species ranges. Kluwer Academic/Plenum Publishers, New York / London. pp 329
- Waltz AEM, Covington WW (2004) Ecological restoration treatments increase butterfly richness and abundance: Mechanisms of response. Restor Ecol 12: 85–96
- Warren MS, Hill JK, Thomas JA, Asher J, Fox R, Huntley B, Roy DB, Telfer MG, Jeffcoate S, Harding P, Jeffcoate G, Willis SG, Greatorex-Davies JN, Moss D, Thomas CD (2001) Rapid responses of British butterflies to opposing forces of climate and habitat change. Nature 414: 65–69
- Wastian L, Unterweger PA, Betz O (2016) Influence of the reduction of urban lawn mowing on wild bee diversity (Hymenoptera, Apoidea). J Hymenopt Res 49:51-63
- Westerkamp C, Gottsberger G (2000) Diversity pays in crop pollination. Crop Sci 40: 1209–1222
- WG (Welsh Government) (2013) The Action Plan for Pollinators in Wales. Welsh Government
- Willmer P (2011) Pollination and floral ecology. Princeton University Press, Princeton
- Wilcock C, Neiland R (2002) Pollination failure in plants: Why it happens and when it matters. Trends in Plant Sc. 7: 270–277
- Williams IH (1994) The dependences of crop production within the European Union on pollination by honeybees. Agric Zool Rev 6: 229–257

- Williams R, Rao GR, Kremen WH (1994) Effects of elevated CO_2 and temperature grown red and sugar maple on gypsy moth performance. Global Change Biol 1: 685–695
- Wilson JRU, Richardson DM, Rouget M, Proches S, Amis MA, Henderson L, Thuiller W (2007) Residence time and potential range: crucial considerations in modelling plant invasions. Divers Distrib 13: 11–22
- Wilson RJ, Gutiérrez D, Gutiérrez J, Martínez D, Agudo R, Monserrat VJ (2005) Changes to the elevation limits and extent of species ranges associated with climate change. Ecol Lett 8: 1138–1146
- Winfree R (2010) The conservation and restoration of wild bees. Annals of the New York Academy of Sciences 1195: 169–197
- World Population Review. Walnut, United States. Available online: www.worldpopulationreview.com (accessed on 12 May 2020)
- Zanella FCV (2000) The bees of the Caatinga (Hymenoptera, Apoidea, Apiformes): a species list and comparative notes regarding their distribution. Apidologie 31: 579–592
- Ziska LH, Pettis JS, Edwards J, Hancock JE, Tomecek MB, Clark A, Dukes JS, Loladze I, Polley HW (2016) Rising atmospheric CO_2 is reducing the protein concentration of a floral pollen source essential for North American bees. Proc Biol Sci. 13; 283(1828): 20160414

6

Chemical Ecology in Cross-Pollination

K. Nishanthini* and R. Kanagarajan

ABSTRACT

Floral chemistry mediates plant-pollinator interactions through floral scents and reward components. Although improved techniques have increased interest in studying floral volatiles and nectar chemistry, these two foci have generally been studied in isolation. The ecological functions of floral chemistry have been relatively well studied and focused on pollinator behavior. While studies comparing chemistry between plant parts and across phylogenies are increasing, work on the evolution of floral chemistry and the importance of community context in mediating pollinator responses is lacking. Future research should concentrate on more holistic studies that include both signal and reward chemistry to understand the relative contribution of these complex and dynamic floral traits to the ecology and evolution of plants and their pollinators

Keywords *Pollination • Pollinators • Floral scent • Attraction • Cross-pollination • Chemical ecology*

6.1 INTRODUCTION

Chemical ecology refers to chemically mediated interactions between organisms and their biotic and abiotic environment. It covers a broad range of chemical interactions and signaling processes; major facets are:

(i) The chemical communication (chemical language, e.g., pheromones) of animals, particularly expressed in arthropods.

(ii) The mutualistic interactions of organisms, e.g., plants and animals (pollination), plants and fungi (mycorrhiza), and plants and bacteria (symbiotic nitrogen fixation).

(iii) The chemical defenses of organisms, e.g., plant defenses against herbivores and pathogens, animal defenses against predators and parasitoids, and microor-ganism defenses against food competitors.

[1]Department of Entomology, Faculty of Agriculture, Annamalai University, Tamil Nadu-608 002, India.

*Corresponding author E-mail: nishanthee97@gmail.comItemenatqui consulos di con

(iv) Protection against abiotic stress, e.g., plant defenses against damage by UV light, drought, or cold.

Animal-pollinated plants generally attract pollinators by advertising their floral rewards with visual and olfactory signals. While the role of plant chemistry in mediating plant-pollinator interactions has long been recognized, technical advances have made floral chemical analyses more feasible, accurate and accessible to ecologists and evolutionary biologists. Although measuring floral chemistry remains challenging, the field is flourishing. Studies on floral volatiles have dominated the literature, while nectar chemistry has lagged behind until very recently, likely due to additional difficulties in analyzing trace compounds in small volumes of nectar. Beyond technical differences in examining floral volatiles and nectar chemistry, these two chemical aspects of floral biology have remained somewhat distinct fields. Our review focuses on the most recent work addressing how floral chemistry contributes to the ecology and evolution of plant-pollinator interactions (Figure 6.1). Specifically, we highlight the theoretical and empirical links between floral volatiles and nectar chemistry and summarize pollinator responses to plant chemistry with the aim to create an integrated view of floral chemical ecology.

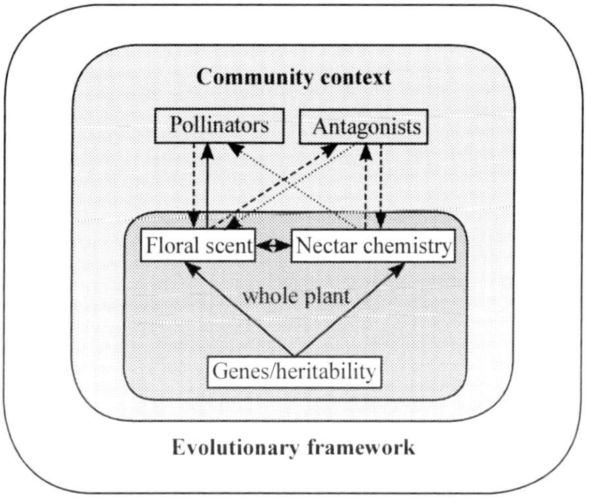

Fig. 6.1 Ecology and evolution of plant-pollinator interactions

A major area of chemical ecology concerns the constant competition between the worlds of autotrophs and heterotrophs or simply plants and animals. During their evolution, plants have evolved sophisticated adaptations to cope with herbivores and pathogens while the latter developed similarly elaborated counter adaptations to overcome plants' defenses. Plants produce a diverse array of metabolites that are not involved in primary metabolism. These secondary metabolites determine our sensory perception of unique characteristics of plants. We see the pretty colors and smell the fragrances of flowers and fruits, and we appreciate the distinctive tastes of spices, vegetables, and fruits. Moreover, all of the biological activities of plants that humans have used for medicinal reasons for hundreds of years can be attributed to secondary metabolites.

6.2 CROSS-POLLINATION

Cross-pollination, also called heterogamy, type of pollination in which sperm-laden pollen grains are transferred from the cones or flowers of one plant to egg-bearing cones or flowers of another. Cross-pollination is found in both angiosperms (flowering plants) and gymnosperms (cone-bearing plants) and facilitates cross-fertilization and outbreeding. This movement of pollen may occur by wind, as in conifers, or via symbiotic relationships with various animals (e.g., bees and certain birds and bats) that carry pollen from plant to plant while feeding on nectar.

6.3 MECHANISMS THAT PREVENT SELF-POLLINATION

Many flowering plants have evolved to limit self-pollination and facilitate cross-pollination. Self-pollination can be reduced or nearly eliminated by the structure of the flower, self-incompatibility, and the timing of the maturation of stamens and pistils of the same flower or plant. Cross-pollination may then be brought about by a number of agents, chiefly insects and wind. Wind-pollinated flowers generally can be recognized by their lack of color, odor, or nectar and their stigmas that are arranged to optimize the capture of airborne pollen. Animal-pollinated flowers are conspicuous by virtue of their structure, color, or the production of scent or nectar, all of which evolved in tandem with animal pollinators.

6.3.1 Structural

Many species of plants have developed physical and structural mechanisms that prevent self-pollination. Some date palms (*Phoenix dactylifera*) and willows (*Salix* species)—have become dioecious; that is, some plants produce only "male" (staminate) flowers, with the rest producing only "female" (pistillate or ovule-producing) ones. In species in which staminate and pistillate flowers are found on the same individual (monoecious plants) and in those with hermaphroditic flowers (flowers possessing both stamens and pistils), a common way of preventing self-fertilization is to have the pollen shed either before or after the period during which the stigmas on the same plant are receptive, a situation known as dichogamy. The more usual form of dichogamy, which is found especially in such insect-pollinated flowers as fireweed (*Epilobium angustifolium*) and salvias (*Salvia* species), is protandry, in which the stamens ripen before the pistils. Protogyny, the situation in which the pistils mature first, occurs in arum lilies and many wind-pollinated plants, such as grasses although several grasses are self-pollinated, including common varieties of wheat, barley, and oats. Avocado has both protogynous and protandrous varieties, and these often are grown together to encourage cross-fertilization

6.3.2 Chemical

Chemical self-incompatibility is another device for preventing self-fertilization. In this phenomenon, which depends on chemical substances within the plant, the pollen may fail to grow on a stigma of the same flower that produced it or, after germination, the pollen tube may not grow normally down the style to effect fertilization. This occurs in white clover (*Trifolium repens*), cabbage (*Brassica oleracea*), and many other species. The process is controlled genetically; it need not be absolute and can change in degree during the flowering season. Not surprisingly, chemical incompatibility usually is not found in those plants that have strong structural or temporal barriers against self-pollination. Formation of one such mechanism during evolution apparently was enough for most plant species.

6.3.3 Floral Chemistry Functions

Floral volatiles act as long distance attractants and provide unique chemical signatures to flowers, thereby influencing pollinator preference and constancy, while nectar chemical composition can affect initial foraging decisions, time pollinators spend foraging and the likelihood of visiting a conspecific flower. Current advances in understanding functional differences of floral scent often compare variation in systems with different pollen vectors (Schiestl and Johnson, 2013). In Mimulus, bumblebee preference is driven by its antennal sensitivity, which may subsequently contribute to reproductive isolation between plant species (Byers et al., 2014); similar preference patterns are detected in pollinating beetles of Protea (Steenhuisen et al., 2012). However, increased attention to repellent/defensive aspects of floral volatiles (Kessler et al., 2013; Theis and Adler, 2012) suggests that limiting our view to pollinator attraction severely reduces our understanding of floral chemical trait function.

Historically, nectar sugars were assumed to drive preferences, however the prevalence of nectar secondary metabolites led to a reassessment in light of these putatively noxious compounds. While nectar compounds can deter pollinators and alter their behavior on flowers, experiments often find these negative effects only with concentrations significantly higher than occur naturally in nectar (Manson et al., 2012; Tiedeken et al., 2014). At low concentrations, nectar secondary metabolites may be neutral or even attractive and their functions may go beyond simple attraction/deterrence. Caffeine-enriched nectar improved honeybees' ability to retain associations between floral scent and reward due to caffeine's effect on adenosine receptors in the insect brain (Wright et al., 2013). Furthermore, plants may seek to manipulate pollinators into consuming less nectar, especially if this leads to higher outcrossing rates (Kessler et al., 2012). Bumblebees and birds consume less nicotine-enriched nectar (Tiedeken et al., 2014; Lerch-Henning and Niclson, 2013) and honeybee nectar storage declines with increased nectar nicotine concentrations (Koehler, 2012). However, sugar can modulate responses; birds and honeybees tolerate higher nicotine concentrations with increased sugar levels (Lerch-Henning and Niclson, 2013; Koehler, 2012).

Thus, nectar chemistry may reflect a complex interplay of pollinator behaviors and tolerances in addition to alternate functions such as protection from antagonists. Similar processes may drive function and evolution of pollen chemistry, however this is a poorly understood aspect of floral chemistry. For many plant species, pollen is simply the vehicle for gamete transportation, however, insects often exploit pollen as a resource. Pollen can therefore play a dual role in many systems, having both sexual and reward functions and its chemical composition may influence how insects interact with flowering plants. Like nectar, pollen contains both nutritional and non-nutritional components, with recent studies finding compounds such as alkaloids (Gosselin et al., 2013) and pollen collection is similar regardless of chemical variation in pollen diet while syrup collection is variable. Micro-colonies fed on *S. aucuparia* and *C. scoparius* pollen produced larger larvae (i.e., better mates and winter survivors) and fed less on nectar compared to the other diets. Pollen from both of these species contains 24-methylenecholesterol and high concentrations of polypeptides/total amino acids. This pollen nutritional "theme" seems therefore to promote worker reproduction in *B. terrestris* micro-colonies and could be linked to high fitness for queen right colonies. As workers are able to selectively forage on pollen of high chemical quality, plants may be evolutionarily selected for their pollen content, which might attract and increase the degree of fidelity of generalist pollinators, such as bumblebees (Vanderplanck et al., 2014). These compounds may have negative (Vanderplanck et al., 2014) or neutral (Sedivy et al., 2012) post-consumption effects and may prompt pollen collecting insects to seek multiple pollen sources in order to reduce costs associated with pollen toxins (Eckhardt et al., 2014).

Despite the importance of pollen protein for many social and solitary bees, studies on pollen chemical ecology are sparse. This area of research is rich in questions, poor on data and deserves significantly more attention. Comprehensive studies of floral advertisements and rewards (including pollen) are required to understand whether chemistry has additive or emergent functions in plant-pollinator interactions. Furthermore, pollinator foraging constancy can be enhanced for multimodal signals such as color and odor combined; in Ipomopsis, the volatile indole is sufficient to attract hawkmoth pollinators but does not elicit nectar probing behavior without the appropriate petal color (Bischoff et al., 2015). Thus in roads into understanding functions of floral chemistry must include the broader context of floral traits.

6.3.4 Whole Plant Chemistry

Early studies on floral chemistry focused on analysing floral headspace or nectar in isolation but more recently, researchers have taken a comparative whole plant approach to assess patterns of production, expression and allocation of compounds and test functional hypotheses. Similar secondary metabolite profiles across tissues suggest that nectar chemistry reflects shared biosynthetic pathways or pleiotropy (Adler et al., 2012).

Whole plant studies show many nectar secondary metabolites are found throughout plants and concentrations are positively correlated across plant parts in many species, although secondary metabolite concentrations are often orders of magnitude smaller in nectar relative to flowers, leaves (Adler et al., 2012; Cook et al., 2013) and phloem (Lohaus and Schwerdtfeger, 2014). However, flowers often possess distinct chemical components or ratios/concentrations (especially in floral scents), which may either signal or shape functional and adaptive differences between flowers and other plant parts. The leaves and flowers of *Brassica rapa* emit nearly identical volatile profiles but compound concentrations were not correlated (Schiestl et al., 2014). In milkweeds (*Asclepias species*), composition and concentrations of individual cardenolides varied across plant parts and two species had a unique nectar compound. Further, leaf and floral volatiles may differ but act synergistically to attract visitors, suggesting that volatiles need to be understood in a whole plant context (Karpati et al., 2013).

Differences in floral chemistry of leaves and flowers are not unexpected, as plants are under selection to attract pollinators to flowers but deter herbivores from leaves. Given that the overall chemical composition of nectar differs significantly from phloem, indicating modification of nectar precursors via metabolism or selective secretion (Lohaus and

Schwerdtfeger, 2014), there is mounting evidence to refute that secondary metabolites simply leak into nectar.

6.4 EVOLUTION OF FLORAL CHEMISTRY

Floral chemistry is hypothesized to be the product of natural selection yet research is only just considering the micro-evolution of these traits (Parachnowitsch, 2014). Selection on volatiles can be detected (Ehrlen et al., 2012) and can be stronger than selection on other advertisements such as flower size or colour (Parachnowitsch et al., 2012). Indirect evidence for the potential for selection on nectar chemistry comes from fitness effects of pollinator responses. For example, nectar alkaloids reduced pollen receipt and subsequent *Gelsemium sempervirens* fruit set (Adler and Irwin, 2012), however, nectar alkaloid enrichment in *Delphinium barbeyi* had no plant fitness effects (Manson et al., 2013). Because floral signals and rewards may be linked, especially if signals are honest (Knauer and Schiestl, 2015), correlational selection may drive their evolution. Nectars themselves can be scented (Parachnowitsch et al., 2013) suggesting a strong possibility for correlated evolution. Rapid evolution of floral scent is suggested by examples of reduced selfing in *Abronia umbellata* populations, where attractive scents are presumably no longer needed (Doubleday et al., 2013), and variation among horticultural varieties of Phlox (Majetic et al., 2014). Further, selection on the eugenol synthesis gene in Gymnadenia suggests changes to single genes can increase floral scent complexity (Gupta et al., 2014). Although some biochemical pathways and genes associated with the production of alkaloids and/or floral volatiles have been characterized, heritability of their variation is largely unknown. Recent advances in understanding the genetic basis of chemistry include mapping the floral scent QTLs of cowpea (Andargie et al., 2014) and correlating floral scent and AFLP data in Sorbus (Feulner et al., 2014).

Use of hybrids has also shown the genetic bases ofscent production can be complex, as hybrid scent can differ from parental scent or hybrids can express the parental mean depending on compounds (Bischoff et al., 2013; Page et al., 2014). There is a striking lack of data on the heritability of nectar chemistry and uncovering the genetic basis of these traits should be a research priority. Evidence for floral chemical trait evolution comes from local adaptation and phylogentic studies, often suggesting that pollinators likely play a role. For example, in the rewarding *Gymnadenia odoratissima* and sexually deceptive *Ophrys sphegodes*, scent likely contributes to local adaptation of pollinator communities (Sun et al., 2014; Breitkopf et al., 2013). In Lithophragma, where *Greya politella* moths act as both pollinators and seed parasites, closely related Lithophragma species exhibit extreme variation in floral scent (Friberg et al., 2013) that seems to drive moth preference for their local host species (Friberg et al., 2014). Broad phylogenetic sampling shows nicotine concentrations in Nicotiana nectar are lower in pollinator dependent species, suggesting that pollinators could drive chemical composition of nectar. Pollinator mediated selection also likely drives floral scent evolution, as suggested by two meta-analyses of oviposition site mimics (Jurgens et al., 2013) and beetle-pollinated plants (Schiestl and Dotter, 2012). Research incorporating the heritability of chemical traits, pollinator responses to trait variation, and pollinator-mediated phenotypic selection will greatly advance our understanding of the evolution of nectar and volatile chemistry.

6.5 CONCLUSION

Studies that bridge spatial and temporal scales to address how floral chemistry affects multispecies interactions, feedback within communities and the co-evolution of plants and their pollinators will push forward the boundaries of our field. Recently, research considering floral chemistry in a changing climate has found that increases in temperature could have species-specific effects on volatile emissions and, coupled with increases in carbon dioxide and nitrogen, can also affect nectar composition. However, without a more comprehensive understanding of floral chemistry in shaping relationships between plants and pollinators, it becomes difficult to predict how shifts in chemical composition might affect the stability and sustainability of key mutualisms. The chemical ecology of plant pollinator interactions is an emerging field ripe for exploration and will no doubt provide incredible insight into the ecology and evolution of species interactions.

REFERENCE

- Adler LS, Irwin RE (2012) Nectar alkaloids decrease pollination and female reproduction in a native plant. Oecologia, 168: 1033–1041
- Adler LS, Seifert MG, Wink M, Morse GE (2012) Reliance on pollinators predicts defensive chemistry across tobacco species. Ecol Lett, 15: 1140–1148
- Andargie M, Knudsen JT, Pasquet RS, Gowda BS, Muluvi GM, Timko MP (2014) Mapping of quantitative trait loci for floral scent compounds in cowpea (*Vigna unguiculata* L.). Plant Breed, 133: 92–100
- Bischoff M, Jurgens A, Campbell DR (2013) Floral scent in natural hybrids of Ipomopsis (Polemoniaceae) and their parental species. Ann Bot http://dx.doi.org/10.1093/aob/mct279
- Bischoff M, Raguso RA, Jurgens A, Campbell DR (2015) Contextdependent reproductive isolation mediated by floral scent and color. Evolution, 69: 1–13 http://dx.doi.org/10.1111/evo.12558
- Breitkopf H, Schluter PM, Xu S, Schiestl FP, Cozzolino S, Scopece G (2013) Pollinator shifts between *Ophrys sphegodes* populations: might adaptation to different pollinators drive population divergence? J Evol Biol, 26: 2197–2208
- Byers KJRP, Bradshaw HD, Riffell JA (2014) Three floral volatiles contribute to differential pollinator attraction in Monkeyflowers (Mimulus). J Exp Biol, 217: 614–623
- Cook D, Manson JS, Gardner DR, Welch KD, Irwin RE (2013) Norditerpene alkaloid concentrations in tissues and floral rewards of larkspurs and impacts on pollinators. Biochem Syst Ecol, 48: 123–131
- Doubleday LAD, Raguso RA, Eckert CG: Dramatic vestigialization of floral fragrance across a transition from outcrossing to selfing in *Abronia umbellata* (Nyctaginaceae). Am J Bot 2013, 100: 2280–2292
- Eckhardt M, Haider M, Dorn S, Mueller A (2014) Pollen mixing in pollen generalist solitary bees: a possible strategy to complement or mitigate unfavorable pollen properties? J Anim Ecol, 83: 588–597

- Ehrlen J, Borg-Karlson AK, Kolb A (2012) Selection on plant optical traits and floral scent: effects via seed development and antagonistic interactions. Basic Appl Ecol, 13: 509–515
- Feulner M, Pointner S, Heuss L, Aas G, Paule J, Dotterl S (2014) Floral scent and its correlation with AFLP data in Sorbus. Org Divers Evol, 14: 339–348 http://dx.doi.org/10.1007/s13127-014-0180-8
- Friberg M, Schwind C, Raguso RA, Thompson JN (2013) Extreme divergence in floral scent among woodland star species (*Lithophragma* spp.) pollinated by floral parasites. Ann Bot, 111: 539–550
- Friberg M, Schwind C, Roark LC, Raguso RA, Thompson JN (2014) Floral scent contributes to interaction specificity in coevolving plants and their insect pollinators. J Chem Ecol, 40: 955–965 http://dx.doi.org/10.1007/s10886-014-0491-4
- Gosselin M, Michez D, Vanderplanck M, Roelants D, Glauser G, Rasmont P (2013) Does Aconitum septentrionale chemically protect floral rewards to the advantage of specialist bumblebees? Ecol Entomol, 38: 400–407
- Gupta AK, Schauvinhold I, Pichersky E, Schiestl FP (2014) Eugenol synthase genes in floral scent variation in Gymnadenia species. Funct Integr Genomics, 14: 779–788 http://dx.doi.org/10.1007/s10142-014-0397-9
- Jurgens A, Wee SL, Shuttleworth A, Johnson SD (2013) Chemical mimicry of insect oviposition sites: a global analysis of convergence in angiosperms. Ecol Lett, 16: 1157–1167
- Karpati Z, Knaden M, Reinecke A, Hansson BS (2013) Intraspecific combinations of flower and leaf volatiles act together in attracting Hawk moth pollinators. PLoS One, 8: e72805
- Kessler D, Bhattacharya S, Diezel C, Rothe E, Gase K, Schottner M, Baldwin IT (2012) Unpredictability of nectar nicotine promotes outcrossing by hummingbirds in *Nicotiana attenuata*: variability in nectar nicotine promotes outcrossing. Plant J, 71: 529–538
- Kessler D, Diezel C, Clark DG, Colquhoun TA, Baldwin IT (2013) Petunia flowers solve the defence/apparency dilemma of pollinator attraction by deploying complex floral blends. Ecol Lett, 16: 299–306
- Knauer AC, Schiestl FP (2015) Bees use honest floral signals as indicators of reward when visiting flowers. Ecol Lett, 18: 135–143
- Koehler A, Pirk CWW, Nicolson SW (2012) Honeybees and nectar nicotine: deterrence and reduced survival versus potential health benefits. J Insect Physiol, 58: 286–292
- Lerch-Henning S, Nicolson SW (2013) Bird pollinators differ in their tolerance of a nectar alkaloid. J Avian Biol, 44: 408–416
- Lohaus G, Schwerdtfeger M (2014) Comparison of sugars, iridoid glycosides and amino acids in nectar and phloem sap of *Maurandya barclayana*, *Lophospermum erubescens*, and *Brassica napus*. PLoS One, 9: e87689
- Majetic CJ, Levin DA, Raguso RA (2014) Divergence in floral scent profiles among and within cultivated species of Phlox. Sci Hortic, 172: 285–291

- Manson JS, Cook D, Gardner DR, Irwin RE (2013) Dose–dependent effects of nectar alkaloids in a montane plant–pollinator community. J Ecol, 101: 1604–1612
- Manson JS, Rasmann S, Halitschke R, Thomson JD, Agrawal AA (2012) Cardenolides in nectar may be more than a consequence of allocation to other plant parts: a phylogenetic study of Asclepias. Funct Ecol, 26: 1100–1110
- Page P, Favre A, Schiestl FP, Karrenberg S (2014) Do flower color and floral scent of *Silene* species affect host preference of *Hadena bicruris*, a seed-eating pollinator, under field conditions? PLoS One, 9: e98755
- Parachnowitsch AL (2014) New synthesis: the evolutionary ecology of floral volatiles. J Chem Ecol, 40: 859 http://dx.doi.org/10.1007/s10886-014-0491-4
- Parachnowitsch AL, Burdon RCF, Raguso RA, Kessler A (2013) Natural selection on floral volatile production in *Penstemon digitalis*: highlighting the role of linalool. Plant Signal Behav, 8: e22704
- Parachnowitsch AL, Raguso RA, Kessler A (2012) Phenotypic selection to increase floral scent emission, but not flower size or color in bee-pollinated *Penstemon digitalis*. New Phytol, 195: 667–675
- Schiestl F P, Johnson SD (2013) Pollinator–mediated evolution of floral signals. Trends Ecol Evol, 28: 307–315
- Schiestl FP (2014) Correlation analyses between volatiles and glucosinolates show no evidence for chemical defense signaling in *Brassica rapa*. Chem Ecol, 2:10
- Schiestl FP, Dotterl S (2012) The evolution of floral scent and olfactory preferences in pollinators: coevolution or pre-existing bias? Evolution, 66: 2042–2055
- Sedivy C, Piskorski R, Mueller A, Dorn S (2012) Too low to kill: concentration of the secondary metabolite Ranunculin in buttercup pollen does not affect bee larval survival. J Chem Ecol, 38: 996–1002
- Steenhuisen SL, Raguso RA, Johnson SD (2012) Floral scent in bird- and beetle-pollinated Protea species (Proteaceae): chemistry, emission rates and function. Phytochemistry, 84: 78–87
- Sun M, Gross K, Schiestl FP (2014) Floral adaptation to local pollinator guilds in a terrestrial orchid. Ann Bot, 113: 289–300
- Theis N, Adler LS (2012) Advertising to the enemy: enhanced floral fragrance increases beetle attraction and reduces plant reproduction. Ecology, 93: 430–435
- Tiedeken EJ, Stout JC, Stevenson PC, Wright GA (2014) Bumblebees are not deterred by ecologically relevant concentrations of nectar toxins. J Exp Biol, 217: 1620–1625
- Vanderplanck M, Moerman R, Rasmont P, Lognay G, Wathelet B, Wattiez R, Michez D (2014) How does pollen chemistry impact development and feeding behavior of polylectic bees? PLoS One, 9: e86209
- Wright GA, Baker DD, Palmer MJ, Stabler D, Mustard JA, Power EF, Borland AM, Stevenson PC (2013) Caffeine in floral nectar enhances a pollinator's memory of reward. Science, 339: 1202–1204

7

Seasonal Management of Honeybees for Their Improved Honey Production

Muhammad Anjum Aqueel[1], Hafiz Khurram Shurjeel[2], Abu Bakar Muhammad Raza[3], Muhammad Yasin[4], Irfan Akram[5], and Muhammad Haseeb Ahsan[6]

ABSTRACT

Honeybees are among the best pollinators on the face of earth. Bees' management method is the normal colonies check up to observe the brood status, providing extra colony super for the colony development throughout the flowering, distinctive and crop of mature honey, space lessening, nourishing and uphold colonies in the food shortage duration and revealing and management of bee pests. The honey production is depends upon the floral availability and proper colonies management throughout the year. Pakistan has a great potential in beekeeping with affluent floral resources, good-quality ecological state of affairs, and continuation of big colonies strength and characterized by high honey production. Bees use plants for a short duration due to the seasonal blooming triggering on the differences among plant species regarding nectar and pollen production for the better honey production. Therefore, the apiarists should be aware of the blooming time and season of plants, environmental extreme hindering honey production and carrying ability of the vicinity, i.e., colonies strength that can be place

[1] Department of Entomology, The Islamia University, Bahawalpur, Pakistan
E-mail: anjum.aqueel@iub.edu.pk

[2] Department of Entomology, University of Sargodha, Pakistan
E-mail: shurjeel_38@yahoo.com

[3] Department of Entomology, University of Sargodha, Pakistan
E-mail: abu.bakar@uos.edu.pk

[4] Department of Entomology, The Islamia University of Bahawalpur, Pakistan
E-mail: yasin_1876@yahoo.com

[5] Department of Entomology, The Islamia University of Bahawalpur, Pakistan
E-mail: yyzuaf@gmail.com

[6] Department of Entomology, The Islamia University of Bahawalpur, Pakistan
e-mail: haseeb.entomologist@gmail.com

for utmost honey yield. Colonies should be inspected internally and externally to avoid the pests access to the colonies. The season-wise apiary management is obligatory to get better honey production.

Keywords: *Honeybees • Seasonal management • Honey production*

7.1 INTRODUCTION

Since ages, apiculture is an imperative enriching heritage of the humans and their farming. Of four species exists in Pakistan, three (*A. cerana, A. dorsata* and *A. florae*) are indigenous whereas on is exotic (*A. mellifera*) brought back in 1979 from Russia and Australia (Jadran, 2011; Hikmat et al., 2016). The honey production, bees' health and successful overwintering is mainly depend upon the honeybee colony management practices. Insect-pests management techniques to control the notorious pests are imperative since they are the main element of colony losses in the summer and winter (Morse and Flottum, 1997).

Honeybees engender various commodities along with honey, royal jelly, pollen, wax and propolis are broadly used in cooking supplies, curative medicines, nutritious and additionally profit-making produces (Wakhal et al., 1999). In Pakistan beekeeping is the source of income of rural population since the 400,00 colonies producing around 10,000 metric tons (Mt) honey yearly, approximately above 4000 beekeeping farmers were occupied in rearing of *A. mellifera* and 27000 family unit were generating wages (PARC, 2010). The bees' larvae devour high-quality nutritious food whereas the adult bees have enormous value serving as pollinators in the agricultural settings and facilitate in keeping up the natural environment. Due to the initiation of framed hives, the bees' management becomes more uncomplicated and practiced in the recent times (ICIMOD, 2012). The lower honey production and other beekeeping foodstuffs resulted from inadequate management methods and shortage of satisfactory beekeeping education (Bhusal and Thapa, 2005; Masuku, 2013). The availability of the floral resources is different according to the different weather seasons. The bee' floral collection ability and particular colony management season-wise tactics has significant function in making over the beekeeping industry (Thomas et al., 2002).

7.2 SEASONAL MANAGEMENT OF HONEYBEE COLONIES

The in-colony and out-colony tricks and tasks of the honeybees diverge and hence their management fluctuates through the varying seasons. The period between September–December probably determine and gauge the foundation of a new-fangled and fresh year for a bee colony. The colony affairs at this time of the year deeply influence its accomplishment for the subsequent year (Thomas et al., 2002). Generally, the floral adequacy is easy to get through certain time around the year and when extra food sources are to be had, famously celebrated as honey flow period. Under drought conditions, there is dearth of flora and food dearth and consequently bees suffers (Tesfaye and Tesfaye, 2007; Workneh and Ranjitha, 2011; Gidey and Mokenen, 2010).

During intense and severe environmental conditions such as extreme winter, intense summer and monsoon, various particular and exacting management approaches are obligatory. Fruitful apicultural practices occupied the perceptive of annual colony sequence and carrying out the recurring bee management functions to construct hives to its highest population during the major floral flow and to endure the shortage duration to guarantee the utmost possible return (Brown, 2001). Appropriate recurring apiary management methods would significantly develop the colony performance and honey production (Haloua, 1989).

7.2.1 Management during Honey Flow Season

This season correspond with spring. Through this season, following points should be practiced

- Provide additional breathing space / extra comb to the bees for honey stores.
- Captive the queen bee to young one chamber using the queen excluder.
- Circumvent from the swarming as much as possible.
- Prior to the floral availability period, afford the sugar syrup to colonies to keep the adequate population.
- Divide the well-built colonies when there is need of the reproduction.
- Keep checking the queen's performance and besides that keep practicing the queen rearing that might be followed in order to make fresh new queens for the subsequent and following colonies.
- Extract the honey precisely from the hive at right time by not disturbing and harming rest of the colony.
- Confiscate protective covering of hives in early spring but in heavily packed colonies, packing is removed when temperature reached 16°C.
- Colonies examined shortly on sunny day once a week for food storage and general colony health and equalize the colonies strength by adding brood frames.
- Unite the brood lacking colonies and if strong then introduce new mated queen or eggs enrich frame and young larvae.
- Provide stimulatory feeding of sugar syrup on the spring commencement and guard them against the robbing since bees will put their efforts during this period for brood rearing.
- In order to avoid the spring dwindling, extra honey and pollen frames should be provided.

The sufficient flora and enough space in the colony in order to store honey in the comb and for other in-colony tasks as well is a good colony strengthen parameter (Amssalu, 2006). Because honeybees spent maximum time almost around the year in building up their honey stores for them to use in the dearth periods. Hence, providing super combs for their correspondence would lead to a strong colony. In the late spring season, some empty frames / combs must be added in the colony. After the swarming season, the colony needs to manage nectar storage space and upon favorable conditions, bees fill the combs with honey–its production depends on the floral availability (Crane, 1990). The colonies should possibly be placed around plant proximity for floral availability and crop pollination as well.

7.2.2 Rainy Season and Monsoon Management

In the hot and humid areas in Pakistan, the monsoon season occur potentially from June to September. Several troubles regarding bees' pests, predators, disproportion of humidity and starvation are faced by the bees (Abrol, Rana & Gupta, 1988). Occasionally because of the incessant rains, bees have to restrain in the hives for an extended duration, consequently bees get lethargic and indolent and may possibly suffer from dysentery. The following approaches can be adopted to manage them timely and appropriately. Hence, proper drainage supply, sufficient food (preferably sugar candy) availability in the colony, queen provision, endow with pollen substitute to avoid bloodlessness, check on pests are the imperative management tasks to must perform. The colonies' position must be sloping towards entrance so that the water could easily be drained out instead inside accumulation.

7.2.3 Summer Management

This season is generally considered the honey flow period. There are hot winds with ambient temperature in the summer season and the bees (*A. cerana* and *A. mellifera*) mostly throw out the old bees and drones and resultantly the population declined. So as to avoid from absconding and other colony weakening parameters, there must be sufficient shadow for bees, increase humidity, sprinkle water on hive, splinter among the supers and brood must be present for ventilation, food substitute and water and seal all the cracks and crevices (Wilson, 2006). Mostly small scale beekeepers extract honey in July and manufacture more honey averagely since they cannot administer more units rigorously to get optimal honey production. In early summer, the brood shell generally takes up minimum two supers. Ultimately, in the late summer flow of honey goes down, and the queens' fecundity decrease and resultantly the brood nest reduced. Habitually, the bees will again fill the brood fractions in the nest with honey, and brood shell can turned limited to one super throughout the late summer.

Being an apiarist, this should timely be decided by you to amount of honey you are going to harvest, i.e., are you going to empties the colony from honey or want to leave some in the colony for their future use during July-August? If you remove all the frames, then be ready to provide them supplemental food for their growth and development and also feed them through drought or dearth periods.

7.2.4 Winter Management

The successful winter management of honeybees primarily means with (a) colony strength (b) sufficient food stores (c) hive ventilation and the (d) colony health. The colonies management in this season is aligned with the floral availability and climatic situations in that particular area however, in some regions, a second honey flow period can be observed (Tesfa et al., 2013). Shorten the space inside the colony for winter preparation and confine the food storage area. Another phenomenon of making queen cells in the colony takes place in this season, so confining food storage area, no extra cells will emerge. By the end of winter, bees' actions also closed in colony and they go for winter sleeping. Near early fall; there is a floral shortage in the colony that affect brood health and leads towards population decline. Generally, the older bees die early than young ones in fall and this decline in the colony could be due to the queen low performance. In the overwintering that is not a productive duration, the colonies could be managed through maintaining diseases and pests free hives, ensure productive queen thoroughly, lessen the comb space by confiscating extra frames, merge weak colonies with of good population, and feed them on sugar syrups in the combs.

7.2.5 Autumn Feeding

Extracting entire honey and leaving nothing behind would affect the bees' health so there must be sufficient amount for the remaining bees in the colony. The right amount of sugar and water solution per hive and its proper application (5-6 times in two days) in winter season must be provide to all the colonies. The preferred autumn syrup preparation contains 1L water + 4 kg honey or $\frac{1}{2}$L water + 1 kg sugar by mixing them in the hot water. The winter foodstuffs' worth is imperative such as fruit juices are not appealing since they contain junk material and can cause dysentery. Merely well-built united bee colonies are capable of overwintering. A storm break is necessary for the affective brood rearing in the early spring season, either man-made or natural (Shah & Shah, 1976).

The colonies in this season must contain adequate honey stores, bees' strength, and queen with a high-quality brood pattern comparatively. There should be controlled structure to resolve the problems due to heavy rains in autumn or winter season. The hive protective covers must be strong to avoid humidity and molds that cause mortality. The other related factors such as colony health, queen's age and performance, honey production, diseases and enemies/pests must be observed on colony cards and calculate the loss of a winter time.

7.2.6 Management during Dearth Period

There are four important practices to adopt when it comes to managing bees in the dearth period. These are (a) removal of the empty frames/combs and lay up in an absolute airtight container, (b) utilization of mannequin partition floorboard to impound bees to an undersized area, (c) merge the weak colonies with strong ones, and (d) provide artificial diet like sugar candy/syrup, pollen mixed supplemental diet and other food substitute.

7.3 SWARMING

Swarming is an innate yearning of honeybees to enlarge their numbers through reproducing and expanding their survival possibilities. In swarming, the mother colony is divided into more different colonies (Miller, 2007).

Factors contributing in the swarming behavior include the short brood area that point out bees' strength and space availability. Swarming also is linked with the creating and releasing of chemicals with those the queen produce. If queen pheromone does not happened sufficiently, bees built queen cells to arrange for swarming (Solomon, 2009). Queen's pheromones are dispersed all over the hive when helper workers touch their sensory organs to queen. Then chemical signals are communicated by antennae through the whole colony particularly during food exchanges and when signals highly dispersed, the queen essence restrains the push for swarm. The queen's health, age, fecundity level and performance must be checked in the spring season. The smooth brood pattern represents the strong colony and when there is scattered pattern, queen starts laying drone eggs in the workers cells (Villa, 2004; Avitabile et al., 1975). Queens' right young colonies are more on the verge to swarm because the old age queen bees generate smaller amount of essence comparatively. Even though autumn requeening is best, however requeening must be continuously be repeating around all the seasons.

The swarming may influenced by the weather particularly when there is great strength and colonies are rapidly growing in the good weather which amplifies the swarming excitement. Additionally, the imbalanced colony with respect to age differences, genetically issues, and deprived ventilation conditions are also contributors in the swarming. However, the first and foremost indication of the swarming preparation is the queen cells present around the bottom bars of frame while supersedure cells could be find on the frame face. Hence to locate them in the colony, look up at the bottom and top chambers and obliterate the existing cells. The bees show a varying behavior since sometimes they do not swarm even over population and other apparent reasons and sometimes it happens with small reasons.

7.3.1 Swarming Period and Causes

It generally happen when the queen has performed well in brood rearing task over the pollen and nectar availability particularly in the late spring or early summer, however can also take place during summer or fall, if there is sufficient floral accessibility. The conditions that shore up the swarming and needs to be managed immediately:

- When there is no/less space to grow offspring and store honey.
- Inadequate ventilation facility
- Extreme colony temperature
- Aged queen with poor performance
- Sudden honey flow

7.3.2 Setbacks due to Swarming

- Its cause colony division that reduces working labor of colony and weakens colony strength.
- It decrease the colony's morale and make it unfavorable for honey harvesting since the bees focus on searching sites and building new cells for queen.

7.3.3 Preparing to Swarm

In order to nurturing more queen bees, hive preparations for swarming contains putting the queen on a diet, nurturing drones more, and dropping the foraging goings-on. When the workers shall be consuming less jelly in this duration, egg laying reduces and abdomen get smaller, permitting her to take off with the primary swarm (an aged queen, some drones, few workers) when it depart the colony. Before leaving the parent colony, the worker bees use honey to swell up themselves and kick-out the queen. Infrequently, further minor swarms pursue with a recently emerge queen. Newly emerged queen start warble/making sounds like "vang vang" 2-3 days before the emergence. These sounds indicate the older queen to start obligatory preparations to leave the colony; on emergence they communicates with each other recognizing that similar sounds and leaves the hive as early as possible.

Generally, the swarming occur before or at noon and reconcile and straighten out on the close proximities like trees, buildings, etc and it could continue to be there for just some minutes or number of days prior to move towards any dark protected enclosure chosen by the scout bees (Seeley, 2009).

7.3.4 Swarm Indications

There are some factors that indicate the potential swarming, such as presence of more drones in the hive, extra queen cells formation, increased fecundity, audible sounds of both queens, bees gathering on smelling queen before flight board, restlessness symptoms in front of the hive, well-built combination among hive and worker bees and exhaustive dashing out of hive opening.

7.3.5 Swarming Time

The swarming time basically depends on the weather conditions; generally it happens in the early timings of sunny days, however of weather situations are not feasible it could even initiate in the early morning and late evening.

When the swarm initiate from colony, bees generally recognize each other and fly mutually with their visage towards the colonies and the entire flock then go for flight (few meters above) that is known as swarm dance which they continue for few minutes. Bees require energy to fly as a swarm so they consume maximum honey and fill their bellies that's why after dance they cannot go far away. Secondly, the old queen in swarm has exhausted wings due to eggs in her belly and hence cannot go far flung and reconcile nearby protected places. It is observed that the worker may join their parent colony again if they don't find sufficient food.

7.3.6 Catching and Hiving a Swarm

The acquiescent swarm can be wedged through the basket designed to catch the swarm. The basket can straightforwardly be positioned above the bunch of swarm and tenderly pushed aloft in order to ascending them in the basket. Following the queen, all members religiously will gather and subsequently it can simply be shifted for hiving in the apiary. In order to reconcile the swarm appropriately, a colony is arranged by providing one frame each of capped brood, pollen and honey and endow with extra frames according to the strength. The bees from the basket is then gently wobbled on top bars of hive and instantly enclosed with burlap cloth, interior cover and upper cover. The supplemental diet like sugar syrup must be provided to these colonies (Doull, 1980a).

7.3.7 Swarm Prevention

Swarm prevention depends on both the in-colony and out-colony factors. A good swarm prevention method decreases the swarming with a lowest colony intrusive and must be well-matched with better colony management for the purpose of delight and income as well. Swarming should be proscribed in order to booming and sustainable beekeeping (Kumar et al., 2013). The swarm affected colonies infrequently recuperate in time to manufacture a honey yield. Regular colony management in spring season typically decreases the swarming. Sufficient space provision to for brood rearing and food storage in obligatory in the early March to April since these are swarm anticipation months. The top brood chamber and bottom brood chamber in the colony shifts the drained comb over rising brood nest, that make space for queen to lay eggs. Honeybees prefer to develop their brood nest upward when the colony is growing.

The process of chambers shifting should be performed when there is a brood nest in one box only and if brood nest is dispersed, it will create anxiety in the bees hence they start thermoregulation in spring nights. During the exchange process of colonies and frames, make sure about the disease-free hives and queenless frames. Keep a newspaper as a separator to keep aggressiveness to a minimum in the hive when putting mature bees in the existing colony.

Note: Avoid turning around the colonies unless the climate has established and there is few possibility of an unexpected temperature fluctuation (big fall).

Equalizing the colony's population also facilitates in swarm prevention and making management practices uncomplicated for the upcoming years. Hence, the weak colonies can be reinforced through changing colonies' position in the same apiary, through addition of sealed brood taken from healthy colonies, adding queenless boosters, merging feeble colonies, avoid overcrowding, remove queen cells, proper ventilation, requeening at least annually, giving sugar syrups to the confined beehives, temporary colony divisions with equal number of frames in later March or April and reconciling prior to the starting of honey flow and combining the queenless with the queen right colonies.

Another important approach to prevent swarming could be the introducing the new queen cell in queenless part when splitting hives and while splitting places the colonies as a minimum at 2 miles away (for better growing and performance). Exclusion of four or six frames form the healthy colony and putting them in the separate hive and afford a queen, supplemental food and the bees could be another approach to avoid swarming. The main hive infrequently swarms subsequent to such cure with sufficient place, and the newly split colony generally develop into a fruitful component when established well in time with a passable nectar flow (Thomas et al., 2002).

7.3.8 Swarm Control

When the queen cells are formed in the colony and bees are devoted to swarm, then there is dire need to manage the swarm. The best technique at that moment should be the division of the colony inside by means of double screen. Introduce the old age queen with 3-5 frames of uncap brood in brood chamber at bottom. Attach an additional hive with unfilled combs and the honey. Put the double screen over the second colony with the entrance facing the hive nurture. On top of it, place 2nd brood chamber holding 5-6 brood frames, mostly caped and 2 pollen combs and honey on every side. Wobble the extra bees from lower colony into the upper section, because the forager bees will homecoming to the lower chamber.

The bees in the lower part of colony eradicate any extra cells; while the bees on top of the double screen develop a fresh queen and through this managing pattern, there are less swarming chances and then reconciles them and takes out the double screen subsequent to the swarming period. This is an exceptional method to requeen the colony or strengthen the weak colony by moving upper body of colony. Utilization of double screen is a wonderful technique for splitting prior to the colony incline towards swarming. The new queen must be introduced in the upper portion, when this technique is employed in the early spring season.

Another method to prevent swarming is the Demaree method, which includes the queen extrication from brood. It takes lot of time and efforts, however, help the colony to grow very fast. As usual, critically check up the colony and obliterate all extra queen cells. Then put the queen in lower chamber and other unsealed frames; full of eggs and larvae, in upper chamber, however, the capped brood could be located in any brood chamber. The original chambers should be providing with the two colonies completely occupied with the empty frames. Prior to the addition of middle supers, put a queen excluder device on upper portion of bottom hive body. Now the colony will be containing three supers; (i) first super; queen, blank frames and unsealed brood (ii) middle super; blank frames and possibly one or two sealed brood (iii) top super; young and unsealed frames of brood.

Through Demaree method, the unsealed larvae in top super magnetize full grown nurse bees from the older brood nest in bottom super, which lighten its population strength. In addition, the blank frames in the bottom body afford sufficient room to queen for egg laying. Additional space bring in as the capped brood hatched in the colony. After at least a week later must examine the colony and remove all the extra queen cells present in the upper portion of the hive body.

7.3.9 Making Space for Honey Crop: Supering

The bees need more space for the brood rearing, food stores, etc. so space must be provided to them through adding up of empty supers to brood nest preferably in the spring season. A hive body is generally required at blooming time of the fruit, and a vigorous and health beehive in late spring could possibly require the comparable of three profound supers exclusively to supply adequate space to adult bees (Alemtsehay, 2011). The suitable time for adding the blank honey super is late spring because bees need space for nectar storage after the swarming time passed. When there are favorable conditions, the bees efficiently load up the hive bodies within two to three days. So timely fashioned supering could efficiently save colony from resources shortage to grow. Good bloomy periods indicate the excellent honey production so in a typical season, there can be an optimal production and similarly unusual weather like drought, dearth period will affect the yield (Gidey et al., 2012). The colonies must be located in close vicinity of sufficient floral availability sources for good honey production and bees' health (BOA, 2010).

The commercial beekeepers place queen excluder device to thwart the brood hatching and production in honey super. The potential reason is that the labor cost for shifting brood frames from honey supers through the extraction procedure probable be too expensive. The excluder device basically assures that queens by no means place brood into honey boxes. Many apiarists not practice excluder since they believe that the disruptive characteristic of these screens increase the nest congestion that impulse the swarming. Generally, these are small beekeepers which cannot afford to invest time and labor costs for keeping brood and honey supers separate.

Alternatively they exercise a filled honey super as an obstruction to the upward queen movement. The queens never wish to traverse the honey frames for egg laying. If queen place any brood on the super frame, those can be substituted with frames of honey and brood shifted to the extra honey super. The brood will hatch out prior to get honey flow over, and bees will load honey in the blank cells. The full honey frames in lower portion will avoid the queen from peripatetic there around. Some beekeepers apply two-thirds full rule; directions for when to place extra super. Of 10, when six frames become full with honey, this is best time to place empty frame as a super in the colony. In the beginning of the summer, some beekeepers put empty frames on the top and some put below, however, here is no published evidence regarding the production difference over to or bottom supering process, either can works. It can be continued form mid to late summer season in anticipation of the floral accessibility ends and nest get smaller in size.

If the queen in the colony dies, the colony itself has no chance of producing fresh queen since around 24% workers start laying unfertilized eggs (Miller and Ratnieks, 2001). The workers' strength seems not to be stable throughout the season (Velthuis, 1970; Hoover et al., 2006). The literature suggested that workers' strength can possibly be increased after the swarming season (Kropa´cova´ and Haslbachova, 1970) that is the only ordinary method of colony multiplication and starts when worker bees commence the new queen rearing (Winston, 1987).

7.3.10 Colony Division

Colony division should only be done when it uses the bees' impulse to swarm. The strong and healthy colonies with a big brood strength (eggs must be present because workers need eggs to make fresh queen) should be divided. During division, the frames are appropriately constructed; interconnected between two beehives. The following process can be carried out;

1. The two or three days before division, examine the hives to make sure the eggs presence since they employed eggs to make fresh queen. Failing to eggs presence, division should not be performed until the eggs are present.
2. Smoke the active hive for two to three minutes.
3. Carefully lower the active colony and relocate it to the new place.
4. Suspend the empty hive in the position used by the active hive. The bees will immediately start joining this hive from the active colony.
5. Gently remove all the bees from comb with brush and place the comb back to the active hive then transfer it to the new hive.
6. Examine the frames, if they are accurate (having lot of eggs and brood), brush all the bees and transfer the frames to the new hive.
7. Carefully one at a time, transfer at least six best frames from active to new colony.
8. All bees should be brushed in active colony(at least with half combs for queen to lay eggs and store food) so that the queen is left behind.
9. Place the combs from inlet and should be sited side by side when placed in the new hive.

10. Sometimes, queen cells are already present in active colony. If unsealed cells exist, colony already plans to divide naturally by swarming and success chances are improved.
11. If sealed queen cells are present, the colony has already swarmed and new queens are forming inside the cells. If queen cells are already present, at least one brood comb with queen cells should be transmitted to new hive. A brood frame with queen cells should be present in active hive in case the queen is missed during colony division process.
12. Restore every removed frame with a new one. Load all blank spaces with fresh top bars to complete two hives, and then restore lids. Reinstate the active hive in its new location.
13. Carefully examine the queens' fecundity and grown up.

7.3.11 Month-wise Colony Management

7.3.11.1 January

Numerous methods are employed in order to efficiently manage the honeybee colonies. During this month, examine the entrances reducer adjustment and also each entrance is unlock and not congested with the dead bees. When the bees' activities are slow and nothing is happening in the yard, this is the right time for any repairing, and makes colonies ready for the next season. Examine each and every colony individually in the day time through visiting and checking the general apiary situation because this time is suitable for any developmental and protective work prior to the next season. Critically check up the inactive or slow performing hives and adopt better solution measures to increase their activity and strength. Inspect the colonies to check the condition of brood diseases, any pest attack and the accumulation of the dead bees for removing them and cleaning the colony. And if the bees are actively flying and continuing in-colony and out-colony tasks and doing transportation of flora; pollen, nectar, water, etc. then that bee colony is developed with strong brood population (Haftom and Tesfay, 2012; Workneh and Ranjitha, 2011; Gidey and Kibrom, 2010).

7.3.11.2 February

According to the environmental conditions, in the late February, generally the first colony inspection is made on a very sunny day having normal wind and a temperature of 60°F. Particularly, the colonies must be monitored the discussed parameters at this time of the year. Examine that colony has proper ventilation system preferably upward. And the significant notion to check ventilation would be the moisture on inner cover; if there is no moisture then that colony has sufficient ventilation (Rao et al., 1998). And if moisture present, then the lager spacer should be provided on the upper side of inner cover. Moisture creates problems when it comes in contact with the winter clusters. The inner cover should contain an unlock center hole for ventilation to work. Examine if there are proper food stores in the colony or not. In brood chamber, if bees are clustered and the super are filled with honey then this is the sign of sufficient food stores in the colony. If bees are clustered in the top super, check the adequacy of pollen and honey. The colony should be fed in case of less pollen and honey availability through transferring honey occupied frames form other colonies to this. Sugar syrup in feeders can also be provided to the colony. Critically examine the presence of bees and brood by removing frame at clusters' edge and moves each frame. In fully covered frame with bees, check the status of brood and or eggs. If broods are marked is the sign of active queen.

7.3.11.3 March

Each honeybee colony should be examined in the early sunny and warm days of this month. The food stores must be checked and of there is less honey and pollen stores available, then colony needs to be feed properly. The brood strength and pattern should also be checked. In the beginning of March, a colony must be vigorous and has good queen with several brood frames and eggs in the cells because the honey production is the result of the presence of these resources and pollens' quantity and cluster size. When frames are replaced inside the colony from other hives, brood pattern must be checked because the queens' effectiveness is observed by the population strength, egg pattern, larvae and pupae on the frame in brood nest. A good brood pattern by an efficient queen seems in a sequence; eggs after eggs, larvae after larvae and pupae after pupae in the cells. Failing to find the filled brood cells with larvae and eggs, the queen immediately be replaced with healthy one. Sometime the queens' performance found slow due to the inappropriate mating, deprived nutrition, disease and pests, and old age factors. The combs should be moved to check their condition properly on upper and lower sides and it must be immediately removed if there are so many drone cells and is in bad condition.

7.3.11.4 April and May

In the early April, there should be two extracting supers and there must be a queen excluder device above second chamber, on the top in every colony. The colony may be observed for; estimating food and honey stores, removal of chemical strips if used and the checking of larval brood and eggs. When there is frequent and sufficient nectar flow in the colony, the next concern is to come up with the potential swarming activity of the bees because this time of the year is so much appealing for the swarming. The protective measure should be adopted accordingly to avoid the swarm otherwise it could have detrimental effects on the honey production and colony strength. Additionally, take out all the fallow supers in the colony when the nectar flow stops, this action will turn away the honeybees from chewing and harming the new established supers. Frames' cells should properly be capped prior to the replacement of any honey frames and delay the removal process until they are not capped appropriately. Undeveloped honey present in the uncapped cells enclosed big moisture level and possibly will grounds honey to ferment if removed gradually.

7.3.11.5 June - July

In Pakistan, the temperature is so hot and may exceed up to 40°C at this time of the year. Hence, the colony strength parameters badly affected losing worker and drone bees, pest attacks and the colony robbing during these months. The colony absconding can potentially be occurred if there is no proper management of the apiary is done (Sande et al., 2009) and food shortage in the colony (Gidey and Mekonen, 2010). So the colony management must be covering the; apiary must be at shady places, brood chamber should be provided with proper upward ventilation other robbing could be take place inside the chamber, cracks and crevices must be sealed/ closed to stop entry of robbers pests, etc., adequate brood and food stores, placement of moisture gunny bags and water on top cover of hives, and availability of the fresh water in order to uphold the colony internal temperature throughout the summer season and spilling it inside and evaporating it through the fanning process (Clauss and Clauss, 1991).

7.3.11.6 August and September

In these months, the bee colonies characteristically initiate collapsing because of the mite infestation. Since this is favorite time of pest infestation, the bee colonies are attacked by wax moth and mites in the early these months. The appropriate control procedures should be performed immediately if the symptoms or wax moth and mite infestation are even slightly observed in and around the hives. Brood chamber correctly be observed frequently regarding the honey and sugar syrup provisions to brood sides and besides that makes sure that cells are on the top. If there is food shortage in the colony, it must be fed immediately with supplemental diet as sugar syrup and pollen substitute diet (Mishra, 1995). The queen existence and performance is checked in the month of August. If she is not present or its effectiveness is deprived then remove and replace the queen instantaneously and these months are considered the best time for queen replacing as compared to the spring season because queens are easily available in this season comparatively.

7.3.11.7 October and November

These are the months to prepare for the winter season so all the colonies must completely be prepared for winter. The honey stores must be checked and fed those are lacking or have less food available. The queen excluder devices should be removed, entrance reducer shout be fitted to avoid predators and mice since guard bees are not fully active due to low temperature, provide the upward ventilation, and check the food (especially honey) for the whole season.

7.3.11.8 December

The hove population begins declining (dying of older bees due to low temperature) badly throughout the month. The queen discontinues the egg laying in the early December and also in the January. Provide the artificial or supplemental diet to the bees and also observe their activities regularly in this month (Standifer et al., 1973b).

7.4 CONCLUSION

In beekeeping, each season gets a hold of new set of tasks and these activities vary depending on honeybee needs and honey flow. Seasonal management of honeybees is imperative because bees necessitate different things right through the year. The activities of a colony vary with the seasons. The lifestyle of bee colonies is intimately associated with the seasons when accessibility of flora, high temperature, and rainfall fluctuate dramatically. Modern beekeeping practices are attuned equivalent to the seasonal colony life cycle by raising honey production, off-putting seasonal behavior (swarming), or monitoring parasites that also pursue seasonal blueprint.

In the early spring, bee queens recommence egg-laying and colony instigates brood rearing. Probably because of lengthy days, warm atmosphere, and availability of early blooming flowering plants, nurse bees begin accumulating honey and pollen for themselves and brood feeding. As flowers bloom, bees start foraging since this is a perilous time of year for colony. Once brood rearing starts, the colony may quickly use all its reserves and run the risk of going hungry. Beekeepers must constantly keep an eye on the colonies, especially in colder locations, to ensure that there is sufficient supply of food and that the colony is kept warm. In times of scarcity, bees can be fed copiously sugar syrup and pollen to keep them alive until nectar flows are stable. The feeble colonies have to be culled to survive disease instead of investing in feeding. Due to copiously spring feeding, strong colonies can be divided and new ones established to avoid the swarming behavior. Late spring brings abundance of floral nutrients, allowing the colony population to grow. However, the swarming can be tricky, so colonies must be checked properly by adding brood or adding honey supers and removing queen cells to lessen overcrowding and postpone swarming.

Colony gathers and saves honey during the summer to use during times of scarcity. If the swarming occurs or colony splits, the newly emerged queens mate and begin laying eggs. In this season mostly beekeepers harvest honey. In the autumn, colonies need extra care with regards to colony weight. Colonies holding huge population and honey stores can excellently stay alive in winter while weak ones can be boosted by nourishing or uniting for their better survival. Proper monitoring of the varroa mites in summer is an imperative task because mite populations increase quickly during brood rearing times.

Beekeepers must address mite infestations early in the season by keeping an eye out for them in colonies. Due to the absence of brood, winter offers a special opportunity to treat the colony for mites. Beekeepers use this time to treat their hives with oxalic acid. The winter mite treatment makes sure that there will not be many, if any, mites when the colony restarts its life cycle. It is recommended to employ a variety of tactics to control pests and diseases, giving preference to methods devoid of chemicals that can harm bees.

REFERENCES

- Abrol DP (1998) Environmental factors influencing flight activities in honeybees, *Apis cerana* Fab and *Apis mellifera* L. (Hymenoptera: Apidae). Indian Bee J 60 (2): 71–75
- Alemtsehay T (2011) Seasonal availability of common bee flora in relation to land use and colony performance in Gergera watershed AtsbiWembwrta district eastern zone of Tigray, Ethiopia, M.Sc. Thesis. Wondo Genet College of Forestry and Natural Resources, School of graduate studies Hawassa University WondoGenet Ethiopia 76.
- Amssalu B (2006) Seasonal intensity of flowering and pollen-forage selectivity by honeybees, *Apis mellifera bandasii* in central highlands of Ethiopia. Proceedings of the 14th annual conference of the Ethiopian Society of Animal Production (ESAP) held in Addis Ababa, Ethiopia, Sep 5-7, 2006. Part II: Technical Papers. Addis Ababa, Ethiopia 3–10
- Avitabile A, Morse RA, Boch R (1975) Swarming honeybees guided by pheromones. Annals of Entomological Society of America 68(6): 1079–1082 Doi:10.1093/aesa/68.6.1079.
- Bhusal SJ, Thapa RB (2005) Comparative study on the adoption of improved beekeeping technology for poverty alleviation. J Inst Agric Anim Sci 26: 117–125

- BOA (2010). Tigrai Beuro of Agriculture. Annual Report.
- Brown C (2001) Responding to deforestation: Productive conservation, the World Bank, and beekeeping in Rondonia, Brazil. Profess Geog 53(1): 106–119
- Clauss B, Clauss R (1991) Zambian beekeeping handbook. A publication of the beekeeping division of the forest department; Mission Press, Ndola, Zambia
- Crane E (1990) Bees and beekeeping: science practice and world resource. Henemann News, Hally Court Jordan Hill OX28Ej.
- Doull KM (1980a). Relationships between consumption of pollen supplement honey production, and brood rearing in colonies of honeybees (*Apis mellifera* L.). Apidologie 11(4): 361–365
- Gidey Y, Bethelhem K, Dawit K, Alem M (2012) Assessment of beekeeping practices in AsgedeTsimbla district, Northern Ethiopia: Absconding, bee forage and bee pests. Afric J Agri Res 7(1)1–5
- Gidey Y, and Mekonen T (2010) Participatory technology and constraints assessment to improve the livelihood of beekeepers in Tigray region, northern Ethiopia. CNCS Mekelle University 2(1): 76–92
- Haftom G, Tesfay B (2012) Determining suitable size of stem cutting for propagating *Becium grandflorum*. Livestock Research for Rural Development 24: 175 http://www.lrrd.org/lrrd24/10/haft24175.htm
- Haloua F (1989) Strategy for development of beekeeping in developing countries. In Proceeding of the fourth international conference on apiculture in tropical climates, Egypt. 176–181.
- Hikmat UK, Syed IA, Nighat S, Baharullah K (2016) Honey production potential of the honeybee (*Apis mellifera*) in Karak and Kohat. Pakistan J Entomol Zool Stud 4(6): 559–564
- Hoover SER, Higo, HA, Winston ML (2006) Worker honeybee ovary development: seasonal variation and the influence of larval and adult nutrition. J Comp Physiol B 176: 55–63
- ICIMOD. (2012) Beekeeping training for farmers in the Himalayas. Kathmandu: ICIMOD.
- Jadran KK (2011) Measurement of economic efficiency for the projects of bee raising inside and outside towns in Wasit Governorate. Al-Anbar J Vet Sci 9(3): 112–121
- Kropa´cova´ S, Haslbachová H (1970) The development of ovaries in worker honeybees in queen right colonies examined before and after swarming. J Apicultural Res 9: 65–70
- Kumar R, Rajput GS, Mishra RC, Agrawal OP (2013). A study on assessment of duration of dearth period for honeybees in Haryana. Entomol Zool 8(1): 434–437 http://www.munisentzool.org/yayin/vol8

- Wakhal MD, Bhujbal M, Pais EVD (1999) Analysis of honey, pollen and royal jelly by high performance liquid chromatography: A review. Apiacta 30(4): 6–11
- Pakistan Agriculture Research Council. (2010) honeybee, Annual Report of PARC. 2010-2011. 98
- Masuku MB (2013) Socioeconomic analysis of beekeeping in Swaziland: A case study of the Manzini Region, Swaziland. J Dev Agri Econ 5(6): 236–241
- Miller DG, Ratnieks FLW (2001) The timing of worker reproduction and breakdown of policing behavior in queenless honeybee (*Apis mellifera* L.) societies. Insectes Soc 48: 178–184
- Miller P (2007) The genius of swarms. National Geographic 212(1): 126–147
- Mishra, R. C. 1995. Social behavior of bees and related management practices, pp. 44-59, In: Honeybees and their management in India. Krishi Anusandhan Bhawan, Pusa, New Delhi, pp. 168
- Mohana R, Singh TSMS, Subba Rao K (1998) Top ventilation helps *A. mellifera* L. colonies to increase honey production during rubber nectar flow. Indian Bee J 60(2): 108–110
- Morse RA, Flottum K (1997) Honeybee pests, predators, and diseases, 3rd ed.; A.I. Root Company: Medina, OH, USA. 718
- Shah FA, Shah TA (1976) A note on the bee activity and bee flora of Kashmir. Indian Bee Journal 38: 29–33
- Sande OS, Crewe RM, Raina SK, Nicolson SW and Gordon I (2009) Proximity to the forest lead to higher honey yields: another reason to conserve. Biological Conservation 142 (11): 2703–2709
- Seeley TD. (2009) The wisdom of the hive: the social physiology of honeybee colonies. Harvard University Press.
- Standifer LN, Waller GD, Levin MD, Haydak MH, Mills J (1973b) Value of three protein ratios in maintaining honeybee colonies in outdoor flight cages. Journal of Apicultural Research 12: 137–43
- Solomon B (2009) Indigenous knowledge and its relevance for sustainable beekeeping development: a case study in the Highlands of Southeast Ethiopia. Livestock Research for Rural Development 21(11) http://www.lrrd.org/lrrd21/11/boga21184.htm.
- Tesfa A, Ejigu K, Kebede A (2013). Assessment of current beekeeping management practice and honeybee floras of Western Amhara, Ethiopia. Int J Agric Biosci 2(5): 196–201
- Thomas Daisy, Pal N., RAO K. (2002). Bee management and productivity of Indian honeybees. Apiacta. 3
- Velthuis, HHW (1970) Ovarian development in *Apis mellifera* worker bees. Entomol Exp Appl 13: 377–394
- Villa José D (2004) Swarming behavior of honeybees (Hymenoptera: Apidae) in Southeastern Louisiana. Annals of Entomological Society of America 97(1): 111–116 doi:10.1603/0013-8746(2004)097[0111:SBOHBH]2.0.CO;2

- Tesfaye K, Tesfaye L (2007) Study of honey production system in Adami Tulu JidoKombolcha district in mid rift valley of Ethiopia. Livestock Research for Rural Development 19(11) http://www.lrrd.org/lrrd19/11/kebe19162.htm
- Wilson RT (2006) Current status and possibilities for improvement of traditional apiculture in sub-Saharan Africa. *Livestock Research for Rural Development 18: 111*
- Winston ML (1987) The biology of the honeybee (Cambridge, MA: Harvard University Press).
- Workneh A, Ranjitha P (2011) Beekeeping sub sector challenges and constraints in AtsbiWemberta District of eastern zone, Tigray Region, Ethiopia. Journal of Agricultural Extension and Rural Development 3(1): 8–12

8

Insect Pollination Technology and Its Economic Significance

Amarjit S. Tanda*

ABSTRACT

Bee pollination contribute significantly to enhancing agriculture yield and works as a fundamental support for crop productivity. Crop plants mostly rely on agents to transfer pollen, such as water, wind, and animal pollinators like bats, moths, hover flies, birds, bees, butterflies, wasps, thrips, and beetles. Field crops are generally cross-pollinated by insect pollinators. Crop pollination depending on animals offer up to 30% of world food production, and bee pollinators contribute to one-third of the total human food demands. Bees are regarded as important pollinators due to their efficiency and availability widely. Crop bee pollination offers great value to crop quality and quantity, enhancing world economic and dietary end results. This chapter presents the contribution of bee pollination, which impacts the economy, and enlists the various kinds of bee species and other insect pollinators related to crop-pollination in sustainable agroindustrial sectors.

Keywords *Crop • Pollination • Economic • Bees • Production • Visitation • Challenges • Impact*

8.1 INTRODUCTION

In the natural resilience of bio-ecosystems and the foundation of agriculture production, insect pollinators contribute significantly interplaying between agriculture and the life cycles. Entpollinatology or insect pollination as proposed by us has a crucial business in boosting the economy by improving the quality and quantity of many field crops [Gill et al., 2016; Breeze et al., 2011; Hristov et al., 2020; Khalifa et al., 2021]. Insect pollination or entpollinatology as recently described (Tanda, 2019a, 2020 a-c,2021a-e; Tanda, 2022 a, b) is defined as the mechanism or process by which pollen is transferred or moved by the insect pollinators from the male anthers to the female stigma, either within the same flower (self-pollination) or between plants (cross-pollination) [Sukumaran et al., 2020; García-Breijo et al., 2020]. Insect pollinators are the chief players of the crop cross-pollination and enhanced yield program as plants totally depend on agents to move their pollen grains in cross-pollination. Employing

* School of Horticulture, University of Western Sydney, Richmond, Australia and CSIRO North Ryde NSW, Australia.

E-mail: drtanda193@gmail.com

both domesticated bee species and native wild bees in an area could boost cross-pollination [MacInnis and Forrest, 2020]. Water, wind, and animal pollinators, as potential agents involve bats, birds, butterflies, hover flies, wasps, thrips, dipterans, and other animals [MacInnis and Forrest, 2020; Lobo et al., 2005; Zeng and Fischer, 2020; Van der Kooi and Ollerton, 2020]. From 200 countries, animal pollinators play a great role in the production of 87 world crops, together with cocoa (*Theobroma cacao*), kiwi (*Actinidia deliciosa* var. *deliciosa*), passion fruit (*Passiflora edulis*), and watermelon (*Citrullus lanatus*). About thirty percent of these crops contribute to the world economy through better quality food production. In 2005, it was estimated that the global economic value of pollination's was EUR 153 billion on an average, which is worth 9.5% of the world agricultural food production. Among the main insect-pollinated crop classes are vegetables and fruit, fetching around EUR 50 billion each, followed by edible petroleum crops, stimulants, nuts, and spices. One ton of crop production that is not reliant on insect pollination is worth about EUR 151, in comparison to an average of EUR 761 for crops dependent on pollinators for seed setting and fruit production [Gallai et al., 2009]. Insect pollination is a major factor in the crop improvement and production of a majority of farm products globally, involving aromatic and medicinal plants, for example, black cumin (*Nigella sativa* Linn), cumin (*Cuminum cyminum* Linn) [Meena et al., 2018], sunflower (*Helianthus* spp.) [Chambó et al., 2011], anise (*Pimpinella anisum* Linn) [Abd El-Wahab et al., 2012], and coriander (*Coriandrum sativum* Linn) [Patil and Pastagia, 2016]. Mainly, honeybees, native bees, and flies every flowering season, cross-pollinate 48 crops of the global most expensive products, enhancing significantly the world economy. [Klein et al., 2007]. In the USA alone, insect pollination add USD 16 billion every year with USD 12 billion accountable only to the availability of honeybees [Calderone, 2012; Rader et al., 2016]. The Western honeybee (*Apis mellifera* L.) is the major species responsible for bee pollination globally and fulfils 34% of pollination needs in the United Kingdom [Stanley et al., 2020]. Though many other species of bees also play role in pollination, research workers have targeted only a few of them at present, especially the bumblebee (*Bombus* spp.) [Rao and Stephen, 2009; Bänsch et al., 2021]. Greenleaf and Kremen found that interactions between wild and honeybees doubled pollination rates and improved the frequency of hybrid sunflowers by fivefolds, as compared to wild bees alone [Greenleaf and Kremen, 2006]. This chapter presents the contribution of bees to crop-pollination and its influence on the world economy. The factors affecting bee foraging in addition to a comparison of bees and other insect pollinators are illustrated.

8.2 ECONOMY AND BEE POLLINATION

There is a growing demand for human food security in spite of threats, for example, changes in climate and land use, habitat transformation, and the increasing human population. Appropriate pollination can make better the quantity and quality of fruits, nuts, oils, and other crop commodities raised [Giannini et al., 2015]. Animal pollination improves the world crop production by an additional USD 235–577 billion annually, with the highest quality economic advantages having been observed in Europe, the Mediterranean, Southern, and Eastern Asia, as per market prices [Potts et al., 2016]. However, higher production also results in enhanced demand for crop pollination jobs [Lautenbach et al., 2012; Tanda, 2022 a, b]. About 5–8% of crop production around the globe, would disappear without animal pollination [Aizen et al., 2009], and pollination also offers different assistance to ecosystems, for example, increasing biodiversity and improving food production without menacing the environment [Montoya et al., 2020]. For crops, bees are the chief pollination players. Gallai et al. (2009) described that insect pollination contributed EUR 153 billion, constituting 9.5% of the total economic value of agricultural production utilized for the human diet [Gallai et al., 2009]. Accordingly, cash crops such as coffee (*Coffea* spp.), cocoa, almond (*Prunus dulcis* ((Mill.)), and soybeans (*Glycine max* L.) are highly dependent upon bee pollination broadly in agriculture [Geeraert et al. 2020; Luo et al., 2020; Sáez et al., 2020]. Research workers have explored many techniques to assess the annual benefit of some ecological costs incurred by native bee pollinators in the USA, which have been demonstrated to amount to more than USD 57 billion, USD 3.07 billion of which is a consequence of bee pollination [Losey and Vaughan, 2006]. In the USA, the value of non-*Apis* pollinator's services was estimated at USD 3.44 billion, however, honeybees added USD 11.68 billion by 2009 [Calderone, 2012]. In North America, honeybees are accountable for pollinating more than 100 profit-oriented crops [Hristov et al., 2020]. For sunflower seed production, both honeybees and wild bees are also economically significant which is an uprising industry assessed at about USD 10.4 million yearly [Greenleaf and Kremen, 2006]. Bee pollination also enhances the crop yield in farmlands. In sub-Saharan Africa, which is the major producer of cotton [Sáez et al., 2020; Losey and Vaughan, 2006; Esquivel et al., 2020], bee pollination improves the cotton yield to 62% in comparison to a calculated 37% without bee pollination [Stein et al., 2017]. Furthermore, economic gains from bee pollination have been documented in smallholder farming systems in Kakamega (western Kenya), where many crops profit from pollination, comprising green gram (*Vigna radiata*), beans, cowpea (*Vigna unguiculata* L. Walp), sunflower, tomato (*Solanum lycopersicum* Linn), bambara groundnut (*Voandzeia subterranean* L.), passion fruit, and capsicum, with pollination boosting the production dramatically and being responsible for 40% of the annual crop production [Kasina et al., 2009]. In 2016, the calculated yearly value of pollination services provided by bees in Brazil was about USD 564,000 in the north (Serra da Bocaina, Pará) and USD 246,000 in the southeastern protected area (Mata do Jambreiro) [Hipólito et al., 2019]. In 2016, it was recorded that out of 36 crops grown in the state of Pará, 20 (55%) are anticipated by insect pollinators, and the overall worth of pollination was USD 983.2 million, similar to 33% of the total value of crop production (USD 2.95 billion). Describing the pollination service value rendered in four groups were as 96% of Pará's consists of, for example, cocoa (USD 187.6 million), watermelon (USD 26.1 million), Acaí palm (USD 635.6 million), and soybean (USD 98.4 million) [Orges et al., 2020]. Wild bees and honeybees have rendered significant degrees of pollination services for several field crops, in agriculturally-intensive regions in the USA. For seven crops the annual agriculture production value of wild pollinators is more than USD 1.5 billion. It is assessed that the value of wild pollinators is highest in apples, with a value of USD 1.06 billion, however the approximate values of watermelon (USD 146 million), sweet cherry (USD 145 million), pumpkin (USD 101 million) blueberry (USD 50 million), and art cherry (USD 32 million), are indicatively great. Across these crops, the economic value of honeybees on yield is approximately USD 6.4 billion [Reilly et al., 2020] which exhibits the great contribution of these pollinators and needs to be protected (Tanda, 2019a, 2020 a-c, 2021a-e; Tanda, 2022 a, b).

8.3 CROP PRODUCTION AND BEE POLLINATION CONTRIBUTION

The flower visitation numbers and the additive impacts of a different bee pollinating species influence not only the number and yield of crops produced but also their fruit quality, which is significant mostly from an economic outlook [Hall et al., 2020]. Crop-pollination services by more than one bee species, involving honeybees, carpenter bees, stingless bees, bumblebees, long-tongued bees, feral bees, social bees, and solitary bees, lead to a great pollination process.

8.3.1 Role of Honeybees

Since the commercial bee pollination services started, the main managed species globally for both honey production and crop pollination, the western honeybee have been extensively engaged as crop pollinators [Valido et al., 2019]. In fact, the western honeybee classifed as the most important pollinator bee species for crops worldwide [Garantonakis et al., 2016], and is the most efficient forager globally, performing about 13% of floral visits to 5% of crops across all plant web [Hung et al., 2018]. Nonetheless, there are at least eight other honeybees in the *Apis* genus, for example, *A. florea* Fabr., *A. cerana* Fabr., *A. andreniformis*, and *A. dorsata* Fabr [Streinzer et al., 2013]. It was assessed that honeybees provided USD 11.68 billion to agroindustry in the USA, in 2009 [Calderone, 2012]. Honeybees are important pollinators due to their pollinating efficiency and broad accessibility [Calderone, 2012]. The mutualistic associations between crop plants and honeybee pollinators arising from the exchange of nutritious food of nectar and pollen (Tanda, 2022 a, b). Bees are a chief player in the crop pollination [Hung et al., 2018], and plants produce a sugar rich liquid similar to nectar from their glands to allure pollinators to their different color flowers so that the pollen grains can stick to the body of bee foragers [Shin et al., 2019]. Apiculturists have studied that honeybees (*A. mellifera* L.) seems to like crops high amounts of nectar and pollen which is beneficial to high stocks of this food, for the colony sustainable development and improvement of foragers efficiency in the flowering field crop [Ghosh et al., 2020; Seo et al., 2019]. Several nations have employed honeybee services and obtained excellent outputs in improvement of the quality and quantity of crops. The pollination jobs of honeybee abundance is well appreciated in three crops species, such as cucumber (*Cucumis sativus* Linn), in the USA, for which there has been a 10% improvement in production and in the number of hives from 40,000 to 45,000; cranberry (*Vaccinium oxycoccos* Linn), which accomplished as an enhancement of crop yield from 3.7 million in 1989 and 5.4 million in 1998 [Morse and Calderone, 2000], and pear (*Pyrus communis* Linn), which showed a 7% additional improvement in fruit size and a net income boost of $400 per hectare [Naumann et al., 1994]. The honeybee pollinators enhanced the fruit quality of guava (*Psidium guajava* Linn), in India, as well as the fruit length and girth of coconut (*Cocos nucifera* Linn) and citrus (*Citrus* spp.) in comparison to the control treatments [Rajagopal et al., 2005; Meléndez-Ramírez et al., 2004]. Honeybees crop services, in Egypt, have significantly improved the percent seed set and seed yield in onion (*Allium cepa* Linn) as compared to other insect pollinatorss [Ewies and EL-Sahhar 1977]. Moreover, in Burkina Faso, the yield of sesame (*Sesamum indicum* Linn) seeds tripled after engaging honeybee services in flower pollination [Stein et al., 2017]. The crop pollination by honeybees, of oilseed rape (*Brassica napus* Linn), strawberry (*Fragaria × ananassa* (Duchesne ex Weston) Duchesne ex Rozier) and buckwheat (*Fagopyrum esculentum* Moench), have significantly increase increased their production and quality [Bartomeus et al., 2014]. Likewise, flowers of black cumin are alluring to a number of insect pollinators, for example, Hemiptera (true bugs), Diptera (flies), Coleoptera (beetles), and Hymenoptera (bees) [Abd El-Wahab and Ebadah, 2011]. Nonetheless, honeybees are the most plentiful pollinators influencing its yield and crop quality, with their pollination effectiveness by improving the number of seeds per fruit and increasing the total crop yield, which has resulted to the recommendation that beekeepers move beehives near black cumin fields for effective pollination and enhanced productivity [Saboor et al., 2018]. The high significant production of anise also depends on bee pollinator action. An investigation described that honeybees showed a daily peak in anise pollination act between 12 noon and 2 p.m., and enhanced the production above heights observed with insect exclusion, though levels were below those achieved with natural bee pollination. Honeybees and six species of Andrenidae are the chief coriander pollinators, with 63% of honeybee flower visits and 100% of the trips by three species of Andrenidae leading in pollinating task [Bendifallah et al., 2013]. Maximum flower visit rates by high-quality honeybee colonies enhanced fruit set by 15%, in addition to the fruit sugar content and seed set compared with flower trips by conventional colonies, in apple (*Malus domestica* Borkh), leading in the farmer's profits boosting by 70%. Powerful colonies also boosted fruit weight by 20% by bee pollination [Geslin et al., 2017]. The Western honeybee pollination increased the equatorial diameter by a mean of 13.3%, fruit mass by 30.3%, seed variety by 7%, and seed mass by 8.4% compared with self-pollination, in the fruit of cape gooseberry (*Physalis peruviana* Linn) [Chautá-mellizo et al., 2012], while the utilization of honeybees for almond pollination enhanced fruit set by 60% compared with bee-remote trees, which converted into a 20% boost in productivity. In the presence of wild bees (*Bombus* spp., *Halictids* bees, *Andrenids* bees, and *Xylocopa virginica*) and managed honeybees in small isolated and big fields, studies on blueberry (*Vaccinium corymbosum* Linn) pollination in Michigan, USA, established that wild bees were the crucial pollinators in the small fields, considering for 58% of flower foraging, whereas honeybees were the major pollinators in the big field area, with 97% of visitations. Likewise, it was observed that flowers in the large fields were foraged by four times as many bees as flowers in the small study fields. The weight of the fruit was pretentious by the degree of bee pollination and the population of bees, and the weight of berries was twice as bigger in the large fields in comparison to the small fields [Isaacs and Kirk, 2010].

8.3.2 Crop Pollination and Bumblebees

Bumblebees (Apidae: Bombini) are indispensable pollinators for farm and wild plants globally, and their pollination aids in food security [Crowther et al., 2019]. Mostly five species of bumblebees are engaged in commercial crop pollination services:

I. *Bombus terrestris* Linn (in Europe, North Africa, Asia, and Australasia),

II. *B. occidentalis* Greene (in western North America),

III. *B. ignitus*,

IV. *B. lucorum* Linn (in East Asia), and

V. *B. impatiens* Cresson (in North America) [Velthuis and Van Doorn, 2006].

To continue visiting surprisingly in high and low temperatures, the aggressive converting to various climates and environments of bumblebees describes their capability [Peat et al., 2005]. Bumblebees have donated a lot to the global crop pollination through improving the production and fruit quality [Fijen et al., 2018]. In fact, horticulturists gain many benefits from pollination by bumblebees, which are effective pollinators of many crops, such as kiwifruit (*Actinidia deliciosa*) [Nayak et al., 2019], sweet pepper (*Capsicum annum* Linn) [Shipp et al., 1994; Roldán Serrano and Guerra-Sanz, 2006], and red clover (*Trifolium pretense* Linn). They are major pollinators of buzz pollinated crops, for example, blueberry and tomato, including large and small flowering plants, providing them the potential to have adequate pollination in open fields and greenhouses [Desjardins and De Oliveira, 2006; Yankit et al. 2018]. In India, it has also been described that buzz pollination by *Bombus haemorrhoidalis* Smith result in longer, bigger, heavier, and healthier fruits, exceptionally in kiwi fruit [Nayak et al., 2019]. Pollination by bumblebees increases the quality and quantity of tomato fruit. The number of fruit per cluster, the number of fruit per plant, fruit length, fruit freshness, fruit breadth, and fruit production were also improved [Yankit et al., 2018]. Additionally, pollination of sweet pepper by bumblebees lead into a larger number of pollen grains and a greater degree of seed set on the fruit than self-pollination, such that flowers foraged by bumblebees develop larger and heavier fruit than non-foraged flowers [Roldán Serrano and Guerra-Sanz, 2006]. eventually, bumblebees have offered greater pollination services to hybrid leek (*Allium porrum* Linn), leading in a 25% increase in plant quality, which helped the plant quality and crop price value by an estimated USD 18,007 and USD 17,174 per hectare, correspondingly [Fijen et al., 2018]. Wild pollinators provides better pollination than honeybees, in some cases, as observed in apple crops pollinated by bumblebees, because all wild bee species are capable to clasp and place more apple pollen grains on stigmas than honeybees [Normandeau Bonneau et al., 2020].

8.3.3 Role of Stingless Bees

In tropical and subtropical regions of the world, stingless bees (Apidae: Meliponini) are usual flower visitors of crops. They show higher food diversity and intensity in their visitation behavior than honeybees and thus impact the evolution of pollination solutions that are best applicable to the requirements of specific crops and environments [Quezada-Euán, 2018]. They are a large, diverse category of eusocial bees, qualifying them as good pollinators. They are small- to medium-sized, with rudimentary stings [Quezada-Euán, 2018]. Few species tend to be large and smooth, with long hairs that assist to bring pollen and other plant products to the bee colony [Ramírez et al., 2018]. The physiology of stingless bees is favorable to flower pollination [Eickwort and Ginsberg, 1980] because they have satisfactory structures for gathering pollen, nectar. The absence of stinging nature, make them to handle easily than the honeybees. Stingless bees, belonging to the genus *Melipona*, show vibration behavior to take out the pollen, which is required in tomato and pepper crops with poricidal anthers [De Luca and Vallejo-Marín, 2013]. The neotropical stingless bee *Melipona quadrifasciata* Lepeletier is employed to pollinate greenhouse tomatoes, and has increased the fruit yield with minimum mechanical injury [Del Sarto et al., 2005]. They also play a vital role in the cucumber crops of greenhouse pollination, by ameliorating the fruit weight and productivity [Solange et al., 2008]. The pollination of cucumbers by the stingless bee *Heterotrigona itama* and hand pollination enhanced crop quantity and fruit quality, resulting in heavier, longer, and wider fruits

[Azmi et al., 2017]. Likewise, the pollination of rockmelon (*Cucumis melo* var. *reticulatus*) by stingless bees and manual cross-pollination had a useful effect on fruit set and the number of seeds per fruit in comparison to self-pollination [Azmi et al., 2019]. Pollination of strawberries in greenhouses by stingless bees also improved the quality and commercial fruit value [Roselino et al., 2009]. Additionally, the pollination of eggplant (*Solanum melongena* Linn) by *Melipona fasciculata* Smith in greenhouses improved fruit set by 29.5% and enhanced fruit quality when compared with self-pollinated plants [Ilva et al., 2013].

8.3.4 Role of Carpenter Bees

A class of large carpenter bees that are found in tropical and subtropical regions and related to the genus *Xylocopa* in the tribe *Xylocopini* (Apidae: Xylocopinae) [Leys et al., 2000]. They have many benefits in crop pollination, as they feed on a wide variety of plants during their long active seasons, comparing with other non-*Apis* bees. They are also capable to perform buzz-pollinate flowers, establishing them among diverse plant pollinators [Somanathan et al., 2019]. Nevertheless, there is a high demand for an adequate breeding process to be designed that involves the selection of new genotypes, under control mating, and nesting [Keasar, 2010]. They are familiar for their ability to construct their nests in hard wood, logs, stumps, or the dead tree branches by making tunnel patterns [Richards, 2020]. During the day and sometimes even work through moonlit nights, carpenter bees remain working the year around and visit a number of flowers in India. It has been observed that flowers foraged by carpenter bees secret nectar that is odoriferous, so it is possible that these bees consume this odor as a cue to forage the right flowers [Raju and Rao, 2006]. The employment of carpenter bees for pollination jobs is essential to guarantee sufficient pollination for many crops, alongwith passion fruit (*Passiflora edulis* f. *flavicarpa*), cucurbits, and other vegetables and fruits, as found in the Philippines, Brazil, USA, and Malaysia [Halder et al., 2019; Barrera et al., 2020]. Yellow passion fruit is adequately pollinated when the flowers are only foraged by local bees, specifically carpenter bees [Silveira et al., 2012]. Likewise, when native carpenter bees (*Xylocopa* (*Lestis*)) were engaged as different to bumblebees for tomato pollination in a glasshouse, the females buzz-pollinated the flowers and the fruits developed were heavier and contained more seeds than those that were not pollinated by these pollinators [Hogendoorn et al., 2000]. The carpenter bee *Xylocopa pubescens* Spinola is also employed to pollinate greenhouse raised honeydew melons (*Cucumis melo* Inodorus Group), as it was seen that while this species had shorter foraging timing per flower than the honeybee. However, pollination by both bees led in a similar fruit mass and seed numbers, and *X. pubescens* pollination improved fruit set threefold when compared with honeybee crop pollination [Adi et al., 2007].

8.3.5 Role of Solitary Bees

Globally, solitary bees involve the majority of bee species. They are responsible for 85% of pollination services of all bee species [Batra, 1984]. Most of solitary bees are polylectic (gather pollen from several plants), while a smaller number are oligolectic (those who forage a narrow range of plant species) and very few are monolithic (use only a single plant species) and in recent decades, in Britain there has been a decrease in monolithic and oligolectic species [Biesmeijer et al., 2006]. They are a major pollination players, and it has been described that wild bees contribute USD 3251/hectare for their pollination services globally, and seven out of ten of which are solitary [Kleijn et al., 2015]. They are more efficient pollinators than honeybees for few crops that rely on pollinators for their reproduction, as in apple. In fact, the economic benefits of utilizing solitary bees for apple production were calculated to be € 51.4 million compared to honeybees of € 21.4 million in the United Kingdom [Garratt et al., 2016].

8.4 BEE FLOWER FORAGING

Flower cross-pollination is accomplished by bees and other insects for a number of crop plants. As flower pollination is so indespensible to plants, they modify themselves to be more attractive to different insect pollinators [Bloch et al., 2006; Pashte and Kulkarni, 2015]. Plants have various techniques to allure bees, such as flower color [Papiorek et al., 2016; Bauer et al., 2017], motion of flowers as in bumblebees [Hennessy et al., 2020], plant cell type (e.g., conical epidermal cells) as in bumblebees [Whitney et al., 2009], visual and olfactory signals as in honeybees and apple pollination [Rachersberger et al., 2019] and nectar and pollen production [Varassin et al., 2001; Gresty et al., 2018], as shown in Fig. 8.1.

Fig. 8.1 Bee flower foraging influenced by various factors

Similarly, plants contribute a lot in affecting the rate of visitation in bee pollinators [Prado et al., 2021]. One of the most significant attributes of plants that allures bees is the flower color [Bradshaw and Schemske, 2003; Handelman and Kohn, 2014]. Bees possess a trichromatic visual system that is very delicate to green, ultraviolet, white, and blue wavelengths, permitting them to observe many colors. Bees mostly forage blue or purple flowers but generally attracted to blue flowers [Giurfa et al., 1995; Hill et al., 1997]. Relatively, bees are rarely attracted to red flowers, although they visit red flowers sometimes reflecting ultraviolet light [Chen et al., 2020]. The flower movement is also important for more attractiveness to bees, as this is essential to their three-dimensional vision system. The comparative movement of the flowers enhances the number of opposite stimuli on the bee‘s vision. Conical epidermal cells possess numerous functions, for example encouraging the attention of flower color and decreasing the wettability of the petal, which improves the efficiency of the cross-pollination procedure. As a consequence, bees mostly like to forage conically shaped *Petunia* flowers, especially those that are most attractive with movement [Alcorn et al., 2012; Whitney et al., 2011]. In fact, maximum numbers of bees appear on the moving part of a flower bed due to their constructive reaction to optical stimulation [Alcorn et al., 2012]. Additionally, in bumblebees, flower volatile substances influence the bee orientation [Giuliani et al., 2020], pollen gathering and the behavior [Solís-Montero et al., 2018]. Flower size has an exceptional function for attracting pollinators [Giuliani et al., 2020], as bumblebees (*Bombus diversus* Smith) like large floral exhibits [Makino et al., 2007]. The quantity and/or quality of pollen and nectar secreted finally, may influence the flower foraging by bee pollinators. For bee and other pollinators, nectar and pollen are main sources of energy, protein, and lipids, [Shrestha et al., 2017] and bees are attracted to plants to offer food for their brood [Perera and Karunaratne, 2019]. Solitary bees forage for pollen but rarely nectar [Gresty et al., 2018], whereas honeybees visit for flowers with big quantities of nectar [Mallinger and Prasifka, 2017]. It has been described previously that the make-up of wild bee populations that forage several plants can best be elucidated by differences in the flowering regions, flowers height, and pollen amount moved in the flowers [Rowe et al., 2020]. For instance, British bumblebees distinguish between *Mimulus guttatus* plants on the basis of their pollen amount and quality, with a great relationship between foraging and pollen quantity [Robertson et al., 1999]. Moreover, both wild bees and honeybees visit sunflowers for their nectar sugar, with the number of flower trips enlarging sufficiently with an increasing nectar sugar level and lowering during the corolla time but appearing to be unaffected by nectar sugar composition. Wild bees make maximum trips to sunflowers that offer pollen (male-fertile plants) and honeybees prefer pollen-free flowers (male sterile plants) [Mallinger and Prasifka, 2017]. Cover crops, especially low-diversity mixes that involve buckwheat and *Phacelia* spp., offer a great abundance of flowers throughout the summer season, leading in maximum bee visitation rates, with *Phacelia* spp. being more attractive to honeybees and bumblebees, while sunflowers and native wildflowers are more alluring to solitary bees [Mallinger et al., 2019]. Honeybees forage native wild and grown crop plants at comparable average rates that are autonomous floral abundance, thus improving their foraging rates for the greatly plentiful plants, whereas the visit rate of wild pollinators is maximum for raised plants than for wild plants. For instance, knapweed (*Centaurea* spp.) is a widely distributed and often locally significant plant for honeybees, as it is liked and regularly visited for its pollen and nectar requirements [Urbanowicz et al., 2020]. The visitation ratio of large solitary bees, wild beetle pollinators, and bumblebees was negatively influenced by the abundance of honeybees hives, in France [Ropars et al., 2019]. A behavior that few bees have evolved during their foraging is buzz pollination, whereby the bees produce vibrations to remove and gather the pollen from the fruit set, fruit mass, and flowers during seed fertilization [Pritchard and Vallejo-Marín, 2020]. About half of bee species can carry out buzz behavior, for example, large carpenter bees, minute sweat bees and bumblebees, but they vary in their buzzing behavior [Pritchard and Vallejo-Marín, 2020; De Luca et al., 2014]. This buzzing vibration is the best method for collecting pollen from plant species that have little pores on their anthers [Vallejo-Marín, 2019]. During this behavior, the bee bites the anther of the flower and produces vibrations with its thoracic muscles while in direct touch with the flower, bringing about the vibrations to be conveyed the flower parts [Russell et al., 2016]. There are many ways of carrying out this buzzing behavior. For instance, the flowers of *Pedicularis* spp. possess long anthers and narrow corolla tubes, so bees produce their buzzing shaking on these corolla tubes for fast pollen collections [Corbet and Huang, 2014; Cardinal et al., 2018]. Consequently, this behavior is associated to the functional uniqueness of flowers, specifically in those flowers in which pollen release needs the stamens modifications. Buzz pollination is known from the audible sound produced during the vibrations and is also generally termed sonication [Corbet and Huang, 2014]. Recently it is trusted that this buzzing sound for pollen gathering is not made by any other insect. Nevertheless, further studies is required to know whether some other flies also exploring the vibration behavior to extract pollen. Several bees forage during the daylight and only five of nine bee families look for flowers in dim light conditions [Wcislo and Tierney, 2009]. Nocturnal bees have various factors that influence their flower visits, counting light intensity and temperature. Different temperature and light levels are lower at night than during the day, and this can influence, specifically, the nocturnal bee's behavior.

8.5 BEE POLLINATION OPPORTUNITIES AND THREATS

Bees are encircled by different parameters that influence their contributions in pollination, for example disease pathogens, shortages in nutrition, changing climate, and deforestation (Figure 8.2) [Ma et al., 2019; Dolezal and Toth, 2018; Huang and Giray, 2012; Christopher Brown and Albrecht, 2001]. Viral and bacterial infections have a negative effect on bee health and its longevity, threatening crop and wild plant pollination needs [Yang et al., 2018]. Viral diseases influence immune systems of bees, infecting the whole colonies [O'Neal et al., 2017]. Colony collapse disorder (CCD) is a phenomenon by which there are mysterious, rapid losses of adult workers in domesticated beehives in the USA, resulting in only the queen and a few nursing bees left behind [van Engelsdorp et al., 2009] and this problem is confronted by several beekeepers [Van Dooremalen and van Langevelde, 2021]. In the USA, the parasites *Nosema ceranae* and *Nosema apis* also have maximum disease effects, resulting in big honeybee losses [Snow et al., 2019]. Conroy et al. [2016] observed that both nutritional scarcity and diseases have a huge effect on bumblebee colonies, with a dearth of pollen and low nectar sugar levels causing lower pollination and, thus, a decline in yield. In addition to the natural factors affecting bee pollination, the over use of pesticides, such as acetamiprid and ergosterol-inhibiting fungicides, challenge crop pollination jobs [Han et al., 2019]. Pesticide residues and other synthetic products persist in the nectar and pollen gathered by bees, resulting in neurotoxicity, changes in behavior, immune deficiency, and chronic ailments [Christen et al., 2018; Arce et al., 2018]. The use of neonicotinoid insecticides, which are systemic insecticides that are transmitted into the pollen and nectar of several pollinated crops, is one of the main co-factors related with bee damages [Woodcock et al., 2016; Jiang et al., 2018]. Applying fungicides, insecticides, and pesticides lead in contamination, toxicity, and declines in the quality and quantity of nutrients in the pollen and nectar, resulting in poor colony health and consequently menacing the bee survival [Tosi et al., 2017; Tomé et al., 2017].

The Environmental Protection Agency (EPA) has been assessing the influence of pesticides on the immune system of bees [Vanegas, 2017], while the European Food Safety Authority assessment offers detail on the chronic toxicity of pesticides on bees [Rortais et al., 2017]. The influence of neem-based insecticide (botanical) and pyrethroid insecticides, deltamethrin and the fungicides thiophanate-methyl and chlorothalonil (synthetic) insecticides on the melon (*Cucumis melo* L.) has been studied. Insecticides and pesticides not only decreased the visitation intensity of bees but also reduced melon production [Tschoeke et al., 2019]. Consequently, good nutrition has a direct impact on the immune function and an indirect effect on availability of energy [Dolezal et al., 2018]. Numerous elements also affect the development, reproduction, and survival of bees, for instance high temperatures, and humidity, decreasing not only the bee biodiversity and other pollinators but also farm yield [Ma et al., 2019]. Most bees forage flowers during the daylight, but some bee families look for flowers in dim light [Wcislo and Tierney, 2009]. Nocturnal bees resolve their flower visits by light intensity and temperature [Liporoni et al., 2020].

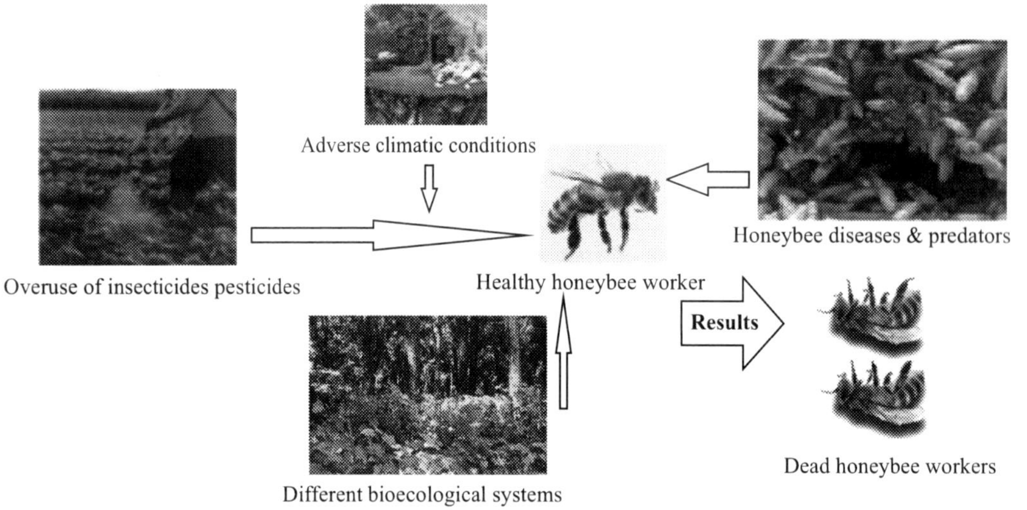

Fig. 8.2 Schematic model of various threats to bee pollinators

Deforestation can also affect bee populations (Fig. 8.2) [Christopher Brown and Albrecht, 2001]. For instance, the population of bumblebees in the tropical agricultural highlands of Guatemala enhanced with the increase in forests and semi-natural vegetation in native areas, but was not influenced by any seasonal changes [Kenefic et al., 2020]. Loss in environment and climate change also influence honeybees globally, resulting in pollinator losses [Grassl et al. 2018; Dalsgaard 2020; Jacques et al., 2017].

8.6 BEE POLLINATION VS. NON-BEE POLLINATION

Bees are contemplated as the most efficient pollinators; however, the role of other insect pollinators cannot be ignored or is negligible, as they contribute to boost and sustain crop pollination and are dependent upon these plants for the pollen and nectar supplies [Jacques et al., 2017]. The major categories of other insect pollinators are butterflies, moths (Lepidoptera), beetles (Coleoptera), and few flies (Diptera) [Cook et al., 2020].

8.6.1 Hover Flies vs. Bees

Hover flies (Diptera: Syrphidae) are regarded as the most anthophilous family belonging to Diptera order [Larson et al., 2001]. *Episyrphus balteatus* DeGeer is one of the most common Hover fly which is mostly observed in crop flowering areas, and many recent reports have established its role in many crops as a pollinator globally [Hodgkiss et al., 2018]. One major crop that is cross pollinated by hover flies *E. balteatus* is oilseed rape, which is an important crop grown in temperate areas. It has been described that when *E. balteatus* is engaged in the pollination services of this crop, sufficiently high numbers of seeds per pod are developed, explaining the efficiency of *E. balteatus* as an edible crop pollinator [Jauker and Wolters, 2008]. The drone fly, *Eristalis tenax* L., has also been observed as a profitable pollinator of many managed crops, for example, pak choi (*Brassica rapa* subsp. *chinensis*) and onion in New Zealand [Howlett and Gee, 2019], onion, spring turnip rape (*Brassica rapa* L. subsp. *oleifera*), and sweet pepper (*C. annuum*) in Canada [Jarlan et al., 1997], kiwifruit in Italy [Barbattini et al., 1994], and carrot (*Daucus carota* L. subsp. *sativus*) in Germany [Schittenhelm et al., 1992]. Accordingly, Brad and Megan, reported when *E. tenax* L. is often employed in large numbers in fields during the flowering period for pollination [Howlett and Gee, 2019]. In a research study in Germany aimed at the influence of pollination by the solitary mason bee *O. rufa* and two Hover fly species (*E. tenax* and *E. balteatus*) were kept in a field of oilseed rape. Studies reported that the fruit yield and number of seeds per pod were better with an additional bee population compared to hover fly abundance, and that fivefold higher density of hover flies than red mason bees were needed to obtain the same fruit sets and production. So, reports established that mason bees were more efficient pollinators than hover flies on this crop trials [Jauker et al., 2012].

8.6.2 Butterflies vs. Bees

About 180,000 species of butterfly and moth (Lepidoptera) are observed and are approximately 10% of all identified as insect pollinators. Butterflies describe about 10% of Lepidoptera [Hahn and Brühl, 2016] and visit psychrophilic flowers that offer small to medium amounts of dilute nectar. Flowers defined by brightly colored petals with pleasant aroma and a flat platform that allows the butterflies to alight in the inflorescence. Large- to medium-sized butterflies function as pollinators through their wings [Santos et al., 2020]. A dry land floral plant in Kenya *Gloriosa* minor *Rendle* (Colchicaceae) that rely on butterflies for pollination, and the seed production, which are a origin of colchicine [Martins, 2014]. *Caesalpinia pulcherrima* was pollinated mostly through butterflies which convey the pollen grains on their wings [Cruden and Hermann-Parker, 1979]. *Angadenia berteroi* is an endangered flowering plant that possess large, showy, yellow, and tubular flowers without fragrance. The tubular flowers provides them a complex structure, so any pollinators allured to such plants should possess mouthparts that are long sufficient to suck the nectar [Pinto et al., 2008; Barrios and Koptur, 2011]. Bees and butterflies forage *A. berteroi*, permitting a comparison of their effectiveness. The long-tongued bees having wider heads than the apical portion of the pollen chamber, compelling them to tap the reproductive parts of the flower, and these bees seldom reforge the same flowers. Long-tongued bees seem to be effective in collecting nectar and moving pollen [Gann et al., 2020]. Conversely, the two categories of butterflies that forage the flowers skippers (Hesperiidae and non-skippers) convey very small amounts of pollen on their proboscides, do not move this pollen on the stigmas of the flowers, and regularly forage the same flowers, thus seem to work as nectar robbers. The reforaging of the same flower by an insect is familiar to have negative influences. Flower re-visitation can result in to fruit abortion and ovule due to self-pollen transfer on the stigma, defining why the mouthparts are related with the efficiency of the pollination procedure.

8.6.3 Moths vs. Bees

In different environments around the globe, moths can be classify as possessing a crepuscular or nocturnal lifestyle and are identified as one of the major pollinators of a large number of crop species [Macgregor et al., 2015]. The studies on the contribution of moth pollination in natural environments is accessible as about 227 flowers have been crossed by moths [Hahn and Brühl, 2016]. Moths are regular flower visitors, and meet plants many times. They play role in pollination of about 40% of plants in rural landscapes, including meadows, field edges, pastures, old farms, and roadsides. Accordingly, the services of moths in farm habitats is mostly ascribed to their pollination of non-crop plants, which helps in sustaining the biodiversity in farm-environments, providing an enormously admirable ecological service [Ricketts et al., 2008; Power, 2010]. They also play a good role in the pollination of some orchid unique plants. Due to few reports on this subject their jobs in pollination are being ignored [Alarcón et al., 2008; Devoto et al., 2011; Sime and Baldwin, 2003]. In the central Darling Downs insect pollination studies on 42 sunflower crops found that Western honeybees were the most frequent foragers, and on average 65.3 bees per 100 flower heads were seen visiting in the mid-morning [Radford et al., 1979]. Conversely, *Helicoverpa armigera* Hübner moths were reported foraging the flowers in 33 crops between 7 and 8 pm, and the population's density was observed to be on average 3.9 moths per 100 flower heads. Consequently, moths were active for less than 2 h per night compared to the 9 h activity of bees. The poor moth's performance and abundance showed that bees contributed significantly in pollination of sunflower in this region [Radford et al., 1979].

8.6.4 Beetles vs. Bees

Beetles (Coleoptera) relate to one of the most heterogeneous insect orders and their services in crop pollination process are more and more being appreciated. Few angiosperms rely on pollination by particular categories of beetles. In the tropics, for instance, members of Cetoniinae (Scarabaeidae) subfamily are frequent flower visitors [Armstrong and Irvine, 1989]. Exclusively, over 184 flowering plants belonging to Magnolia in Magnoliaceae are cross-pollinated by beetles [Bernhardt, 2000; Dieringer et al., 1999]. Few beetles utilize flowers as meeting place besides their utilization in their food which adds to their activities as flower pollinators. As opposed to tropical pollinators, beetles rely on odor to search flowers, but Hopliine beetles utilize visual signals alone and, even without food or smell, are allured to colors looking bright [Mayer et al., 2006]. Mostly they are related to the pollination of flowers having open bowl-shapes [Bernhardt, 2000]. They were reported to be the second most significant insect class pollination players in Lambir (27%) Malaysia and Kakachi (17%) in India. Bees and beetles constitute over 60% of the crop pollination of tree species in Lambir and 34% in Kakachi [Devy and Davidar, 2003]. Bees are the major pollination players performing 32% in Lambir whereas beetles contribute upto 20% inpollination services [Momose et al., 1998]. However, in the southwestern area of Cape Province and Namaqualand, the beetle *Hopliini* sp. (Scarabaeidae) has been observed as the most significant and effective pollinator of trees [Colville et al., 2002; Van Kleunen et al., 2007]. To determine the ecological function of beetles and their efficiency in in pollination, require further investigations [Sayers et al., 2019].

8.6.5 Thrips vs. Bees

Thrips (Thysanoptera) play a role in pollination of plants, however still need further information to establish their pollination services [Suetsugu, 2019]. They possess piercing-sucking mouthparts and are generally seen on flowers covered with pollen, and rely on nectar, pollen, or the plant tissue contents in their nutrition [Ananthakrishnan, 1982]. In the southern state of Chiapas, Mexico in three flowering seasons, these tiny insects have been observed in the blooming period of coffee species *C. arabica* L. and *C. canephora* L. during the years 2013–2015. Many thrips species on coffee flowers were seen to be moving a few pollen grains on their bodies [Infante et al., 2017]. In shaded coffee agro-forests of South India, bees have enhanced coffee fruit yield by 50% through pollination more than wind contribution. The role of other flower pollinators with bees including Thysanoptera in coffee was insignificant. Because they did not touch the flower anther or stigma sufficiently in addition to infrequent visits.

8.6.6 Wasps vs. Bees

In the Neotropical area, social wasps (Hymenoptera) are observed as the flower pollinators. These wasps act as flower foragers as predators [Resende et al., 2001; Beggs, 2001]. There are several attributes that allure wasps to blooms including color and shape of flowers. Reddish brown, dirty purple, and dirty brown flowers attract them from a distance. Schremmer (1962) has described that wasps are allured to tiny flowers with bulbous shape, broad openings and sucrose-rich nectar [Schremmer, 1962]. To these wasps, flower aroma is one of the major allurement. Social wasps are the chief pollinators of *Epipactis helleborine* L. due to the attractive flower fragrance [Brodmann et al., 2008; Jakubska-Busse et al., 2005]. Similar research has been carried out on the coconut flower exploring wasps and honeybees. Because of their inability in picking sufficient quantities of pollen grains and their behavior in discouraging the honeybees, the wasps (*Polistes crinita* Felt) could not perform as an effective pollinator. Contrastingly, honeybees were found to be efficient coconut pollinators.

8.7 CONCLUSION

By an increasing world human population, there is a growing demand for food sustainability ostensibly threats caused. Bees are great in crop pollination alongside other animal pollinators constituting birds, bats, moths, hover flies, beetles, wasps, thrips, and butterflies and other agents for example wind and water. To the world food security, bees add through pollinating numerous fruits, vegetables, oilseeds, legume crops (Hristov et al., 2020; Khalifa et al., 2021; Tanda, 2022 a, b). Better economic gains of bees in enhancing the food productivity per year was found along with cash crops of coffee, cocoa, almond and soybean, when compared to the crop self-pollination. The quality and quantity of fruits, nuts, and oils was also improved using bee pollination services. Beehives are encountered with several threats that affect their growth, reproduction, and sustainability, especially change in climate, overuse of pesticides, land use, and bee management clout, as these factors help in profitable crop pollination. Bee pollination offers a various types of benefits to mankind, adding to food processing industry, raw materials, medicines, fibers, social, cultural values, and the sustainability of bio-diverse ecological systems. Further, bee pollination directly influences the world crop cultivars through profitability and productivity such as improving most vegetables, seeds, and nuts, and increasing high-value farm products, including coffee, cocoa, and rapeseed. Presently, approximately 5–8% of all world crop yield would be gone in the absence of bee pollination services, demanding alterations in the human diet and the enlargement of crop lands to settle shortages in food production. Bee populations are fronted with various challenges that can deform their growth, such as shifting in farm land usage, climatic dynamics, over use of pesticides, varietal genetics and crop cultivation management system. Anxieties about the decline of managed and wild bees have expanded the demand to promote the utilization of the wild bees on farm lands for pollination. Noticing wild bee foraging have advanced with the growth of high-diversity bee environments in the landscape around, the rehabilitation of high-diversity bee habitats is must to encourage free pollination strengths. A safe bee environment should be offered for healthy crop production. The application of pesticides is harming to human health as both food and bee products got intoxicated with sprays that people must in the end consume (Hristov et al., 2020; Khalifa et al., 2021). Though the contribution of non-bee pollinators cannot be disregarded, bee pollination still remains a golden strength that should be safe guarded. Not only to develop habitat balance but also to sustain food security globally, bee populations for pollination must be secured. The contributions of bees are significant for crop production globally alongwith improving the yield and quality of some medicinal plants (Hristov et al., 2020; Khalifa et al., 2021;Tanda, 2022 a, b). Apiculturists should concentrate their attention on investigating the influences that bee pollination put in fruit quality, which should offer better observations concerning how bees can change the biochemical make up through pollination in some agricultural cultivars.

REFERENCES

- Abd El-Wahab, TE, Ebadah IMA (2011) Impact of honeybee and other insect pollinators on the seed setting and yield production of black cumin *Nigella sativa* L. J. Basic. Appl. Sci. Res, 1, 622–626.
- Abd El-Wahab, TE, Ebadah IMA, Mahmoud YA (2012) Insect pollinators of anise plants (*Pimpinella anisum* L.) and the important role of honeybees (*Apis mellifera* L.) on their yield productivity. Arch. Phytopathol. Plant Prot., 45, 677–685.
- Adi S, Avi S, Tamar K (2007) The carpenter bee *Xylocopa pubescens* as an agricultural pollinator in greenhouses. Apidologie, 38, 508–517.
- Aizen MA, Garibaldi LA, Cunningham SA, Klein AM (2009) How much does agriculture depend on pollinators? Lessons from long-term trends in crop production. Ann. Bot., 103, 1579–1588.
- Alarcón R, Davidowitz G, Bronstein JL (2008) Nectar usage in a southern Arizona hawk moth community. Ecol. Entomol., 33, 503–509.
- Alcorn K, Whitney H, Glover B (2012) Flower movement increases pollinator preference for flowers with better grip. Funct. Ecol., 26, 941–947.
- Ananthakrishnan TN (1982) Thrips and pollination biology. Curr. Sci., 51, 168–172.
- Arce AN, Ramos Rodrigues A, Yu J, Colgan TJ, Wurm, Y, Gill RJ (2018) Foraging bumblebees acquire a preference for neonicotinoid-treated food with prolonged exposure. Proceedings. Biol. Sci., 285, 8–11.

- Armstrong JE, Irvine AK (1989) Floral biology of *Myristica insipida* (Myristicaceae), a distinctive beetle pollination syndrome. Am. J. Bot., 76, 86–94.
- Asiwe JAN (2009) Insect mediated outcrossing and geneflow in cowpea (*Vigna unguiculata* (L.) Walp): Implication for seed production and provision of containment structures for genetically transformed cowpea. African J. Biotechnol., 8, 226–230.
- Azmi WA, Ghazi R, Sultan U, Abidin Z, Chuah T, Mara UT (2017) Effects of stingless bee (*Heterotrigona itama*) pollination on greenhouse cucumber (*Cucumis sativus*). Malaysian Appl. Biol., 46, 51–55.
- Azmi, WA, Sembok, WZW, Yusuf N, Hatta MFM, Salleh AF, Hamzah MAH, Ramli SN (2019) Effects of pollination by the Indo-Malaya stingless bee (Hymenoptera: Apidae) on the quality of greenhouse-produced rockmelon. J. Econ. Entomol., 112, 20–24.
- Bänsch S, Tscharntke T, Gabriel D, Westphal C (2021) Crop pollination services: Complementary resource use by social vs solitary bees facing crops with contrasting flower supply. J. Appl. Ecol., 58, 476–485.
- Barbattini R, Greatti M, Zandigiacomo P, Costa G, Testolin R, Vizzotto G. Insect pollinators of kiwifruit and their role in crop pollination. In Proceedings of the Atti XVII Congresso Nazionale Italiano di Entomologia, Udine, Italy, 13–18 June 1994; pp. 13–18.
- Wilfredo B. Barrera Jr., Krizz Audrie D. Trinidad, Jeric A. Presas (2021) Hand pollination and natural pollination by carpenter bees (*Xylocopa* spp.) in *Passiflora edulis* Sims. f. *flavicarpa* Deg. (yellow passion fruit), Journal of Apicultural Research, 60(5): 845–852.
- Barrios B, Koptur S (2011) Floral biology and breeding system of *Angadenia berteroi* (Apocynaceae): Why do flowers of the pineland golden trumpet produce few fruits? Int. J. Plant Sci., 172, 378–385.
- Barrios B, Pena SR, Salas A, Koptur S (2016) Butterflies visit more frequently, but bees are better pollinators: The importance of mouthpart dimensions in effective pollen removal and deposition. AoB Plants 8: plw001.
- Bartomeus I, Potts SG, Steffan-Dewenter I, Vaissière BE, Woyciechowski M, Krewenka KM, Tscheulin T, Roberts SPM, Szentgyörgyi H, Westphal C, et al. (2014) Contribution of insect pollinators to crop yield and quality varies with agricultural intensification. PeerJ, e328–e347.
- Batra SW (1984) Solitary bees. Sci. Am., 250, 120–127.
- Bauer AA, Clayton MK, Brunet J (2017) Floral traits influencing plant attractiveness to three bee species: Consequences for plant reproductive success. Am. J. Bot., 104, 772–781.
- Beggs J (2001) The ecological consequences of social wasps (*Vespula* spp.) invading an ecosystem that has an abundant carbohydrate resource. Biol. Conserv., 99, 17–28.
- Bendifallah L, Louadi K, Doumandji S (2013) Bee fauna potential visitors of coriander flowers *Coriandrum sativum* L. (Apiaceae) in the Mitidja area (Algeria). J. Apic. Sci., 57, 59–70.

- Bernhardt P (2000) Convergent evolution and adaptive radiation of beetle-pollinated angiosperms. Pl Syst. Evol., 222, 293–320.
- Biesmeijer JC, Roberts SPM, Reemer M, Ohlemüller R, Edwards M, Peeters T, Schaffers AP, Potts SG, Kleukers R, Thomas CD (2006) Parallel declines in pollinators and insect-pollinated plants in Britain and the Netherlands. Science, 313, 351–354.
- Blettler DC, Fagúndez GA, Caviglia OP (2018) Contribution of honeybees to soybean yield. Apidologie, 49, 101–111.
- Bloch D, Werdenberg N, Erhardt A (2006) Pollination crisis in the butterfly-pollinated wild carnation *Dianthus carthusianorum*? New Phytol., 169, 699–706.
- Bosch J, Osorio-Canadas S, Sgolastra F, Vicens N (2021) Use of a managed solitary bee to pollinate almonds: Population sustainability and increased fruit set. Insects, 12, 56.
- Boyle NK, Pitts-Singer TL (2019) Assessing blue orchard bee (*Osmia lignaria*) propagation and pollination services in the presence of honeybees (*Apis mellifera*) in Utah tart cherries. PeerJ, 7, e7639.
- Bradshaw HD, Schemske DW (2003) Allele substitution at a flower color locus produces a pollinator shift in monkeyflowers. Nature, 426, 176–178.
- Breeze TD, Bailey AP, Balcombe KG, Potts SG (2011) Pollination services in the UK: How important are honeybees? Agric. Ecosyst. Environ., 142, 137–143.
- Brodmann J, Twele R, Francke W, Hölzler G, Zhang QH, Ayasse M (2008) Orchids mimic green-leaf volatiles to attract prey-hunting wasps for pollination. Curr. Biol., 18, 740–744.
- Calderone NW (2012) Insect pollinated crops, insect pollinators and US agriculture: Trend analysis of aggregate data for the period 1992–2009. PLoS One, 7, e37235. 17.
- Rader R, Bartomeus I, Garibaldi LA, Garratt MPD, Howlett BG, Winfree R, Cunningham SA, Mayfield MM, Arthur AD, Andersson GKS, et al. (2016) Non-bee insects are important contributors to global crop pollination. Proc. Natl. Acad. Sci. USA, 113, 146–151.
- Cardinal S, Buchmann SL, Russell AL (2018) The evolution of floral sonication, a pollen foraging behavior used by bees (Anthophila). Evolution, 72, 590–600.
- Chambó ED, Garcia RC, de Oliveira NTE, Duarte-Júnior JB (2011) Honeybee visitation to sunflower: Effects on pollination and plant genotype. Sci. Agric., 68, 647–651.
- Chautá-mellizo A, Campbell SA, Argenis M, Thaler JS, Poveda K (2012) Effects of natural and artificial pollination on fruit and offspring quality. Basic Appl. Ecol., 13, 524–532.
- Chen Z, Liu CQ, Sun H, Niu Y (2020) The ultraviolet color component enhances the attractiveness of red flowers of a bee-pollinated plant. J. Plant Ecol., 13, 354–360.
- Christen V, Kunz PY, Fent K (2018) Endocrine disruption and chronic effects of plant protection products in bees: Can we better protect our pollinators? Environ. Pollut., 243, 1588–1601.
- Christopher Brown J, Albrecht C (2001) The effect of tropical deforestation on stingless bees of the genus *Melipona* (Insecta: Hymenoptera: Apidae: Meliponini) in central Rondonia, Brazil. J. Biogeogr., 28, 623–634.

- Colville J, Picker MD, Cowling RM (2002) Species turnover of monkey beetles (Scarabaeidae: Hopliini) along environmental and disturbance gradients in the Namaqualand region of the succulent Karoo, South Africa. Biodivers. Conserv., 11, 243–264.
- Conroy TJ, Palmer-Young EC, Irwin RE, Adler LS (2016) Food limitation affects parasite load and survival of *Bombus impatiens* (Hymenoptera: Apidae) infected with *Crithidia* (Trypanosomatida: Trypanosomatidae). Environ. Entomol., 45, 1212–1219.
- Cook DF, Voss SC, Finch JTD, Rader RC, Cook JM, Spurr CJ (2020) The role of flies as pollinators of horticultural crops: An Australian case study with worldwide relevance. Insects, 11, 341.
- Corbet SA, Huang SQ (2014) Buzz pollination in eight bumblebee-pollinated *Pedicularis* species: Does it involve vibration-induced triboelectric charging of pollen grains? Ann. Bot., 114, 1665–1674.
- Crowther LP, Wright DJ, Richardson, DS, Carvell C, Bourke AFG (2019) Spatial ecology of a range-expanding bumblebee pollinator. Ecol. Evol., 9, 986–997.
- Dalsgaard B (2020) Land-use and climate impacts on plant-pollinator interactions and pollination services. Diversity, 12, 168.
- De Luca PA, Cox DA, Vallejo-marín M (2014) Comparison of pollination and defensive buzzes in bumblebees indicates species-specific and context-dependent vibrations. Naturwissenschaften, 101, 331–338.
- De Luca PA, Vallejo-Marín M (2013) What's the "buzz" about? The ecology and evolutionary significance of buzz-pollination. Curr. Opin. Plant Biol., 16, 429–435.
- De Milfont MO, Rocha EEM, Lima AON, Freitas BM (2013) Higher soybean production using honeybee and wild pollinators, a sustainable alternative to pesticides and autopollination. Environ. Chem. Lett., 11, 335–341.
- Del Sarto MCL, Peruquetti RC, Campos LAO (2005) Evaluation of the neotropical stingless bee *Melipona quadrifasciata* (Hymenoptera: Apidae) as pollinator of greenhouse tomatoes. J. Econ. Entomol., 98, 260–266.
- Desjardins ÈC, De Oliveira D (2006) Commercial bumblebee *Bombus impatiens* (Hymenoptera: Apidae) as a pollinator in lowbush blueberry (Ericale: Ericaceae) fields. J. Econ. Entomol., 99, 443–449.
- Deuri A, Rahman A, Gogoi J, Borah P, Bathari M (2018) Pollinator diversity and effect of *Apis cerana* F. pollination on yield of mango (*Mangifera indica* L.). J. Entomol. Zool. Stud., 6, 957–961.
- Devoto M, Bailey S, Memmott J (2011) The "night shift": Nocturnal pollen-transport networks in a boreal pine forest. Ecol. Entomol., 36, 25–35.
- Devy MS, Davidar P (2003) Pollination systems of trees in Karachi, a mid-elevation wet evergreen forest in Western Ghats, India. Am. J. Bot., 90, 650–657.
- Dieringer G, Cabrera RL, Lara M, Loya L, Reyes P, Journal I (1999) Beetle pollination and floral thermogenicity in *Magnolia tamaulipana* (Magnoliaceae). Int. J. Plant Sci., 160, 64–71.

- Dolezal AG, Toth AL (2018) Feedbacks between nutrition and disease in honeybee health. Curr. Opin. Insect Sci., 26, 114–119.
- Eickwort GC, Ginsberg HS (1980) Foraging and mating behavior in Apoidea. Annu. Rev. Entomol., 25, 421–446.
- Esquivel IL, Coulson RN, Brewer MJ (2020) A native bee, *Melissodes tepaneca* (Hymenoptera: Apidae), benefits cotton production. Insects, 11, 487.
- Ewies MA, EL-Sahhar KF (1977) Observations on the behavior of honeybees on onion and their effects on seed yield. J. Apic. Res., 16, 194–196.
- Fijen TPM, Scheper JA, Boom TM, Janssen N, Raemakers I, Kleijn D (2018) Insect pollination is at least as important for marketable crop yield as plant quality in a seed crop. Ecol. Lett., 21, 1704–1713.
- Free JB, Raw A, Williams IH (1975) Pollination of coconut (*Cocos nucifera* L.) in Jamaica by honeybees and wasps. Appl. Anim. Ethol., 1, 213–223.
- Gallai N, Salles JM, Settele J, Vaissière BE (2009) Economic valuation of the vulnerability of world agriculture confronted with pollinator decline. Ecol. Econ. 68, 810–821.
- Gann GD, Woodmansee SW, Bradley KA (2020) Conserving rare plants in locally-protected urban forest fragments: A case study from Miami-Dade County, Florida. Urban For. Urban Greening., 20, 1–11.
- Garantonakis N, Varikou K, Birouraki A, Edwards M, Kalliakaki V, Andrinopoulos F (2016) Comparing the pollination services of honeybees and wild bees in a watermelon field. Sci. Hortic. (Amsterdam), 204, 138–144.
- García-Breijo F, Armiñana J.R, Garmendia A, Cebrián N, Beltrán R, Merle H (2020) In vivo pollen tube growth and evidence of self-pollination and prefloral anthesis in cv. Macabeo (*Vitis vinifera* L.). Agriculture, 10, 647.
- Garratt MPD, Breeze TD, Boreux V, Fountain MT, McKerchar M, Webber SM, Coston DJ, Jenner N, Dean R, Westbury DB, et al. (2016) Apple pollination: Demand depends on variety and supply depends on pollinator identity. PLoS One, 11, e0153889.
- Geeraert L, Aerts R, Berecha G, Daba G, De Fruyt N, D'hollander J, Helsen K, Stynen H, Honnay O (2020) Effects of landscape composition on bee communities and coffee pollination in *Coffea arabica* production forests in southwestern Ethiopia. Agric. Ecosyst. Environ., 288, 106706–106717.
- Geslin B, Aizen MA, Garcia N, Pereira A, Vaissière BE (2017) The impact of honeybee colony quality on crop yield and farmers' profit in apples and pears. Agric. Ecosyst. Environ., 248, 153–161.
- Ghosh S, Jeon H, Jung C (2020) Foraging behavior and preference of pollen sources by honeybee (*Apis mellifera*) relative to protein contents. J. Ecol. Environ., 44, 4–10.
- Giannini TC, Cordeiro GD, Freitas BM, Saraiva AM, Imperatriz-Fonseca VL (2015) The dependence of crops for pollinators and the economic value of pollination in Brazil. J. Econ. Entomol., 108, 849–857.
- Gill RJ, Baldock KC, Brown MJ, Cresswell JE, Dicks LV, Fountain MT, Garratt MP, Gough LA, Heard MS, Holland JMOJ (2016) Protecting an ecosystem service:

Approaches to understanding and mitigating threats to wild insect pollinators. Adv. Ecol. Res., 54, 135–206.

- Giuliani C, Giovanetti M, Lupi D, Mesiano MP, Barilli R, Ascrizzi R, Flamini G, Fico G (2020) Tools to tie: Flower characteristics, voc emission profile, and glandular trichomes of two Mexican salvia species to attract bees. Plants, 9, 1645.
- Giurfa M, Núñez J, Chittka L, Menzel R (1995) Colour preferences of flower-naive honeybees. J. Comp. Physiol. A, 177, 247–259.
- Grassl J, Holt S, Cremen N, Peso M, Hahne D, Baer B (2018) Synergistic effects of pathogen and pesticide exposure on honeybee (*Apis mellifera*) survival and immunity. J. Invertebr. Pathol., 159, 78–86.
- Greenleaf SS, Kremen C (2006) Wild bee species increase tomato production and respond differently to surrounding land use in Northern California. Biol. Conserv., 133, 81–87.
- Greenleaf SS, Kremen C (2006) Wild bees enhance honeybees' pollination of hybrid sunflower. Proc. Natl. Acad. Sci. USA, 103, 13890–13895.
- Gresty CEA, Clare E, Devey DS, Cowan RS, Csiba L, Malakasi P, Lewis OT, Willis KJ (2018) Flower preferences and pollen transport networks for cavity-nesting solitary bees: Implications for the design of agri-environment schemes. Ecol. Evol., 8, 7574–7587.
- Hahn M, Brühl CA (2016). The secret pollinators: An overview of moth pollination with a focus on Europe and North America. Arthropod. Plant. Interact., 10, 21–28.
- Halder S, Ghosh S, Khan R, Khan AA, Perween T, Hasan MA (2019) Role of pollination in fruit crops: A review. Pharma Innov. J., 8, 695–705.
- Hall MA, Jones J, Rocchetti M, Wright D, Rader R (2020) Bee visitation and fruit quality in berries under protected cropping vary along the length of polytunnels. J. Econ. Entomol., 113, 1337–1346.
- Han W, Yang Y, Gao J, Zhao D, Ren C, Wang S, Zhao S, Zhong Y (2019) Chronic toxicity and biochemical response of *Apis cerana* (Hymenoptera: Apidae) exposed to acetamiprid and propiconazole alone or combined. Ecotoxicology, 28, 399–411.
- Handelman C, Kohn JR (2014) Hummingbird color preference within a natural hybrid population of *Mimulus aurantiacus* (Phrymaceae). Plant Species Biol., 29, 65–72.
- Hennessy G, Harris C, Eaton C, Wright P, Jackson E, Goulson D, Ratnieks FFLW (2020) Gone with the wind: Effects of wind on honeybee visit rate and foraging behavior. Anim. Behav., 161, 23–31.
- Herrmann JD, Beye H, de la Broise C, Hartlep H, Diekötter T (2019) Positive effects of the pollinators *Osmia cornuta* (Megachilidae) and *Lucilia sericata* (Calliphoridae) on strawberry quality. Arthropod. Plant. Interact., 13, 71–77.
- Hill PSM, Wells PH, Wells H (1997) Spontaneous flower constancy and learning in honeybees as a function of colour. Anim. Behav., 54, 615–627.
- Hipólito J, Sousa BdSB, Borges RC, de Brito RM, Jaffé R, Dias S, Imperatriz Fonseca VL, Giannini TC (2019) Valuing nature's contribution to people: The pollination services provided by two protected areas in Brazil. Glob. Ecol. Conserv., 20, e00782.

- Hodgkiss D, Brown MJF, Fountain MT (2018) Syrphine hover flies are effective pollinators of commercial strawberry. J. Pollinat. Ecol., 22, 55–66.
- Hogendoorn K, Steen Z, Schwarz MP (2000) Native Australian carpenter bees as a potential alternative to introducing bumblebees for tomato pollination in greenhouses. J. Apic. Res., 39, 67–74.
- Holzschuh A, Dudenhöffer JH, Tscharntke T (2012) Landscapes with wild bee habitats enhance pollination, fruit set and yield of sweet cherry. Biol. Conserv., 153, 101–107.
- Howlett BG, Gee M (2019) The potential management of the drone fly (*Eristalis tenax*) as a crop pollinator in New Zealand. N. Z. Plant Prot., 72, 221–230.
- Howpage D, Spooner-Hart RN, Vithanage V (2001) Influence of honeybee (*Apis mellifera*) on kiwifruit pollination and fruit quality under Australian conditions. New Zeal. J. Crop Hortic. Sci., 29, 51–59.
- Hristov P, Neov B, Shumkova R, Palova N (2020) Significance of apoidea as main pollinators. Ecological and economic impact and implications for human nutrition. Diversity, 12, 280.
- Huang ZY, Giray T (2012) Factors affecting pollinators and pollination. Psyche, A Journal of Entomology, Article ID 302409, 3 pages.
- Hung KLJ, Kingston JM, Albrecht M, Holway DA, Kohn JR (2018) The worldwide importance of honeybees as pollinators in natural habitats. Proc. R. Soc. B Biol. Sci., 285, 20172140–20172147.
- Ilva PNU, Rncir MH, Inês C, Ilva S (2013) Stingless bees, *Melipona fasciculata*, as efficient pollinators of eggplant (*Solanum melongena*) in greenhouses. Apidology, 44, 537–546.
- Infante F, Ortíz JA, Solis-Montero L, Mound LA, Vega FE (2017) Thrips (Thysanoptera) of coffee flowers. Ann. Entomol. Soc. Am., 110, 329–336.
- Isaacs R, Kirk AK (2010) Pollination services provided to small and large highbush blueberry fields by wild and managed bees. J. Appl. Ecol., 47, 841–849.
- Jacques A, Laurent M, Consortium E, Ribière-Chabert M, Saussac M, Bougeard S, Budge GE, Hendrikx P, Chauzat MP (2017) A pan-European epidemiological study reveals honeybee colony survival depends on beekeeper education and disease control. PLoS One, 12, e0172591.
- Jakubska-Busse A, Przado D, Steininger M, Anioł-Kwiatkowska J, Kadej M (2005) Why do pollinators become "sluggish"? Nectar chemical constituents from *Epipactis helleborine* (L.) Crantz (Orchidaceae). Appl. Ecol. Environ. Res., 3, 29–38.
- Jarlan A, De Oliveira D, Gingras J (1997) Pollination of sweet pepper (*Capsicum annuum* L.) in greenhouse by the syrphid fly *Eristalis tenax* (L.). Acta Hortic., 437, 335–339.
- Jauker F, Bondarenko B, Becker HC, Steffan-Dewenter I (2012) Pollination efficiency of wild bees and hover flies provided to oilseed rape. Agric. For. Entomol., 14, 81–87.
- Jauker F, Wolters V (2008) Hover flies are efficient pollinators of oilseed rape. Oecologia, 156, 819–823.

- Jiang J, Ma D, Zou N, Yu X, Zhang Z, Liu F, Mu W (2018) Concentrations of Imidacloprid and Thiamethoxam in Pollen, Nectar and Leaves from Seed-Dressed Cotton Crops and their Potential Risk to Honeybees (*Apis mellifera* L.); Elsevier Ltd.: Amsterdam, The Netherlands; Volume 201, ISBN 8605388242.
- Kasina JM, Mburu J, Kraemer M, Holm-Mueller K (2009) Economic benefit of crop pollination by bees: A case of kakamega small-holder farming in Western Kenya. J. Econ. Entomol., 102, 467–473.
- Keasar T (2010) Large carpenter bees as agricultural pollinators. Psyche A J. Entomol. 2010, 1–7.
- Kenefic NE, González PL, Theodorou P, Cardona E (2020) Disentangling the effects of local resources, landscape heterogeneity and climatic seasonality on bee diversity and plant-pollinator networks in tropical highlands. Oecologia, 194, 333–344.
- Khan MR, Khan MR (2004) The role of honeybees *Apis mellifera* L. (Hymenoptera: Apidae) in pollination of apple. Pakistan J. Biol. Sci., 7, 359–362. Insects 2021, 12.
- Khalifa S, Elshafiey EH, Shetaia AA, El-Wahed A, Algethami AF, Musharraf SG, Al Ajmi MF, Zhao C, Masry S, Abdel-Daim MM, Halabi MF, Kai G, Al Naggar Y, Bishr M, Diab M, & El-Seedi HR. (2021) Overview of bee pollination and its economic value for crop production. Insects, 12(8), 688. https://doi.org/10.3390/insects12080688.
- Klatt BK, Holzschuh A, Westphal C, Clough Y, Smit I, Pawelzik E, Tscharntke T (2014) Bee pollination improves crop quality, shelf life and commercial value. Proc. Biol. Sci., 281, 20132440–20132447.
- Kleijn D, Winfree R, Bartomeus I, Carvalheiro LG, Henry M, Isaacs R, Klein AM, Kremen C, M'Gonigle LK, Rader R, et al. (2015) Delivery of crop pollination services is an insufficient argument for wild pollinator conservation. Nat. Commun., 6, 1–9.
- Klein AM, Steffan-Dewenter I, Tscharntke T (2003) Fruit set of highland coffee increases with the diversity of pollinating bees. Proc. R. Soc. B Biol. Sci., 270, 955–961.
- Klein AM, Vaissière BE, Cane JH, Steffan-Dewenter I, Cunningham SA, Kremen C, Tscharntke T (2007) Importance of pollinators in changing landscapes for world crops. Proc. R. Soc. B Biol. Sci., 274, 303–313.
- Krishnan S, Kushalappa CG, Shaanker RU, Ghazoul J (2012) Status of pollinators and their efficiency in coffee fruit set in a fragmented landscape mosaic in South India. Basic Appl. Ecol., 13, 277–285.
- Larson BMH, Kevan PG, Inouye DW (2001) Flies and flowers: Taxonomic diversity of anthophiles and pollinators. Can. Entomol., 133, 439–465.
- Lautenbach S, Seppelt R, Liebscher J, Dormann CF (2012) Spatial and temporal trends of global pollination benefit. PLoS One, 7, e35954.
- Leys R, Cooper SJB, Schwarz MP (2000) Molecular phylogeny of the large carpenter bees, genus *Xylocopa* (Hymenoptera: Apidae), based on mitochondrial DNA sequences. Mol. Phylogenet. Evol., 17, 407–418.
- Liporoni R, Cordeiro GD, Prado PI, Schlindwein C, Warrant EJ, Alves-dos-Santos I (2020) Light intensity regulates flower visitation in *Neotropical nocturnal* bees. Sci. Rep., 10, 1–11.

- Lobo JA, Quesada M, Stoner KE (2005) Effects of pollination by bats on the mating system of *Ceiba pentandra* (Bombacaceae) populations in two tropical life zones in Costa Rica. Am. J. Bot., 92, 370–376.
- Losey JE, Vaughan M (2006) The economic value of ecological services provided by insects. Bioscience, 56, 311–323.
- Luo D, Silva DP, De Marco Júnior P, Pimenta M, Caldas MM (2020) Model approaches to estimate spatial distribution of bee species richness and soybean production in the Brazilian Cerrado during 2000 to 2015. Sci. Total Environ., 737, 139674.
- Ma W, Li X, Shen J, Du Y, Xu K, Jiang Y (2019) Transcriptomic analysis reveals *Apis mellifera* adaptations to high temperature and high humidity. Ecotoxicol. Environ. Saf., 184, 109599.
- Macgregor CJ, Pocock MJO, Fox R, Evans DM (2015) Pollination by nocturnal Lepidoptera, and the effects of light pollution: A review. Ecol. Entomol., 40, 187–198.
- MacInnis G, Forrest JRK (2020) Field design can affect cross-pollination and crop yield in strawberry (*Fragaria* × *ananassa* D.). Agric. Ecosyst. Environ., 289, 106738–106745.
- Makino TT, Ohashi K, Sakai S (2007) How do floral display size and the density of surrounding flowers influence the likelihood of bumblebee revisitation to a plant? Funct. Ecol., 21, 87–95.
- Malerbo-Souza DT, Nogueira-Couto RH, Couto LA (2004) Honeybee attractants and pollination in sweet orange, *Citrus sinensis* (L.) Osbeck, var. Pera-Rio. J. Venom. Anim. Toxins Incl. Trop. Dis., 10, 144–153.
- Mallinger RE, Franco JG, Prischmann-Voldseth DA, Prasifka JR (2019) Annual cover crops for managed and wild bees: Optimal plant mixtures depend on pollinator enhancement goals. Agric. Ecosyst. Environ., 273, 107–116.
- Mallinger RE, Gratton C (2015) Species richness of wild bees, but not the use of managed honeybees, increases fruit set of a pollinator-dependent crop. J. Appl. Ecol., 52, 323–330.
- Mallinger RE, Prasifka JR (2017) Bee visitation rates to cultivated sunflowers increase with the amount and accessibility of nectar sugars. J. Appl. Entomol., 141, 561–573.
- Mandal E, Amin MR, Rahman H, Akanda AM (2018) Abundance and foraging behavior of native insect pollinators and their effect on mustard (*Brassica juncea* L.). Bangladesh J. Zool, 46, 117–123.
- Martins DJ (2014) Butterfly pollination of the dryland wildflower gloriosa minor. J. East African Nat. Hist., 103, 25–30.
- Mayer C, Soka G, Picker M (2006) The importance of monkey beetle (Scarabaeidae: Hopliini) pollination for Aizoaceae and Asteraceae in grazed and ungrazed areas at Paulshoek, Succulent Karoo, South Africa. J. Insect Conserv., 10, 323–333.
- Meena NK, Lal G, Meena RS, Meena BM, Meena RD (2018) Pollinator's diversity and abundance on cumin (*Cuminum cyminum* L.) and their impact on yield enhancement at semi-arid regions. J. Entomol. Zool. Stud., 6, 1017–1021.

- Meléndez-Ramírez V, Parra-Tabla V, Kevan PG, Ramírez-Morillo I, Harries H, Fernández-Barrera M, Zizumbo-Villareal D (2004) Mixed mating strategies and pollination by insects and wind in coconut palm (*Cocos nucifera* L. (Arecaceae)): Importance in production and selection. Agric. For. Entomol., 6, 155–163.
- Momose K, Yumoto T, Nagamitsu T, Kato M, Nagamasu H, Sakai S, Harrison RD, Itioka T, Hamid AA, Inoue T (1998) Pollination biology in a lowland dipterocarp forest in Sarawak, Malaysia. I. Characteristics of the plant-pollinator community in a lowland dipterocarp forest. Am. J. Bot., 85, 1477–1501.
- Montoya D, Gaba S, de Mazancourt C, Bretagnolle V, Loreau M (2020) Reconciling biodiversity conservation, food production and farmers' demand in agricultural landscapes. Ecol. Modell., 416, 108889–108909.
- Morse RA and Calderone NW (2000) The value of honeybees as pollinators of U.S. crops in 2000. Bee Cult., 128, 1–15.
- Munawar MS, Sarwar G, Raja S, Waghchoure ES, Iftikhar F, Mahmood R (2009) Pollination by honeybee (*Apis mellifera*) increases seed setting and yield in black seed (*Nigella sativa*). Int. J. Agric. Biol., 11, 611–615.
- Muto NA, Leite RO de S, Pereira DS, Rogez HLG, Venturieri GC (2020) Impact of the introduction of stingless bee colonies (*Scaptotrigona* aff. *postica*) on the productivity of acai (*Euterpe oleracea*). Rev. Verde Agroecol. Desenvolv. Sustentável, 15, 265–273.
- Naumann K, Winston ML, Slessor KN, Smirle MJ (1994) Synthetic honeybee (Hymenoptera: Apidae) queen mandibular gland pheromone applications affect pear and sweet cherry pollination. J. Econ. Entomol., 87, 1595–1599.
- Nayak RK, Rana K, Sharma HK, Rana VS, Thakur M (2019) Influence of bumblebee pollination on quantitative and qualitative parameters of kiwifruit. Indian J. Hortic., 76, 294–299.
- Neto S, Lima FG, Gonçalves BB, Lima Bergamini L, Araújo B, Bergamini R, Antônio M, Elias S, Franceschinelli EV (2013) Native bees pollinate tomato flowers and increase fruit production. J. Pollinat. Ecol., 11, 41–45.
- Nicodemo D, Couto RHN, Malheiros EB, de Jong D (2009) Honeybee as an effective pollinating agent of pumpkin. Sci. Agric., 66, 476–480.
- Normandeau Bonneau M, Samson-Robert O, Fournier V, Chouinard G (2020) Commercial bumblebee (*Bombus impatiens*) hives under exclusion netting systems for apple pollination in orchards. Renew. Agric. Food Syst., 36, 234–244.
- O'Neal ST, Swale DR, Anderson TD (2017) ATP-sensitive inwardly rectifying potassium channel regulation of viral infections in honeybees. Sci. Rep., 7, 8668–8676.
- Orges RCB, Rito RMB, Onseca VLIM, Iannini TCG (2020) The value of crop production and pollination services in the Eastern Amazon. Neotrop. Entomol., 49, 545–556.
- Papiorek S, Junker RR, Alves-dos-Santos I, Melo GAR, Amaral-Neto LP, Sazima M, Wolowski M, Freitas L, Lunau K (2016) Bees, birds and yellow flowers: Pollinator-dependent convergent evolution of UV patterns. Plant Biol., 18, 46–55.

- Pashte VV, Kulkarni SR (2015) Role of pollinators in qualitative fruit crop production: A review. Trends Biosci., 8, 3743–3749.
- Patil PN, Pastagia JJ (2016) Effect of bee pollination on yield of coriander. *Coriandrum sativum* Linnaeus. Int. J. Plant Prot., 9, 79–83.
- Peat J, Darvill B, Ellis J, Goulson D (2005) Effects of climate on intra- and interspecific size variation in bumblebees. Funct. Ecol., 19, 145–151.
- Perera RASN, Karunaratne WAIP (2019) Floral visits of the wild bee, *Lithurgus atratus*, impact yield and seed germinability of okra, *Abelmoschus esculentus*, in Sri Lanka. Pollinat. Ecol., 25, 1–6.
- Perrot T, Gaba S, Roncoroni M, Gautier JL, Bretagnolle V (2018) Bees increase oilseed rape yield under real field conditions. Agric. Ecosyst. Environ., 266, 39–48.
- Pinto CE, Oliveira R, Schlindwein C (2008) Do consecutive flower visits within a crown diminish fruit set in mass-flowering *Hancornia speciosa* (Apocynaceae)? Plant Biol., 10, 408–412.
- Pires VC, Silveira FA, Sujii ER, Torezani KRS, Rodrigues WA, Albuquerque FA, Rodrigues SMM, Salomao AN, Soares Pires CS (2014) Importance of bee pollination for cotton production in conventional and organic farms in Brazil. J. Pollinat. Ecol., 13, 151–160.
- Popak AE, Markwith SH, Strange J (2019) Economic valuation of bee pollination services for passion fruit (Malpighiales: Passifloraceae) cultivation on smallholding farms in São Paulo, Brazil, using the avoided cost method. J. Econ. Entomol., 112, 2049–2054.
- Potts SG, Imperatriz-Fonseca V, Ngo HT, Aizen MA, Biesmeijer JC, Breeze TD, Dicks LV, Garibaldi LA, Hill R, Settele J, et al. (2016) Safeguarding pollinators and their values to human well-being. Nature, 540, 220–229.
- Power AG (2010) Ecosystem services and agriculture: Tradeoffs and synergies. Philos. Trans. R. Soc. B Biol. Sci., 365, 2959–2971.
- Prado SG, Collazo JA, Marand MH, Irwin RE (2021) The influence of floral resources and microclimate on pollinator visitation in an agroecosystem. Agric. Ecosyst. Environ., 307, 107196–107204.
- Pritchard DJ, Vallejo-Marín M (2020) Floral vibrations by buzz-pollinating bees achieve higher frequency, velocity and acceleration than flight and defence vibrations. J. Exp. Biol., 223, jeb220541.
- Pritchard DJ, Vallejo-marín M (2020) Quick guide buzz pollination ll. Curr. Biol., 30, R858–R860.
- Quezada-Euán JJG (2018) Stingless Bees of Mexico. In Stingless Bees of Mexico: The Biology, Management and Conservation of an Ancient Heritage; Springer: New York, NY, USA, pp. 1–37. ISBN 9783319777849.
- Rachersberger M, Cordeiro GD, Schäffler I, Dötterl S (2019) Honeybee pollinators use visual and floral scent cues to find apple (*Malus domestica*) flowers. J. Agric. Food Chem., 67, 13221–13227.

- Radford BJ, Nielsen RGH, Rhodes JW (1979) Agents of pollination in sunflower crops on the central Darling Downs, Queensland. Aust. J. Exp. Agric., 19, 565–569.
- Rajagopal D, Eswarappa G, Raju AJS (2005) Pollination Potentiality of Honeybees in Increasing Productivity of Guava in Karnataka. In Changing Trends in Pollen Spore Research; Today and Tomorrow's Printers & Publishers: New Delhi, India, pp. 131–141. ISBN 8173715173.
- Raju AJS, Rao SP (2006) Nesting habits, floral resources and foraging ecology of large carpenter bees (*Xylocopa latipes* and *Xylocopa pubescens*) in India. Curr. Sci., Vol. 90, No. 9 (10 May 2006), pp. 1210-1217.
- Ramírez VM, Ayala R, González HD (2018) Crop pollination by stingless bees. In Pot-Pollen in Stingless Bee Melittology; Springer: Cham, Switzerland, pp. 139–153
- Rao S, Stephen WP (2009) Bumblebee pollinators in red clover seed production. Crop Sci., 49, 2207–2214.
- Reilly JR, Artz DR, Biddinger D, Bobiwash K, Boyle NK, Brittain C, Brokaw J, Campbell JW, Daniels J, Elle E, et al. (2020) Crop production in the USA is frequently limited by a lack of pollinators. Proc. R. Soc. B Biol. Sci., 287, 20200922–20200930.
- Resende JJ, de M Santos GM, Bichara Filho CC, Gimenes M (2001) Atividade diária de busca de recursos pela vespa social *Polybia occidentalis* (Olivier, 1791) (Hymenoptera, Vespidae). Rev. Bras. Zoociências, 3, 105–115.
- Richards MH (2020) Built environments influence carpenter bee sociality and vice versa. Insectes Soc., 67, 201–202.
- Ricketts TH, Regetz J, Steffan-Dewenter I, Cunningham SA, Kremen C, Bogdanski A, Gemmill-Herren B, Greenleaf SS, Klein AM, Mayfield MM, et al. (2008) Landscape effects on crop pollination services: Are there general patterns? Ecol. Lett., 11, 499–515.
- Robertson AW, Mountjoy C, Faulkner BE, Roberts MV, Macnair MR (1999) Bumblebee selection of *Mimulus guttatus* flowers: The effects of pollen quality and reward depletion. Ecology, 80, 2594–2606.
- Roldán Serrano A, Guerra-Sanz JM (2006) Quality fruit improvement in sweet pepper culture by bumblebee pollination. Sci. Hortic. (Amsterdam), 110, 160–166.
- Ropars L, Dajoz I, Fontaine C, Muratet A, Geslin B (2019) Wild pollinator activity negatively related to honeybee colony densities in urban context. PLoS One, 14, e0222316.
- Rortais A, Arnold G, Dorne JL, More SJ, Sperandio G, Streissl F, Szentes C, Verdonck F (2017) Risk assessment of pesticides and other stressors in bees: Principles, data gaps and perspectives from the European Food Safety Authority. Sci. Total Environ., 587–588, 524–537.
- Roselino AC, Santos SB, Hrncir M, Bego LR (2009) Differences between the quality of strawberries (*Fragaria × ananassa*) pollinated by the stingless bees *Scaptotrigona* aff. *depilis* and *Nannotrigona testaceicornis*. Genet. Mol. Res., 8, 539–545.
- Rowe L, Gibson D, Bahlai CA, Gibbs J, Landis DA, Isaacs R (2020) Flower traits associated with the visitation patterns of bees. Oecologia, 193, 511–522.

- Russell AL, Leonard AS, Gillette HD, Papaj DR (2016) Concealed floral rewards and the role of experience in floral sonication by bees. Anim. Behav., 120, 83–91.
- Saboor N, Muhammad A, Muhammad AR, Younisand A (2018) Role of pollinators in recommended and densely grown black cumin (*Nigella sativa* L.) yield at Dera Ismail Khan. J. Entomol. Zool. Stud., 6, 983–986.
- Sáez A, Aizen MA, Medici S, Viel M, Villalobos E, Negri P (2020) Bees increase crop yield in an alleged pollinator-independent almond variety. Sci. Rep., 10, 3177–3183.
- Santos RdS, Milfont MdO, Silva MM, Carneiro LT, Castro CC (2020) Butterflies provide pollination services to macadamia in northeastern Brazil. Sci. Hortic., 259, 108818–108825.
- Sayers TDJ, Steinbauer MJ, Miller RE (2019) Visitor or vector? The extent of rove beetle (Coleoptera: Staphylinidae) pollination and floral interactions. Arthropod. Plant. Interact., 13, 685–701.
- Schittenhelm S, Gladis T, Rao VR (1997) Efficiency of various insects in germplasm regeneration of carrot, onion and turnip rape accessions. Plant Breed., 116, 369–375.
- Schremmer F (1962) Wespen und Hornissen. Die einheimischen sozialen Faltenwespen. A. Ziemsen Verlag, Wittenberg Lutherstadt: 104 pp
- Seo HJ, Song J, Yoon HJ, Lee KY (2019) Effects of nectar contents on the foraging activity of honeybee (*Apis mellifera*) on Asian pear (*Pyrus pyrifolia* Nakai). Sci. Hortic. (Amsterdam), 245, 185–192.
- Shin D, Choi WT, Lin H, Qu Z, Breedveld V, Meredith JC (2019) Humidity-tolerant rate-dependent capillary viscous adhesion of bee-collected pollen fluids. Nat. Commun., 10, 1–9.
- Shipp JL, Whitfield GH, Papadopoulos AP (1994) Effectiveness of the bumblebee, *Bombus impatiens* Cr. (Hymenoptera: Apidae), as a pollinator of greenhouse sweet pepper. Sci. Hortic. (Amsterdam), 57, 29–39.
- Shrestha M, Garcia JE, Burd M, Dyer AG (2020) Australian native flower colours: Does nectar reward drive bee pollinator flower preferences? PLoS One, 15, e0226469.
- Silveira MV, Abot AR, Nascimento JN, Rodrigues ET (2012) Is manual pollination of yellow passion fruit completely dispensable? Sci. Hortic. (Amsterdam), 146, 99–103.
- Sime KR, Baldwin IT (2003) Predominantly self-fertilizing native tobacco. BMC Ecol., 3, 6–14.
- Snow JW, Ceylan Koydemir H, Karinca DK, Liang K, Tseng D, Ozcan A. (2019) Rapid imaging, detection, and quantification of *Nosema ceranae* spores in honeybees using mobile phone-based fluorescence microscopy. Lab Chip, 19, 789–797.
- Solange ABS, Ana CR, Luci RB (2008) Pollination of cucumber *Cucumis sativus* L. (Cucurbitales: Cucurbitaceae), by the stingless bees *Scaptotrigona* aff. *depili* Moure and *Nannotrigona testaceicornis* Lepeletier (Hymenoptera: Meliponini) in greenhouses. Neotrop. Entomol., 37, 506–512.
- Solís-Montero L, Cáceres-García S, Alavez-Rosas D, García-Crisóstomo JF, Vega-Polanco M, Grajales-Conesa J, Cruz-López L (2018) Pollinator preferences for floral

volatiles emitted by dimorphic anthers of a buzz-pollinated herb. J. Chem. Ecol., 44, 1058–1067.

- Somanathan H, Saryan P, Balamurali GS (2019) Foraging strategies and physiological adaptations in large carpenter bees. J. Comp. Physiol. A Neuroethol. Sensory, Neural, Behav. Physiol., 205, 387–398.
- Stanley DA, Msweli SM, Johnson SD (2020) Native honeybees as flower visitors and pollinators in wild plant communities in a biodiversity hotspot. Ecosphere, 11, e02957.
- Stanley J, Sah K, Subbanna ARNS (2017) How efficient is the Asian honeybee, Apis cerana in pollinating mustard, *Brassica campestris* var. toria? Pollination behavior, pollinator efficiency, pollinator requirements and impact of pollination. J. Apic. Res., 56, 439–451. Insects 2021, 12, 688.
- Stein K, Coulibaly D, Stenchly K, Goetze D, Porembski S, Lindner A, Konaté S, Linsenmair EK (2017) Bee pollination increases yield quantity and quality of cash crops in Burkina Faso, West Africa. Sci. Rep., 7, 17691–17700.
- Streinzer M, Brockmann A, Nagaraja N, Spaethe J (2013) Sex and caste-specific variation in compound eye norphology of five honeybee species. PLoS One, 8, e57702.
- Suetsugu K (2019) Social wasps, crickets and cockroaches contribute to pollination of the holoparasitic plant *Mitrastemon yamamotoi* (Mitrastemonaceae) in southern Japan. Plant Biol., 21, 176–182.
- Sukumaran A, Khanduri VP, Sharma CM (2020) Pollinator-mediated self-pollination and reproductive assurance in an isolated tree of *Magnolia grandiflora* L. Ecol. Process., 9, 45–53.
- Tanda AS (2022a) Chapter 11 Pest management technology and bee pollinators' integration. In: Tanda A S (ed.) Advances in Integrated Pest Management Technology: Innovative and Applied Aspects. Springer Nature Switzerland (in press).
- Tanda AS (2022b) Chapter 1: Mutulistic Plant Associations related to Insect Resistance. In: Tanda AS. (ed.) Molecular Advances in Insect Resistance of Field Crops: Modern and Applied Approaches. Springer Nature Switzerland (in press).
- Tanda AS (2021a) Biofloralphenology, Foraging Behavior and entpollinatological effect of honeybees in pomegranate (*Punica granatum*) fruit quality and yield. Journal of Horticulture 08 2.
- Tanda AS (2021b) Bee pollinators matter in crop improvement need immediate conservation strategies. Journal of Insect Science (Accepted).
- Tanda AS (2021c) Comparative pollination efficacies of *Apis mellifera* L. and *Tetragonula carbonaria* (Smith) on peach (*Prunus persica* L.). Indian Journal Entomology (Accepted).
- Tanda AS (2021d) Why insect pollinators important in crop improvement?. Indian Journal Entomology (Accepted).
- Tanda AS (2021e) Insect pollinators matter in sustainable world food production. Indian Journal Entomology (Accepted).

- Tanda AS (2020a) Biogenetic engineering in developing insect resistant crops: constraints and applications. *5th Edition of Global Congress on Plant Biology and Biotechnology* (*GPB 2020*) Valencia, Spain.
- Tanda AS (2020b) An Assessment of honeybee (*Apis mellifera*) foraging activity and pollination efficacy in Australian raspberry *Rubus parvifolius* at Rosehill. Australian Journal of Entomology (Submitted).
- Tanda AS (2020c) Entpollinatology is for sustainable agroecosystems and global agroindustry. Indian Journal of Entomology (Submitted).
- Tanda AS (2020d) Entpollinatology: A strong relationship between plants and insects for crop improvement. *6th Edition of Global Conference on Plant Sciences and Molecular Biology* (*GPMB 2020*) Paris, France.
- Tanda AS (2019a) Entomofauna enhance the quality and quantity in Okra. Indian Journal Entomology 81(1): 16–17.
- Tanda AS (2019b) Entomophilous crops get better fruit quality and yield: An appraisal. Indian Journal Entomology 81(2): 227–234.
- Tanda AS (2019c) Floral biology, foraging behavior and efficiency of European honeybee (*Apis mellifera*) in bitter gourd *(Momordica charantia* L.) pollination at Sydney Australia. Bee World (Submitted).
- Tanda AS (1984) Bee pollination increases yield of 2 interplanted varieties of Asiatic cotton (*Gossypium arboreum* L.). The Amer. Bee Jour. 124: 539-540.
- Tanda AS (1983) Assessing the role of honeybees in a field of Asiatic cotton (*Gossypium arboreum* L.). American Bee Journal 123: 593–594.
- Tanda AS and Goyal NP (1979) Insect pollination in Asiatic cotton (*Gossypium arboreum*). J Apic Res 18(1): 64–72.
- Tomé HVV, Ramos GS, Araújo MF, Santana WC, Santos GR, Guedes RNC, Maciel CD, Newland PL, Oliveira EE (2017) Agrochemical synergism imposes higher risk to neotropical bees than to honeybees. R. Soc. Open Sci., 4, 160866.
- Tosi S, Nieh JC, Sgolastra F, Cabbri R, Medrzycki P (2017) Neonicotinoid pesticides and nutritional stress synergistically reduce survival in honeybees. Proc. R. Soc. B Biol. Sci., 284, 20171711–20171719.
- Tschoeke PH, Oliveira EE, Dalcin MS, Silveira-Tschoeke MCAC, Sarmento RA, Santos GR (2019) Botanical and synthetic pesticides alter the flower visitation rates of pollinator bees in Neotropical melon fields. Environ. Pollut., 251, 591–599.
- Urbanowicz C, Muñiz PA, McArt SH (2020) Honeybees and wild pollinators differ in their preference for and use of introduced floral resources. Ecol. Evol., 10, 6741–6751.
- Valido A, Rodríguez-Rodríguez MC, Jordano P (2019) Honeybees disrupt the structure and functionality of plant-pollinator networks. Sci. Rep., 9, 4711–4721.
- Vallejo-Marín M (2019) Buzz pollination: Studying bee vibrations on flowers. New Phytol., 224, 1068–1074.
- Van der Kooi CJ, Ollerton J (2020) The origins of flowering plants and pollinators. Science (80-.), 368, 1306–1308.

- Van Dooremalen C, van Langevelde F (2021) Can colony size of honeybees (*Apis mellifera*) be used as predictor for colony losses due to *Varroa destructor* during winter? Agriculture1, 11, 529.
- van Engelsdorp D, Evans JD, Saegerman C, Mullin C, Haubruge E, Nguyen BK, Frazier M, Frazier J, Cox-Foster D, Chen Y, et al. (2009) Colony collapse disorder: A descriptive study. PLoS One, 4, e6481.
- Van Kleunen M, Nänni I, Donaldson JS, Manning JC (2007) The role of beetle marks and flower color on visitation by monkey beetles (Hopliini) in the greater c\ape Floral Region, South Africa. Ann. Bot., 100, 1483–1489.
- Vanegas M (2017) The silent beehive: How the decline of honeybee populations shifted the environmental protection agency's pesticide policy towards pollinators. Ecol. Law Q., 44, 311–342.
- Varassin IG, Trigo JR, Sazima M (2001) The role of nectar production, flower pigments and odor in the pollination of four species of *Passiflora* (Passifloraceae) in south-eastern Brazil. Bot. J. Linn. Soc., 136, 139–152.
- Velthuis HHW, Van Doorn A (2006) A century of advances in bumblebee domestication and the economic and environmental aspects of its commercialization for pollination. Apidologie, 37, 421–451.
- Viana BF, da Encarnacao Coutinho JG, Garibaldi LA, Braganca Castagnino GL, Gramacho KP, Oliveira Silva F (2014) Stingless bees further improve apple pollination and production. J. Pollinat. Ecol., 14, 261–269.
- Walters SA (2005) Honeybee pollination requirements for triploid watermelon. HortScience, 40, 1268–1270.
- Wcislo WT, Tierney SM (2009) Behavioral environments and niche construction: The evolution of dim-light foraging in bees. Biol. Rev., 84, 19–37.
- Whitney HM, Chittka L, Bruce TJA, Glover BJ (2009) Conical epidermal cells allow bees to grip flowers and increase foraging efficiency. Curr. Biol., 19, 948–953.
- Whitney HM, Poetes R, Steiner U, Chittka L, Glover BJ (2011) Determining the contribution of epidermal cell shape to petal wettability using isogenic antirrhinum lines. PLoS One, 6, e17576.
- Woodcock BA, Isaac NJB, Bullock JM, Roy DB, Garthwaite DG, Crowe A, Pywell RF (2016) Impacts of neonicotinoid use on long-term population changes in wild bees in England. Nat. Commun., 7, 12459.
- Wu P, Tscharntke T, Westphal C, Wang M, Olhnuud A, Xu H, Yu Z, van der Werf W, Liu Y (2021) Bee abundance and soil nitrogen availability interactively modulate apple quality and quantity in intensive agricultural landscapes of China. Agric. Ecosyst. Environ., 305, 107168–107176.
- Yankit P, Rana K, Kumar Sharma H, Thakur M, Thakur RK (2018) Effect of bumblebee pollination on quality and yield of Tomato (*Solanum lycopersicum* Mill.) grown under protected conditions. Int. J. Curr. Microbiol. Appl. Sci., 7, 257–263.

- Zeng X, Fischer GA (2020) Wind pollination over 70 years reduces the negative genetic effects of severe forest fragmentation in the tropical oak *Quercus bambusifolia*. Heredity (Edinb). 124, 156–169.
- Peña JF, Carabalí A (2018) Effect of honeybee (*Apis mellifera* L.) density on pollination and fruit set of avocado (*Persea americana* Mill.) cv. Hass. J. Apic. Sci., 62, 5–14.
- Yang D, Xu X, Zhao H, Yang S, Wang X, Zhao D, Diao Q, Hou C (2018) Diverse factors affecting efficiency of RNAi in honeybee viruses. Front. Genet., 9, 384–392.
- Cruden RW, Hermann-parker SM (1979) Butterfly pollination of *Caesalpinia pulcherrima*, with observations on a psychophilous syndrome. Br. Ecol. Soc., 67, 155–168.

9

Wild Bees are Important in Sustainable Agriculture

Amarjit S. Tanda*

ABSTRACT

Major field crops chiefly dependent upon managed honeybees, however wild bees are also of great significance in sustainable agro environment and food security worldwide. Bee pollinated crops being a great part of the bio-diverse system, 4,000 native bees, and offer over US$1.5 billion each year in North America. In US, the value of wild bees in food production was estimated to be over $1.5 billion annually. However, the value of wild bee pollination services in insect cross-pollinated crops may be much greater. These important wild players are now in declining phase quickly or possibly extinct due to disturbed habitats by the humans. More studies are needed in different topics of wild bee fauna, for example, basic research in their population biology, bee dynamics, abundance, bee protection measures, suitable habitat, and nesting sites especially their immediate conservation strategies. After realizing the importance of wild bees in pollination, in the present chapter, we highlight the various methods and actions to conserve the wild bee populations so that they can serve the growers as co-players to managed honeybees in sustainable agricultural food production globally.

Keywords *Wild bees • Pollinators • Bio-diversity • Habitat • Crop • Improvement • Yield • Buzz pollination • Conservation • Ecosystem • Management • Sustainability*

9.1 INTRODUCTION

Wild bees are the chief insect pollinators worldwide, being responsible for the flower pollination, seed set, and fruit production of different crops. Being neglected part of the bio-diverse system for a long time, they have been badly declining by the avoidable human disturbances and farming practices. Beekeepers and scientists are now worried about the immediate and effective measures to save the native bees for use in crop cross-pollination, enhanced agriculture production, and sustainability. Due to many new research findings presented in the newspapers radio and TV, people's interest in wild bee conservation has increased greatly (Schatz, 2020; Tanda, 2019, 2021a-i). Some insects remain unnoticed from people acceptance which is less

* School of Horticulture, University of Western Sydney, Richmond, Australia and CSIRO North Ryde NSW, Australia.

E-mail: drtanda193@gmail.com

charming (Hart and Sumner, 2020), encouraging bee pollinator protection programs is an interesting subject to educate the majority of the people groups. Though this topic is familiar to many, a lack of knowledge and understanding of the subject is flagrant and has a space between information and its comprehension (Wilson et al., 2017). This pause is due to the absence of knowledge about the importance of hover flies and butterflies in the bio-diverse ecosystem. About 80% and 99% of the reports in Great Britain and the United States described that bees are the key crop pollinators (Wilson et al., 2017). Mostly, people understood the honeybee *Apis mellifera*, as the main crop pollinators. Further, the crop pollination mechanism is now known as entpollinatogy (Tanda, 2021c), the significance of flower assets and nesting site environment, major groups of pollinators with specific needs has become more significant. The insufficient reports may result in irrelevant activities for the conservation of endangered bee pollinators (Wilson et al., 2017; Schatz, 2020; Didham, 2020; Penn, 2021; Tanda, 2019, 2021a-i). Harvey et al. (2017) designed a program for pollinator's conservation and their retrieval as a base for bee scientist studies. In the last 10 years, the population of several insect species has disappeared by 9% and including the native bees (van Klink et al., 2020). Observations on the ill-effects of ecosystem on the population dynamics of native bees are abundant (Potts et al., 2010; Goulson et al., 2015; Tanda, 2019, 2021a-i) and several bees are endangered locally and facing disappearance, but still survive in their geographical habitat (Primack et al., 2012).

9.2 BEES

Bees are insects with wings closely related to wasps and ants, known for their role in pollination and, in the case of the best-known bee species, the western honeybee, for producing honey. Bees are a monophyletic lineage within the superfamily Apoidea. They are presently considered a clade, called Anthophila. There are over 16,000 known species of bees in seven recognized biological families. Some species – including honeybees, bumblebees, and stingless bees – live socially in colonies while most species (> 90%) – including mason bees, carpenter bees, leafcutter bees, and sweat bees – are solitary.

9.3 COMPARATIVE SIGNIFICANCE OF HONEYBEES AND WILD BEES

Honeybees are domesticated and kept in beehives – those white square boxes that everyone is familiar with. Sometimes these domesticated bees escape and form wild colonies which are then called feral honeybees (Fig. 9.1).

Fig. 9.1 Comparative images of a wild bee and honeybee pollinators

This unique event can be found by looking at European and IUCN Red Lists studies (IUCN 2016). A number of native bees are under threat in Belgium, or even disappeared locally but their population is not absent at the European scale (Drossart et al., 2018) showing the protection activities in various regions. There is a space between scientific knowledge and awareness (Wilson et al., 2017). This gap is generally described by ignorance of knowledge about the prominent abundance of bees, hover flies, and butterflies in the living environment. In Great Britain and the United States, studies identified that bees are important, but only 3% and 14% can assess the pollinator's distribution in different regions (Wilson et al., 2017). Many people are capable to recognize honeybees and bumblebees crop pollinators, however, the other native bees are poorly identified as the main species. Less than 50% of the participants were even unable to name at least one bee worker (Wilson et al., 2017). Through large presentations, mainly focused on the honeybee *Apis mellifera*, the general public was understanding the value of crop pollination, but not the distribution and its population size. Flowers' importance, sites of bee nesting, favorable environment, and the existence of main bee pollinators with their specific requirements were also not known to the people. This incomprehension of scientific reports could therefore support unrelated activities, wrong, or even ineffective measures for the conservation of endangered bees (Wilson et al., 2017; Drossart and Gerard, 2020). This shows that with scientific understanding, public awareness has not advanced. For workers, bee identification and ranking endangered species are challenging and a principal project to ongoing studies and an informative program for wild bee conservation and their improvement system to be framed. About 9% of terrestrial insect populations per decade have perished and the wild bee abundance is also not an anomaly (van Klink et al., 2020). On wild bee populations, studies on the harmful impact of pressures are sufficient (Potts et al., 2010; Goulson et al., 2015; Tanda, 2019, 2021a-i). Knowing the harmful impacts of climate, decrease in the bee population, national and regional actions associated with the wild bee fauna conservation, and their protection methods should be implemented urgently. All the possible conservation measures will be of great importance in the preservation of native bees and their bio-diverse habitat. As many native bees are threatened and vanishing, they may still survive quite a long time in their habitats (Schatz, 2020; Penn, 2021; Tanda, 2021h). Such changes can be reported in IUCN Red Lists and European scale studies (IUCN 2016). Many bee species are disappeared in Belgium, as their abundance is not good at the European level (Drossart et al., 2018). Undoubtedly, the bee population degree, as well as the declining factors, are greatly investigated, still, reports on the preservation methods of wild bees are not well described. A general and up-to-date evaluation of the conservation actions, technology as well as their effectiveness and efficiency, is still wanting. Keeping this in mind, firstly, we reviewed the significance of preserving native bees, bee population's assessments at risk along the factors associated with declining their abundance. Secondly, we emphasized the conservation measures, associated factors, and the effectiveness of these methods. We have highlighted the conservation actions that improve the bee-friendly environment such as semi-natural landscapes to urban and crop habitats naming the required floral resources and nesting sites, an alien species with the habitat control process. To fill many gaps in native bee protection studies, some new projects and reports have been presented in this chapter.

9.4 HOW TO CONSERVE WILD BEES

In many bio-environments, bees being the main flower pollinators are involved in the propagation of 80% of angiosperms (IPBES, 2016). Due to this breeding process, cross-pollination is considered one of the most important activities in the functioning of the bio-diverse systems and enhancing crop production worldwide (IPBES, 2016; Potts et al., 2016). About 85% of field crops gain from insect pollination service that directly influences the quality and quantity of yield (IPBES, 2016) which is worth 100 to 500 billion euros annually globally (Lautenbach et al., 2012). The decline in wild bees is interesting to move in humans to non-pollinator dependent crops for food and facing scarcity in essential nutrients, causing economic and health issues (IPBES, 2016; Potts et al., 2016; Bauer and Wings, 2016). They can be the chief pollinators in some field crop productions (IPBES, 2016), while other crops are pollinated by honeybees. Only a little bee population visits the crop plants, thus, pollination management alone is not sufficient to confirm true alone for the protection of native bees. Wild bees are also of specific importance due to their ability to forage flowers in different types of dynamic climatic conditions (Brittain et al., 2016). Bumblebees can visit flowers even in cold weather and forage at high sonication. In wild flora, the high strength and diversity of local plant pollinators increase a complementary and synergistic activity with managed honeybees (Fründ et al., 2013; Garibaldi et al., 2014; Isaacs et al., 2017). Grab et al. (2019) have mentioned this in the phylogenetic diverse studies and bee abundance related to crop pollination. The engaged bee population also acts as the main operator to boost the crop yield (Martins et al., 2015). In the urban area (Säumel et al., 2016), forests (Cummings et al., 2016), and natural bio-ecosystem (Cummings et al., 2016; Massaro et al., 2013), the contribution of native bees have also been reported significant and their abundance is associated to the density wildflowers (Ollerton, 2017). In bees and flowers, the ecological interaction disappearance is disregarded frequently instead of the bee extinction reports (Valiente-Banuet et al., 2015). This loss of bio-diverse factor still occurs at the same time or may result in the disappearance of the bee species (Valiente-Banuet et al., 2015; Jacquemin et al., 2020). Many bees are related to a specific habitat and food resources, which make them resist that environment and agitating food. Specialist bees forage on a few flowers whereas generalist pollinators have great foraging alternatives (Jacquemin et al., 2020). Thus, the all-rounders are more resistant to the different climatic environments associated with human actions as they are capable to manage other food resources (Roger et al., 2017). Damage of these biotic interplay affects rapid species disappearances and negatively influences the working of the eco-bio-structure (Diaz et al., 2017). To prevent the collapse in environmental services to humans, the importance of the interplay of biotic agents be contemplated (Jacquemin et al., 2020) to assess the fitness of the environment and to determine the possible environmental matters (Aizen et al., 2012; Dirzo et al., 2014). The mission of the Andalucía BeeFun project is the increasing knowledge and understanding of the impact of the environment, bee crop-pollination process, and social groups (Underwood et al., 2020; Drossart and Gerard, 2020).

9.5 DECLINE IN BEE POPULATION

Assessment of bee population technology and studies requires a basic action for worldwide fluctuations in the decline and benefits of the bees. In this global change, a few species can survive or vanish as observed in the United Kingdom (Powney et al., 2020). IUCN Red List technology advises assessing the possible extinction of a bee species at regional, national, and global levels (IUCN, 2016) and there may be variations in the assessment techniques in various habitats. To establish a fundamental base for the conservation measures and assessment of the species to execute monitoring, protection plan, management, and policy formation, this is the most powerful technology. For wild bees, Red List, at the European level, and many more countries are also preparing their Red Lists at national and regional levels (Drossart et al., 2018; Reemer et al., 2018). In North America, IUCN has also started the evaluation for the *Bombus* spp. efficiency (Hatfield et al., 2020). This finding for wild bees helps in indicating the bee abundance, species, and areas under threat. Bee protection activities also show to record the factors influencing bee abundance (Harvey et al., 2020; Primack et al., 2012; IPBES, 2012). Scientists are focusing on the degree of decline and factors responsible for population abolition (Forister et al., 2019). Fire, drought, hydrological and geophysical events have a non-negligible effect on the native pollinators (Nicholson et al., 2019), however, global warming, crop growth, environmental and bee diseases greatly influence the populations of bees (Potts et al., 2010; IPBES, 2016). The influence of a single factor (Potts et al., 2010) and feasible interactions (Goulson et al., 2015; Meeus et al., 2018) have been observed in many countries, still, a scientific knowledge gap remains in the assessment of the spatial and temporal influences from the various threats along with the historical information independently (Bartomeus et al., 2019). The knowledge of the genetic variation in species, population dynamics, and speciation is also significant and more used in conservative biological features (Epps and Keyghobadi, 2015). New reports in genetic conservation have emphasized that scientists utilize molecular biotechnology to know better about bee bio-diverse systems (Epps and Keyghobadi, 2015; Lopez-Uribe et al., 2017), with new gene biotechnology (Woodard et al., 2016). This is evident in cryptic bumblebee species having a high degree of morphological conjunction. Less genetic diversity which can lead to inbreeding depression and lessened health is also alarming (Packer et al., 2016). Based on the historical reports and distance assessments (IBD), the assessment of the relation among bee population, their effective part and biotic and abiotic operators, workers can classify the disappearing bee species (Cerna et al., 2017; Lecocq et al., 2017, 2018). Using different techniques and the assessment of the threatening components, through population assessments workers are now able to design conservation strategies by making policies, plans, and measures with ongoing research (van Klink et al., 2020; Forister et al., 2019; Drossart and Gerard, 2020).

9.6 VARIOUS STRATEGIES OF BEE CONSERVATION

The important measure is to protect the suitable habitats after knowing the propensity and decline in the bee population. There is great wild bee heterogeneity and population diversity depending upon the scale and the landscape (Belsky and Joshi, 2019). Native bee structure is made by landscape strength globally, from mountainous tropical habitats of Colombia

(Cely-Santos and Philpott, 2019) to dry grasslands in Missouri, USA (Grover et al., 2017). Grasslands with flowers offer a great bee abundance and habitats than crops provided to livestock and are full of blooms. To protect such favorable environments can be achieved by the evolution of safe areas using legal actions to avoid any changes, and by maintaining such bio-eco-habitats. Flexible management of bees and the environment, in such protected areas, have to be established to explain the best programs and adaptive actions for the success and failures of the bee management. Nevertheless, many anthropogenic urban habitats can never provide the same degree of bee shelter as the semi-natural ecosystem but still need conservation activities. To attain similar targets, many joint programs aiming at the habitat, intensification of the bio-diverse ecosystem, least entry of alien bees, or the communication about diversity damage have been carried out by the World Bee Project. To combine cloud computing with wild bee research worldwide the World Bee plan aims to serve all new intuitions and information to design modern programs for bee decline worldwide, various environmental conditions, and enhanced food sustainability and subsistence to intercontinental extent merging other strategies undertaken by the European Union (EU, 2011) implemented nationally (Belgian NFP-CBD, 2020) and in the USA at other different scales to manage the bee disappearance (Heinz Center, 2013); in France, (Gadoum and Roux-Fouillet, 2016); in Ireland, (PPSG, 2015). Bee and Butterfly Habitat Fund, Seeds for Bees, and The Dutch Bee Strategy, the English National Pollinator Strategy are helping stakeholders to exchange their expertise and information collected from various cultures (Saunders et al., 2018; Turo and Gardiner, 2019). The public suggestions and the participation of youth are significant for running such activities (Turo and Gardiner, 2019). The major efforts should be targeted on the safety, preservation, and restoration of native bee habitats, concentration on the urban and agricultural fields, and execution of man-made tools to provide nests, potential invasions by alien species, and the training of people groups with the efficient transmission.

9.7 SEMI-NATURAL ENVIRONMENT PROTECTION

The conservation of wild bees can be initiated by the protection of semi-natural environments to rebuild huge natural habitats for the establishment of a bio-diverse ecosystem. In the disappearance and protection of terrestrial insects, the safety of suitable bee habitat is most important (van Klink et al., 2020). These habitats are bio-diverse activities to explain the important action to consider for the distinct diverse habitat (Sobral-Souza et al., 2018). Similar new and standardized models can be tested by the Ecolgical Niche Models (Krechemer and Marchioro, 2020) as assessed in South America for bumblebees. Protected eco-environments can increase native bee populations in landscapes and geographic regions (Tonietto and Larkin, 2018). Improvement of the environment by repairing techniques implicit an investigation of the habitat of the target species. They evaluated the preferences of bumblebee likings in the crops and suggested natural landscapes and field boundaries for the bee population survival. Wild bees can react differently in changing environments. Carrié et al. (2017) described that hedgerows, grasslands, and forest edges with potential attributes, for example, in solitary and ground dwelling bees foraging a large number of flowering plants, but the social and above-ground nesting bees visit a few blooms in such ecosystem. Rainforest could improve above-ground nesting for bees as found in Brazil, conversely (Ferreira et al., 2015) showing the conservation activities importance in a protected bee-habitat. The restoration devices can be made double-edged and depending upon the demand and situation. As grazing and burning are followed in grasslands to aid floral blooming, however, they can also destroy wild bees hibernating in the plants (Tonietto and Larkin, 2018). These restoration measures are conducted in the framework of LIFE plans funded by the European Union. They focus to renovate bee habitats directed in Natura 2000 bee sites mentioned in "Birds" and "Fauna-Flora-Habitats". If bee protection is at its inception compared to birds and mammal's safety, aiming at butterflies and LIFE in Quarries will indirectly help to local bee fauna with the bee conservation and the environment renovation similar to quarries, peat bogs, biological meadows, and hedgerows (Folschweiller et al., 2019). As LIFE projects failed, similar programs such as Urban bees LIFE are in action (www.urbanbees.eu) and are targeting management to increase the bee population and bio-diverse environment of local bees in potential urban and peri-urban areas. Hall and Steiner (2019) described that US state projects did not consider the significance of bees in comparison to vertebrates leading to less understanding of their needs, and restoration actions. The conservation focusing on semi-natural environments are believed in schemes at the national/sub-national level.

9.8 MEASURES IN BEE CONSERVATION

In Urban and Agricultural Areas Conservation measures are crucial and in anthropogenic areas, more than 55% of bees live in the peri-urban regions. Under such important programs, hedgerows, parks, roadsides, and urban gardens can constitute crucial environments for bee multiplication in quality and quantity, and transitional areas as the beneficial environment (Hall et al., 2017; Nowakowski and Pywell, 2016; Gosselin et al., 2018; Crone et al., 2019). Such ecosystems can assist in building big specific diversity and a boosting factor for rare bees (Senapathi et al., 2017) and often pillar a great bee diverse ecosystem (Fortel et al., 2014). Bee-friendly plans using roof-top gardens, parks, and roadsides have increased the populations of native bees in Amsterdam (Givetash, 2018) as this environmental interconnection proved important ecological segmentation could be harmful to little bee species. Tiny bees in segmented habitats, can be incompetent to approach the favorable landscapes and face difficulties in anthropogenic environments (Warzecha et al., 2016; Gérard et al., 2020; La Vie Sauvage, 2020; François and Féon, 2017).

It has been revealed largely that the strength of floral wealth is a key framework for native bee diversity when renovating prairies as in Minnesota, USA (Lane et al., 2020; Ritchie and Roser, 2020). Agro-environment actions, as flower strips have been accepted in Europe to increase biodiversity in assiduously managed agro-landscapes (Grass et al., 2016; Cole et al., 2020) and beneficial for bumblebees, honeybees, and hover flies in Germany, Belgium, and in England (Wood et al., 2015). More flower supplies mostly helped in the improvement of the bumblebee population, size, density, and young one's number (Wood et al., 2015; Vaudo et al., 2018). The impact of AES was determined rarely (Batáry et al., 2015) and Geppert et al. (2020) observed the effect of organic practices and floral strips for bee population survival and development. Both actions were positively corresponding to pollinators' strength and population and growth of bumblebee hives, but the efficiency of these actions relied firmly on the landscape around them (Geppert et al., 2020). In England, (Wood et al., 2017) also assessed a Higher

Level Stewardship farms—HLS to experiment if grown flowers helped the native solitary bee populations. For instance, as honeybees and bumblebees were positively influenced by the affair of *Phacelia* sp., solitary bees mostly foraged sunflower and seed mixes of wild blooms (Mallinger et al., 2020; Nichols et al., 2019). Still, a move in flower resources happened among the most environments of native bees facing stress and could result in the exhaustion of flower assets and alterations in the crop-pollinator web (Gérard et al., 2020; DEFRA, 2020). The restructuring of flower mixtures could also be effective in bee environments with maximum flower density in areas under crop and uncultivated land (Quinet et al., 2016; Moquet et al., 2017). Angiosperms should offer the flow of floral gifts throughout the flowering season for sufficient food needs (Vaudo et al., 2015; Filipiak et al., 2018) based on the floral phenology and specific behavior of bee species. Gresty et al. (2018) demonstrated that plants like *Rosa canina*, *Malva sylvestris*, and *Ranunculus acris* allured specifically those solitary bees living in cavities.

This framework is pivotal for bee multiplication and abundance but is often neglected during the selection of bee-friendly flowering plant cultivars. This is accurate for the brood nutritional needs, which vary from adult requirements (Filipiak et al., 2018). The strength and nutritional power of crops like *Brassica napus* for *Osmia bicornis*) can positively help in the development rate of bee abundance (Bukovinszky et al., 2017; Filipiak, 2019). In fact, the variety of proteins and essential amino acids required is important for the growth and development of the bee population. This can help bee species health through nutritional needs using floral assets available in the habitats, particularly in flower resource-exhausted environments. Different agrotechnology could also be tested such as friendly planting, which can enhance the quantity and quality as in the strawberry *Fragaria* × *ananassa* and the borage *Borago officinalis* (Griffiths et al., 2020) but the effect on pollinator abundance was not studied. So it is difficult to describe the chemical toxicity taken from reports on honeybees and generalize about bumblebees and solitary bee taxa, furthermore, solitary bee's sensitivity will be highly different. POSHBEE, European strategies could be beneficial in comprehending how pesticides can influence this native bee fauna and synergistic effect with other decline elements. Still, it has been demonstrated to also influence bee pollinators and is not a particular reaction to safeguard the bee world (Dicks et al., 2016; Egan et al., 2020). Egan et al. (2020) demonstrated a newly designed strategy that is known as Integrated Pest and Pollinators Management (IPPM; Biddinger and Rajotte, 2015) the same we propose to be called, "Pollinator and integrated pest management technology (PIPMT)", as this is a technology or holistic strategy to amalgamate crop pollinators, bio-control agents, minimum use of pesticides and various other IPM procedures, aiming at the management of the pests below economic injury levels for enhancing the agriculture food production (Tanda, 2019, 2021a-i). Applied management actions are also recommended, such as the choice of varieties with great bee pollinator allurement and maximum pest resistance. In Ireland, The Protecting Farmland Pollinators plan is reliant upon the designing of a crop bee pollinator assessing campaign which can permit growers to determine simply which control processes on agro lands help pollinators in decreased pesticides usage, offer little environments and floral resources, for example, flowering pattern at the farmland level. Additionally, the Interreg-Sudoe Poll-Ole-GI project (2016-2019) focused on recognizing and suggesting efficient practices as green infrastructures (GI) to positively affect bee pollinator abundance and bee ecosystem amenity in the two most significant Mediterranean crops of arable farmland in the South-western European Space (SUDOE) which comprises Southern France, Spain, and Portugal, including sunflower and oilseed rape.

9.9 BEE HOTELS ACCOMMODATION

To increase the accessibility of flower assets for pollinators in research, however, barely a few studies have targeted the bee nesting sites (Fortel et al., 2016). Endangered bees could be settled in soil-nesting and other wild bees with special nesting behavior, nesting in cavities, subterranean, carder bees, and in snail's shells. We recommend copious nesting provisions, including the development of Wild Bee Inns for little bee species for wide distribution areas (Fortel et al., 2016). Maclvor and Packer (2015) also established the newly protected environmental enhancement strategies for wild bee needs. They highlighted that about 50% of the bee population housed in the bee hotels were newly introduced alien bees, however, 75% of them were lived in by wasps. Alarmingly, they found a negative interrelationship between the wild bees and the species present in bee hotels (Geslin et al., 2020). This study described the positive reaction of artificial devices for native bees' nesting and exhibits the role of developing nesting sites to multiply various bee species for crop pollination (Maclvor and Packer, 2015; Fortel et al., 2016). In fact, the aim should be on the present and bee hotels cavities hole diameter as needed by various bees for nesting, as small diameters could help many wild species and prevent bigger exotic bees like *M. sculturalis* to adopt bee hotel lodging (Geslin et al., 2020). May this be the little patches of bare soil installations, bee inns, or the bee hotels, the empirical use of their positive contribution should be thoroughly studied.

9.10 ALIEN PLANTS

Local plants can be managed by professionals using protection strategies, many alien species can invade and populate bee-friendly habitats, or even they are grown voluntarily. However, the effect of alien plants can depend upon various ecological factors or life history characteristics, resulting in some species to suffer from the invasions whereas for others it may be favorable (Davis et al., 2018; Drossart and Gerard, 2020). Incursion only happens when an exotic species is having invasive power and a sufficiently suitable environment to survive. Generally, such plant species have been developed in habitats closely associated with human existence. Pretentious habitats related to humans, alien plants can offer food resources for some bee pollinators and improve crop production via cross-pollination (Hobbs et al., 2013; Drossart et al., 2017; Jachuła et al., 2020). They may have detrimental effects on specific bees with small diet flexibility. However, some unwanted alien plants contend with local vegetation and can oust plants that are visited by specific pollinators. Few non-indigenous species also show resources that are inaccessible to most wild bees such as, Petunia ssp., (Lowenstein and Minor, 2016). Majewska and Altizer (2018) described in a meta-analysis that no organized positive or negative effect on pollinator population could be attributed to exotic species and that their influence is case-specific. It can be hypothesized that all exotic species are harmful and thus as detrimental to the knowledge magnification as is the principle that these species are fully protected (Boltovskoy et al., 2018). In the future, the Asian hornet *Vespa velutina*, for example, could be a threat for wild or managed bee pollinators (Arca et al., 2014; Keeling et al., 2017;

Laurino et al., 2020). In importation programs, planned international measures to prevent bio-invasions of alien species should thus be a main important matter (Sutherland et al. 2016). Observing invasive species should be kept on top of the activities, it could allow acquiring the data required to carry out the population dynamics research and test their potential effect on native species and eco-environments (Le Féon et al., 2018; Aizen et al., 2019). A law to prevent and alleviate the harmful effects of alien species in the European Union was entered in 2015 (EU Regulation (EU). Assessing the IUCN Red Lists, a parallel program called "Black List" has been recommended for invasive alien species and contemplates the extent of climatic effect for the species assessed. The IUCN Invasive Species Specialist Group (ISSG) gathers specialists in minimizing challenges developed by aliens to native habitats and the species they restrain by increasing awareness, elimination, management, or removal pertaining to them. Each group is accountable and adapted nationally, making and improving invasive species lists in their respective country.

9.11 PROTECTION APPROACHES IN WILD BEE ABUNDANCE

From the scientific knowledge and liaising more efficiently with scientific data, the efficiency of preservation actions to a large audience is a base to set up a solid conservation starting point. In education, training and communication few precedents can be explained to strengthen bee conservation technology (Saunders et al., 2019). Maximum use of citizen sciences in bee conservation procedures in various taxa, encouraged monitoring population studies over a long time period (Gardiner and Didham, 2020; Ubach et al., 2020). Many people have paid attention to wild bee pollinator conservation in the last decade. Inspite notable biases in records, such as observing the most striking and colourful species, cryptic species incorrect identification, using this tool, permits to effectively note a large number of specimens in a large area, showing an important part of the species in an environment and assessing the species dynamics for scientific research (Duchenne et al., 2020) and Red Lists program. Silva and Minor (2017) described that the degree of education and knowledge about wild bees were directly related to the positive attitudes of the respondents towards the native bee abundance, encouraging the pivotal role of people's awareness. This knowledge should be exchanged largely among scholars, and social media groups and also true in demonstrating the geographic and taxonomic extent of reports (Saunders et al., 2019). Favouring different concepts prior to the pesticides, mowing/pruning agenda constitutes the main action to efficiently integrate bee groups pollinators in the management system of all parks, gardens, and flowering patches (Schonfelder and Bogner, 2017; Folschweiller et al., 2019; Turo and Gardiner, 2019). Austria designed many research projects targeting the positive impact on wild pollinators for the renovation of flowering areas in 20ha using 18 local officials by monitoring native bee populations (Underwood et al., 2017). Apart from this pilot project, they also established many acceptable green habitats in orchards, road edges, schools, nurseries, and along waterways for bee conservation (Underwood et al., 2017). An alliance was established between an NGO and a supermarket chain for bee conservation actions. Similar collaborative programs are initiated between fruit growers and local city municipalities in Flanders (Belgium) or between a beer brewer, NGO, and public officials to boost bee conservation schemes (Underwood et al., 2017).

Populations of native bees are vanishing globally for the last few decades, generally by human interferences. The people's interest in their protection increased greatly, through scientific publications communicated through various media sources. Despite this great interest, due to the lack of knowledge and subject comprehension is flagrant and describes a space between consciousness and apprehension. As bee disappearance is investigated, knowledge of conservation measures is still scattered in the literature. For enhancing native bee populations and the improvement of bee environment, we lack bee preventative concepts and scientists are working for effective methods.

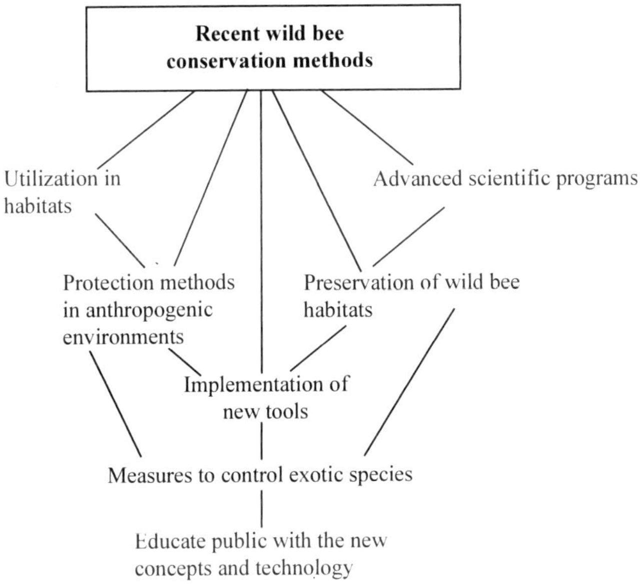

Fig. 9.2 Schematic representation of conservation strategies for wild bee pollinators

We present a recent new analysis of the wild bee conservation methods and their utilization in habitats. Further the advanced scientific programs that fill up the gaps in the literature and reframe new applied conservation strategies in this chapter has been described. We have applied our knowledge focusing on wild bee pollinators such as (i) new conservation methods in anthropogenic environments, (ii) protection and renovation of native bee habitats, (iii) implementation of newly designed tools, (iv) measures to control exotic species, and (v) education to public with the new concepts efficiently for maximum bee benefits in crop pollination (Fig. 9.2). This review can be beneficial to implement a factual and experimental native bee protection strategy for enhanced and sustainable food production globally.

9.12 CONCLUSION

To transfer new ideas, concepts and guidance based on the research findings in bee diversity protection and tools used, the importance of wild bee conservation needs apiculturists (Primack et al., 2012). Bee conservation actions are carried out research on honeybees or bumblebees, but, conclude such reports and can rarely be considered an alternative for local solitary bees

(Schmitt et al., 2020; Wood et al., 2020). The bees observed recently foraged flowering plants (Gresty et al., 2018) liking the foraging period of selected wild bees and necessities of adults and immature stages for food (Filipiak, 2019). At the time of designing new strategies for bee protection, this topic is unceremoniously investigated. In addition to this, there is an imbalance between the investigations regarding the wild bee flora, favorable environments, and safe nesting sites. Such studies have to be handled in redesigning the bee protection plans, further investigations are requireed to study such important subjects more thoroughly, especially in assessing a large area of bee dwellings than only bee resorts, and bee hotels. The planet which is moving forward anthropogenically progressively, various bee flora habitats could be, promptly, an answer to the deficit world food issues. The habitats having small native bee-flora, few alien invaded species can help native bee populations to rehabilitate. These subjects should be considered depending upon evidences as in New York a new report described that exotic species could help honeybees better than the native bees (Urbanowicz et al., 2020). In such anthropogenic environments, the program to run bee-friendly habitats has also to be rethink, in particular by discovering options for pesticide applications. Under such circumstances, the influence of pollinator integrated pest management technology (PIPMT) could be enormously explored for the maximum safety of the bee world. In addition, the overwhelming most bee habitat restoration research is carried out in North America and Europe, which constitute only a small portion of worldly environments inhabited by native bees. Such bee re-establishing methods could not be prototypical of what should be adopted in other bee habitats having dissimilar landscapes (Tonietto and Larkin, 2018; Drossart and Gerard, 2020; Tanda, 2019, 2021a-i) also requires attention. Further studies should be undertaken to assess conservation methods extensively in the rainforest and arctic environments. The transmission of knowledge and education methods about wild bees has to be redesigned in collecting the bee facts. It is a reality that we do not preserve birds by offering hen homes, so we will have to enhance our endeavors so that the beautiful world of bees does not go around the hives, for honey and waxes. To solve this, we should not disregard to involve work packages in our research work about the communication of scientific knowledge to different schools and those who are really engaged to implement conservation actions. These strategies are often the ones that are ignored while they should be the acme of scientific projects. Notwithstanding gathering information about bee population threatening components, we should share, import, away from the preventative ideas as they are mostly studied, established, recognized, and at least evaluated in half (Folschweiller et al., 2019; Tanda, 2019, 2021a-i; Penn, 2021). Experts now are inviting for protection measures, which require the urgent attention (Folschweiller et al., 2019). Research programs as the Interreg SAPOLL project, designing a strategy for wild bee conservation as the main target, or the biodivERsA NUTRIB2 project could support to fill the space in studies.

These modern programs and integrating the various bee conservation players would further strengthen the public awareness of bee biodiversity and environmental services. In bio-ecosystems where people are more cut off from the natural bee world. All the housetops, commercial buildings, roadsides, near bridges, and railway track sides should be used for the bee habitats and nesting sites (Fortel et al., 2016; Penn, 2021; Tanda, 2021g-i). Few methods and actions which can conserve the bee populations (Fig. 9.3) are well described below.

We should pledge to conserve the wild bee pollinators and join BEE-SAFE on your piece of land, garden, and the backyard of your company or your rooftops. Towns, schools, corporations, and individuals land pieces can be utilized for providing suitable habitats.

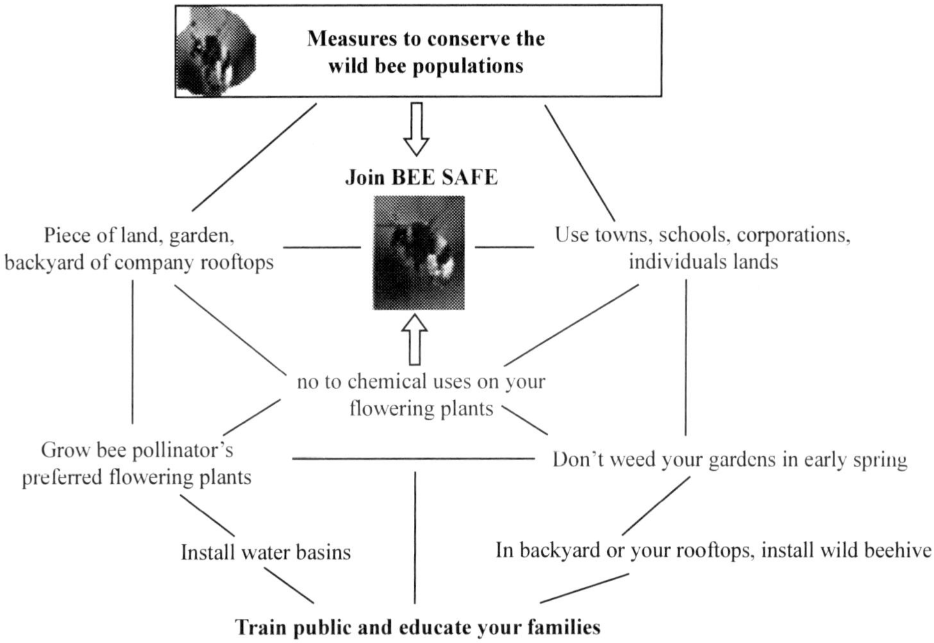

Fig. 9.3 Schematic scientific representation of measures to conserve the wild bee populations

To avoid any chemical contamination to bee populations, stop spraying on your garden flowering plants. Also check plants you buy are not pre-treated with neonics insecticides.

Always grow bee pollinator's preferred flowering plants in your garden as they are a big asset for bees and butterflies providing food as nectar and pollen.

Keep away from lawns as they desert for bee pollinators, plant prairies.

As dandelions are best source of bee food in early spring and medicine for human, so don't weed your gardens.

We can install water basins in summer in every balcony with pebbles or floating corks on water to prevent bees from drowning.

Keep in touch with Facebook page of New York Bee Sanctuary and Instagram account to get enlightened and sign regular petitions to pressurise the world to pass laws to conserve the bees banning neonicotinoids.

By showing bee documentaries, educate and train the public including your family. Native bees are harmless, and visit flowers for pollination. Know about their free services in food production by paying regards as they are preparing food for the whole world.

In your back yard or your rooftops, install a beehive and modified handmade devices for native bees. It's an excellent process to provide home and nesting sites for wild bee populations.

REFERENCES

- Aizen MA, Smith-Ramirez C, Morales CL, Vieli L (2019) Coordinated species importation policies are needed to reduce serious invasions globally: The case of alien bumblebees in South America. Journal of Applied Ecology, 56: 100–106
- Bartomeus I, Stavert JR, Ward D, Aguado O (2019) Historical collections as a tool for assessing the global pollination crisis. Philosophical Transactions of the Royal Society B, 374: 20170389
- Batáry P, Dicks LV, Kleijn D, Sutherland WJ (2015) The role of agri-environment schemes in conservation and environmental management. Conservation Biology, 29: 1006–1016
- Bauer DM, Wings IS (2016) The macroeconomic cost of catastrophic pollinator declines. Ecological Economics, 126: 1–13
- Belgian National Focal Point to the Convention on Biological Diversity (Belgian NFP-CBD). (2020). Biodiversity 2020–Update of Belgium's National Biodiversity Strategy; Royal Belgian Institute of Natural Sciences: Brussels, Belgium, pp. 148
- Belsky J, Joshi NK (2019) Impact of biotic and abiotic stressors on managed and feral bees. Insects 10: 233
- Biddinger D, Rajotte EG, Joshi NK (2018) Integrating pollinator health into tree fruit IPM—A case study of Pennsylvania apple production. In the pollination of cultivated plants: A compendium for practitioners, 2nd ed.; FAO: Rome, Italy 1: 69–83
- Biddinger DJ, Rajotte EG. (2015). Integrated pest and pollinator management–adding a new dimension to an accepted paradigm. Current Opinion in Insect Science 10: 204–209
- Biddinger DJ, Robertson JL, Mullin C, Frazier J, Vaughn M. (2013) Comparative toxicities and synergism of apple orchard pesticides to *Apis mellifera* (L.) and *Osmia cornifrons* (Radoszkowski). PLoS One, 8, e72587.
- Boltovskoy D, Sylvester F, Paolucci EM (2018) Invasive species denialism: Sorting out facts, beliefs, and definitions. Ecology and Evolution 8: 11190–11198
- Bukovinszky T, Rikken I, Evers S, Kleijn D (2017) Effects of pollen species composition on the foraging behavior and offspring performance of the mason bee *Osmia bicornis* (L.). Basic Applied Ecology 18: 21–30
- Carrié R, Andrieu E, Cunningham SA, Ouin A (2017) Relationships among ecological traits of wild bee communities along gradients of habitat amount and fragmentation. Ecography 40: 85–97
- Cely-Santos M, Philpott SM (2019) Local and landscape habitat influences on bee diversity in agricultural landscapes in Anolaima, Colombia. Journal Insect Conservation 23: 133–146.
- Cerna K, Munclinger P, Vereecken NJ, Straka J (2017) Mediterranean lineage endemism, cold-adapted paleodemographic dynamics and recent changes in population size in two solitary bees of the genus Anthophora. Conservation Genetics 18: 521–538
- Cole LJ, Kleijn D, Dicks LV, Bevk D (2020) A critical analysis of the potential for EU Common Agricultural Policy measures to support wild pollinators on farmland. Journal Applied Ecology 57: 681–694

- Crone EE, Brown, LM, Hodgson JA, Lutscher F, Schultz CB (2019) Faster movement in nonhabitat matrix promotes range shifts in heterogeneous landscapes. Ecology, 100, e02701
- Cummings AR, Read JM (2016) Drawing on traditional knowledge to identify and describe ecosystem services associated with Northern Amazon's multiple-use plants. International Journal of Biodiversity Science, Ecosystem Services & Management 12: 39–56
- Davis ES, Kelly R, Maggs CA, Stout JC (2018) Contrasting impacts of highly invasive plant species on flower-visiting insect communities. Biodiversity and Conservation 27: 2069–2085
- DEFRA (2020) The national pollinator strategy: For bees and other pollinators in England. Defra Publ.PB14221
- Dicks LV, Viana B, Bommarco R, Brosi B, Pires C (2016) Ten policies for pollinators. Science 354: 975–976
- Didham RK, Basset Y, Collins CM, Schönrogge K (2020) Interpreting insect declines: Seven challenges and a way forward. Insect Conservation and Diversity 13: 103–114.
- Drossart M, Gerard M (2020) Beyond the decline of wild bees: Optimizing conservation measures and bringing together the actors. Insects 11: 649
- Drossart M, Michez D, Vanderplanck M (2017) Invasive plants as potential food resource for native pollinators: A case study with two invasive species and a generalist bumblebee. Scientific Reports 7: 16242
- Drossart M, Rasmont P, Vanormelingen P, D'Haeseleer J (2018) Belgian Red List of Bees; Belgian Science Policy (BRAIN-be (Belgian Research Action through Interdisciplinary Networks); Presse Universitaire de l'Université de Mons:Mons, Belgium, 2018; p. 140
- Duchenne F, Thébault E, Michez D, Elias M, Fontaine C (2020) Phenological shifts alter the seasonal structure of pollinator assemblages in Europe. Nature Ecology & Evolution 4: 115–121
- Egan PA, Dicks LV, Hokkanen HMT, Stenberg JA (2020) Delivering integrated pest and pollinator management (IPPM). Trends in Plant Science 25: 577–589
- Epps CW, Keyghobadi N (2015) Landscape genetics in a changing world: Disentangling historical and contemporary influences and inferring change. Molecular Ecology 24: 6021–6040
- Ferreira PA, Boscolo D, Carvalheiro LG, Viana BF (2015) Responses of bees to habitat loss in fragmented landscapes of Brazilian atlantic rainforest. Landscape Ecology 30: 2067–2078
- Filipiak M (2018) A better understanding of bee nutritional ecology is needed to optimize conservation strategies for wild bees: the application of ecological stoichiometry. Insects 9: 85
- Filipiak M (2019) Key pollen host plants provide balanced diets for wild bee larvae: A lesson for planting flower strips and hedgerows. Journal of Applied Ecology 56: 1410–1418

- Filipiak M, Kuszewska K, Asselman M, Weiner J (2017) Ecological stoichiometry of the honeybee: Pollen diversity and adequate species composition are needed to mitigate limitations imposed on the growth and development of bees by pollen quality. PLoS One 12, e0183236
- Folschweiller M, Drossart M, D'Haeseleer J, Jacquemin F (2019) Plan d'action transfrontalier en faveur des pollinisateurs sauvages. In Projet Interreg V SAPOLL–Sauvons nos Pollinisateurs–Samenwerken voor Pollinators; Université de Mons: Mons, Belgium p. 147
- Forister ML, Pelton EM, Black SH (2019) Declines in insect abundance and diversity: We know enough to act now. Conservation Science and Practice 1: 80
- Fortel L, Henry M, Guilbaud L (2016) Use of human-made nesting structures by wild bees in an urban environment. Journal of Insect Conservation 20: 239–253
- François D, Le Féon V (2017) Abeilles sauvages et dépendances vertes routières, Pourquoi et comment developer la capacité d'accueil des dépendances vertes routières en faveur des abeilles sauvages. In Ouvrages Scientifiques, OSI2; Ifsttar: Marne-la-Vallée, France p. 120. ISBN 978-2-85782-733-7
- Gardiner T, Didham RK (2020) Glowing, glowing, gone? Monitoring long term trends in glow-worm numbers in south-east England. Insect Conservation and Diversity 13: 162–174
- Geppert C, Hass A, Akter A, Batáry P (2020) Agri-environment schemes enhance pollinator richness and abundance but bumblebee reproduction depends on field size. Journal of Applied Ecology 57: 1818-1828
- Gérard M, Martinet B, Michez D (2020) Shift in size of bumblebee queens over the last century. Global Change Biology 26: 1185–1195
- Gérard M, Martinet B, Michez D, Gérard, Vanderplanck M (2020) Global warming and plant-pollinator mismatches. Emerging Topics in Life Sciences 4(1): 77–86
- Geslin B, Gachet S, Féon VL (2020) Bee hotels host a high abundance of exotic bees in an urban context. Acta Oecologica 105: 103556
- Gibbs J, Elle E, Bobiwash K, Haapalainen T, Isaacs R (2016) Contrasting pollinators and pollination in native and non-native regions of highbush blueberry production. PLoS One 11, e0158937
- Givetash L (2018) Bees are dying at an alarming rate. Amsterdam May Have the Answer. NBC News, 7 September
- Goulson D, Nicholls E, Botias C, Rotheray EL (2015) Bee declines driven by combined stress from parasites pesticides, and lack of flowers. Science 347: 1255957
- Gresty CEA, Gresty E, Willis KJ (2018) Flower preferences and pollen transport networks for cavity-nesting solitary bees: Implications for the design of agri-environment schemes. Ecology and Evolution 15: 7574–7587
- Griffiths J, Nicholls E, Goulson D (2020) Companion planting to attract pollinators increases the yield and quality of strawberry fruit in gardens and allotments. Ecological Entomology 45: 1025–1034

- Grover SN, Miller JED, Damschen E (2017) Indirect effects of landscape spatial structure and plant species richness on pollinator diversity in Ozark glades. Castanea 82: 24–31
- Hall DM, Camilo GR, Tonietto RK, Frankie G, Threlfall CG. (2017). The city as a refuge for insect pollinators. Conservation Biology 31: 24–29
- Hall DM, Steiner R (2019) Insect pollinator conservation policy innovations at subnational levels: Lessons for lawmakers. Environmental Science & Policy 93: 118–128
- Hart AG, Sumner S (2020) Marketing insects: Can exploiting a commercial framework promote undervalued insect species? Insect Conservation and Diversity13: 214–218
- Harvey JA, Heinen R, Armbrecht I, Basset Y, Cardoso P (2020) International scientists formulate a roadmap for insect conservation and recovery. Nature Ecology and Evolution 4: 174–176
- Hatfield R, Colla S, Jepsen S, Richardson LL, Jordan SF (2020) IUCN Assessments for North American *Bombus* sp. for the North American IUCN Bumblebee Specialist Group. The Xerces Society for Invertebrate Conservation. Available online: https://xerces.org/sites/default/files/publications/14-065.pdf
- IPBES (2016) The Assessment Report of the Intergovernmental Science-Policy Platform on Biodiversity and Ecosystem Services on Pollinators, Pollination and Food Production; Potts, S.G., Imperatriz-Fonseca, V.L., Ngo, H.T., Eds.; Secretariat of the Intergovernmental Science-Policy Platform on Biodiversity and Ecosystem Services: Bonn, Germany pp. 552
- Isaacs R, Williams N, Ellis J, Vaughan M (2017) Integrated crop pollination: Combining strategies to ensure stable and sustainable yields of pollination-dependent crops. Basic and Applied Ecology 22: 44–60
- IUCN (2016) Guidelines for Appropriate Uses of IUCN Red List Data. Incorporating, as Annexes, the 1) Guidelines for Reporting on Proportion Threatened Species (ver. 1.1); 2) Guidelines on Scientific Collecting of Threatened Species (ver. 1.0); and 3) Guidelines for the Appropriate Use of the IUCN Red List by Business (ver. 1.0.) Version 3. Adopted by the IUCN Red List Committee, 3rd ed.; IUCN: Gland, Switzerland; p. 32
- Jachuła J, Denisow B, Strzałkowska-Abramek M (2020) Does an invader have a bright side? Floral reward in two Solidago species. Journal of Apicultural Research 59(2): 1–10
- Jacquemin F, Violle C, Munoz F, Mahy G, Dufrêne M (2020) Loss of pollinator specialization revealed by historical opportunistic data: Insights from network-based analysis. PLoS One, 15, e0235890
- Keeling MJ, Franklin DN, Budge GE (2017) Predicting the spread of the Asian hornet (*Vespa velutina*) following its incursion into Great Britain. Scientific Reports 7, 6240
- Krechemer FS, Marchioro CA (2020) Past, present and future distributions of bumblebees in South America: Identifying priority species and areas for conservation. Journal of Applied Ecology 57: 1829-1839
- La Vie Sauvage (2020) Emprunte Aussi nos Routes. Available online: http://biodiversite.wallonie.be/fr/la-viesauvage.html?IDC=3649 (accessed on 14 August)

- Lane I, Portman ZM, Cariveau DP (2020) Floral resource diversity drives bee community diversity in prairie restorations along an agricultural landscape gradient. Journal of Applied Ecology 57: 2010-2018
- Laurino D, Lioy S, Carisio L, Manino A, Porporato M (2020) *Vespa velutina*: An alien driver of honeybee colony losses. Diversity 12: 5
- Le Féon V, Aubert M, Genoud D, Geslin B (2018) Range expansion of the Asian native giant resin bee *Megachile sculpturalis* (Hymenoptera, Apoidea, Megachilidae) in France. Ecology and Evolution 8: 1534–1542
- Lecocq T, Gérard M, Michez D, Dellicour S (2017) Conservation genetics of European bees: New insights from the continental scale. Conservation Genetics 18: 585–596
- Lecocq T, Michez D, Gérard M, Dellicour S (2018) Divergent geographic patterns of genetic diversity among wild bees: Conservation implications. Diversity and Distributions 24: 1860–1868
- Lopez-Uribe MM, Soro A, Jha S (2017) Conservation genetics of bees: Advances in the application of molecular tools to guide bee pollinator conservation. Conservation Genetics 18: 501–506
- Lowenstein DM, Minor ES (2016) Diversity in flowering plants and their characteristics: Integrating humans as a driver of urban floral resources. Urban Ecosystems 19: 1735–1748
- Maclvor JS, Packer L (2015) 'Bee Hotels' as tools for native pollinator conservation: A premature verdict? PLoS One 10, e0122126
- Majewska AA, Altizer S. (2018) Planting gardens to support insect pollinators. Conservation Biology 34: 15–25
- Mallinger RE, Franco JG, Prischmann-Voldseth DA, Prasifka JR. (2019) Annual cover crops for managed and wild bees: Optimal plant mixtures depend on pollinator enhancement goals. Agriculture, Ecosystems & Environment 273: 107–116
- Martins KT, Gonzalez A, Lechowicz MJ (2015) Pollination services are mediated by bee functional diversity and landscape context. Agriculture, Ecosystems & Environment 200: 12–20
- Meeus I, Pisman M, Smagghe G, Piot N (2018) Interaction effects of different drivers of wild bee decline and their influence on host-pathogen dynamics. Current Opinion in Insect Science 26: 136–141
- Moquet L, Vanderplanck M, Moerman R, Jacquemart AL (2017) Bumblebees depend on ericaceous species to survive in temperate heathlands. Insect Conservation and Diversity 10: 78–93
- Nichols RN, Goulson D, Holland JM (2019) The best wildflowers for wild bees. Journal of Insect Conservation 23: 1–12
- Nicholson CC, Egan PA (2019) Natural hazard threats to pollinators and pollination. Global Change Biology 26: 380–391
- Nowakowski M, Pywell RF (2016) Habitat creation and management for pollinators; centre for ecology & hydrology: Wallingford, UK pp. 77

- Packer JG, Delean S, Kueffer C, Prider J, Carthew SM (2016) Native faunal communities depend on habitat from non-native plants in novel but not in natural ecosystems. Biodiversity and Conservation 25: 503–523
- Penn S. (2021) "Climate change reduces the abundance and diversity of wild bees." Science Daily, 12 January 2021
- Pollinator Plan Steering Group. (2015) All-Ireland Pollinator Plan 2015–2020; National Biodiversity Data Centre Series No. 3; National Biodiversity Data Centre: Waterford, Ireland, 48 p, ISSN 2009-6844
- Potts SG, Imperatriz-Fonseca V, Settele J. (2016) Safeguarding pollinators and their values to human well-being. Nature 540: 220–229
- Powney GD, Carvell C, Isaac NJB. (2019) Widespread losses of pollinating insects in Britain. Nature Communications 10: 1018
- Quinet M, Mabeluanga T, Moquet L, Jacquemart AL. (2016) Introduction of new tools to improve pollination in European pear orchards. Scientia Horticulturae 213: 5–12.
- Reemer M. (2018) Basisrapport voor de Rode Lijst Bijen; EIS Kenniscentrum Insecten: Leiden, The Netherlands p. 174
- Ritchie H, Roser M. (2020) "Crop Yields". Published online at OurWorldInData.org. Available online: https://ourworldindata.org/crop-yields
- Roger N, Moerman R, Carvalheiro LG, Rasmont P (2017) Impact of pollen resources drift on common bumblebees in NW Europe. Global Change Biology 23: 68–76
- Säumel I, Weber F, Kowarik I (2016) Toward livable and healthy urban streets: Roadside vegetation provides ecosystem services where people live and more. Environmental Science & Policy 62: 24–33
- Saunders ME, Janes J, O'Hanlon J (2019) Moving on from the insect apocalypse narrative: Engaging with evidence-based insect conservation. BioScience: 1–32
- Saunders ME, Janes J, O'Hanlon J (2020) Semantics of the insect decline narrative; recommendations for communicating insect conservation to peer and public audiences. Insect Conservation and Diversity 13: 211–213
- Saunders ME, Smith TJ, Rader R (2018) Bee conservation: Key role of managed bees. Sciences 360: 381–389
- Schatz B, Drossart M, Michez D (2020) A boom of convergent information about the urgency to conserve pollinators. Acta Oecologica: 105
- Schmitt S, Maréchaux I, Hérault B (2020) Functional diversity improves tropical forest resilience: Insights from a long-term virtual experiment. Journal of Ecology 108: 831–843
- Schonfelder ML, Bogner FX (2017) Individual perception of bees: Between perceived danger and willingness to protect. PLoS One 12, e0180168
- Senapathi D, Goddard MA, Kunin WE, Baldock KCR (2017) Landscape impacts on pollinator communities in temperate systems: Evidence and knowledge gaps. Functional Ecology 31: 26–37
- Silva A, Minor ES. (2017) Adolescents' experience and knowledge of, and attitudes toward, bees: Implications and recommendations for conservation. Anthrozoos 30

- Sutherland WJ, Barnard P, Esmail N (2016) Horizon scan of emerging issues for global conservation and biological diversity. Trends in Ecology & Evolution 32: 31–42
- Tanda AS (2021a) Comparative pollination efficacies of *Apis mellifera* L. and *Tetragonula carbonaria* (Smith) on peach (*Prunus persica* L.) Indian Journal Entomology (Under publication)
- Tanda AS (2021b) Insect pollinators matter in sustainable world food production. Indian Journal Entomology (Under publication)
- Tanda AS (2021c) Why insect pollinators important in crop improvement?. Indian Journal Entomology, Online published Ref. No. e20370)
- Tanda AS (2021d) Biofloral phenology, foraging behavior and entpollinatological effect of honeybees in pomegranate (*Punica granatum*) fruit quality and yield. Journal of Horticulture 8 2: 1–3
- Tanda AS (2021e) Urbanization and Its Impact on Native Pollinators. The 1st International Electronic Conference on Entomology to be held in 1st–15th July 2021 virtually (Abstract accepted)
- Tanda AS (2021f) Native bees are important and need immediate conservation measures: a review. The 1st International Electronic Conference on Entomology to be held in 1st–15th July 2021 virtually (Abstract accepted)
- Tanda AS (2021g) New ideas in pollinator and integrated pest management technology in sustainable agriculture. Journal of Integrated Pest Management (Under publication)
- Tanda AS (2021h) Native bees are important and need immediate conservation measures: a review. MDPI Proceedings 2021, 68, x. https://doi.org/10.3390/xxxxx): 1–19
- Tanda AS (2021i) Wild bees and their conservation. Indian Journal Entomology (Accepted)
- Tanda AS. (2020a) Biogenetic engineering in developing insect resistant crops: constraints and applications. *5th Edition of Global Congress on Plant Biology and Biotechnology* (*GPB 2020*) Valencia, Spain
- Tanda AS (2020 b) Entpollinatology: A strong relationship between plants and insects for Crop Improvement. *6th Edition of Global Conference on Plant Sciences and Molecular Biology* (*GPMB 2020*) Paris, France
- Tanda AS (2020c) Entpollinatology is for sustainable agroecosystems and global agroindustry. Indian Journal of Entomology (Under publication)
- Tanda AS (2019) Entomophilous crops get better fruit quality and yield: An appraisal. Indian Journal Entomology 81(2): 227-234
- Tonietto RK, Larkin DJ (2018) Habitat restoration benefits wild bees: A meta-analysis. Journal of Applied Ecology 55: 582–590
- Tosi S, Nieh JC (2019) Lethal and sublethal synergistic effects of a new systemic pesticide, flupyradifurone (Sivanto), on honeybees. Proceedings of the Royal Society 286: 2019433
- Turo KJ, Gardiner MM (2019) From potential to practical: Conserving bees in urban public green spaces. Frontiers in Ecology and the Environment 17: 167–175

- Ubach A, Paramo F, Gutierrez C, Stefanescu C (2020) Vegetation encroachment drives changes in the composition of butterfly assemblages and species loss in Mediterranean ecosystems. Insect Conservation and Diversity 13: 151–161
- Underwood E, Darwin G, Gerritsen E (2017) Pollinator Initiatives in EU Member States: Success Factors and Gaps. Report for European Commission under Contract for Provision of Technical Support Related to Target 2 of the EU Biodiversity Strategy to 2020–Maintaining and Restoring Ecosystems and Their Services; Institute for European Environmental Policy: Brussels, Belgium, ENV.B.2/SER/2016/0018
- Urbanowicz C, Muniz PA, McArt SH (2020) Honeybees and wild pollinators differ in their preference for and use of introduced floral resources. Ecology and evolution 10: 6741–6751
- Valiente-Banuet A, Aizen MA, Alcantara JM, Arroyo K, Jordano P (2015) Beyond species loss: The extinction of ecological interactions in a changing world. Functional Ecology 29: 299–307
- van Klink R, Bowler DE, Gongalsky KB, Chase JM (2020) Meta-analysis reveals declines in terrestrial but increases in freshwater insect abundances. Science 368: 417–420
- Vaudo AD, Farrell LM, Patch, Tooker JF (2018) Consistent pollen nutritional intake drives bumblebee (*Bombus impatiens*) colony growth and reproduction across different habitats. Ecology and Evolution 8: 5765–5776
- Wilson JS, Forister ML, Carril OM (2017) Interest exceeds understanding in public support of bee conservation. Frontiers in Ecology and Environment. 15: 460–466
- Wood TJ, Holland JM, Goulson D (2015) Pollinator-friendly management does not increase the diversity of farmland bees and wasps. Biological conservation 187: 120–126
- Wood TJ, Holland JM, Goulson D (2017) Providing foraging resources for solitary bees on farmland: Current schemes for pollinators benefit a limited suite of species. Journal of Applied Ecology 54: 323–333
- Wood TJ, Michez, D, Leclercq N (2020) Managed honey bees as a radar for wild bee decline? Apidologie 51: 1100–1116

10

Natural Foraging and Artificial Feeding of Domesticated Honeybees

Muhammad Anjum Aqueel[1], Hafiz Khurram Shurjeel[2], Sajjad Ali[3], Muhammad Abu Bakar[4], Mubasshir Sohail[5], Ziyad Abdul Qadir[6], Amir Manzoor[7], Shams ul Islam[8] and Imran Khalid[9]

ABSTRACT

Honeybees acquire their food through foraging in the field and the sufficient and healthy diet is imperative colony strength parameter alternatively, in the dearth periods in order to accomplish their nutritional needs, colonies are fed with the supplemental diets. The artificial diet must be enclosed with the essential pollen and protein necessary for the growth and development. Since the mostly sites extensively turn in to agricultural monocultures, and pollination activities by the bees frequently take place within an anthropogenic' designed ecosystem, so the dietary

[1]Department of Entomology, Faculty of Agriculture and Environment, The Islamia University, Bahawalpur, Pakistan
E-mail: anjum.aqueel@iub.edu.pk

[2]University of Sargodha, Pakistan
E-mail: shurjeel_38@yahoo.com

[3]Department of Entomology, Faculty of Agriculture and Environment, The Islamia University of Bahawalpur, Pakistan
E-mail: sajjad.ento@gmail.com

[4]Department of Entomology, University of Sargodha, Pakistan
E-mail: mabubakar_26@yahoo.com

[5]Plant Protection Division, Nuclear Institute of Agriculture, NIA, Pakistan.
E-mail: mubasshirsohailroy@gmail.com

[6]HBRI, NARC, Islamabad, Pakistan.
E-mail: qadir.narc@gmail.com

[7]Department of Entomology, The Islamia University, Bahawalpur, Pakistan
E-mail: amir.manzoor@iub.edu.pk

[8]Department of Entomology, The Islamia University, Bahawalpur, Pakistan
E-mail: shams9930@gmail.com

[9]Department of Agri-Extension, The Islamia University, Bahawalpur, Pakistan
E-mail: imran.khalid@iub.edu.pk

requirements of bees may not be properly fulfilled. Consequently, the question crop up that how bees can be fed on supplemental diet in the dearth period. Because the bees are social and super creature therefore, the sustenance can be examined on three stages – colony nutrition, adult feeding and larval feeding – with rising intricacy, because disarray in the preceding level impinge on the following levels, and vice versa. Colony with less pollen stores, possibly will be hampering the adults from nourishing larvae correctly or from becoming larval to successful adult. Consequently, the eminence or the adults' strength in the next cohort perhaps deprived, which could influence colony' food condition and manipulate the following brood nurturing.

Keywords: *Honeybees • Feeding bees • Artificial diet • Natural foraging • Domestic bees*

10.1 INTRODUCTION

The sufficient diet is an imperative colony strength parameter. The pollen particles are male germs of flowers, enriched with premium protein, which contribute in the development, growth and tissue repairment of honeybees in colonies (Somerville, 2000; Alghamdi, 2002; Mishima et al., 2005). They also significantly contribute in the royal jelly formation for the queens' nourishment (Alqarni, 2006). The pollen is also an essential element required at the time of wax manufacturing and honey flow (Somerville, 2000). Generally, a normal honeybee colony devours almost 15-30 kg of pollen in a year (Seeley, 1985). The worker bees instigate to feed on the pollen immediately a few hours subsequent to emergence and their pollen consuming capacity starts increasing gradually and reaches the highest when they are about four to nine days mature (Crailsheim et al., 1992).

During the dearth period or absolute pollen deficiency, or in the occurrence of just deprived and substandard pollen, beekeepers frequently supply substitute and/or supplemental food to the colonies as pollen substitute diet. In this situation, this is a perfect food material that endows with the necessary nutrients for their nourishment (Saffari et al., 2004; Zahra and Talal, 2008). The pollen containing artificial diet holding around 20% of the soybean flour or brewer's yeast are greatly appetizing to the honeybees and have important nutritive supplies for their development and population strength (Mattila and Otis, 2006). This is evidenced that the pollen-alternate food and natural pollen itself are uniformly accepted and consumed by the colony members, thus it can easily be endowed with as sugar patties to the colonies in normal beehives (Saffari et al., 2004). Beekeepers nourish the colonies in order to arouse brood rearing in late winter season or start of spring, or to lighten the nutritional strain through times when there is no sufficient natural pollen supply or is of inconsequential dietary importance (Mattilla and Otis, 2006; Nabors, 2000).

Since mostly sites extensively turn into agricultural monocultures, and pollination activities by the bees frequently take place within an anthropogenic' designed ecosystem, so the dietary requirements of bees may not be properly fulfilled (Naug, 2009). Consequently, the question crop up that how bees can be fed on supplemental diet in the dearth period. Because the bees are social and super creature (Seeley, 1989) therefore, the sustenance can be examined on three stages – colony nutrition, adult feeding and larval feeding – with rising intricacy, because disarray in the preceding level impinge on the following levels, and vice versa. Colony with less pollen stores, possibly will be hampering the adults from nourishing larvae correctly or from becoming larval to successful adult. Consequently the eminence or the adults' strength in the next cohort perhaps deprived, which could influence colony' food condition and manipulate the following brood nurturing.

The nutritional stages are intimately associated through different adult-brood relations and trophallactic acquaintances. The trophallaxis depict the common food exchange between the adult bees mutually in a directed approach, and producing a common stomach that facilitate all bees to acquire the knowledge regarding the colony food condition (Crailsheim, 1991, 1998). Equally the larvae and adults are extremely reliant on the colony food stores, and adult bees become accustomed their brood-care approaches as per their requirement and provision of carbohydrates and proteins (Schmickl and Crailsheim, 2004). The physiological and dietary requirement of all the colony members (workers, queens, drones) vary comparatively (Hrassnigg and Crailsheim, 2005). The bees' colony health and strength is not merely distinct by the nonexistence of diseases, but also by existence of numerous well-fed members proficient to produce offspring and opposing the various stressors.

10.2 NATURAL FORAGING

10.2.1 Foraging Philosophy

Foraging is basically the division of ecology (behavioral) which focuses on the food searching behavior of natural world in return to the ecosystem where the animals live. The term foraging is referred as looking for natural resources for flora and fauna. It impinges on the animal's health and strength since it has significant responsibility in the individual survival and reproduction.

The foraging is uniquely distinctive manners exist in the honeybees that connect the bee colonies with their instant environmental conditions. This particular behavior of the bees facilitates them in reproduction and other health related issues. Since these are social insects having three castes; queen, worker and bees in the colony and workers are also further divided into nurse bees, guard bees, forager bees, etc, they all carry out particular tasks as per their age and colony needs characteristically the queen focus on the egg laying and overall colony care, drones just present there for mating and worker execute all activities inside and outside the colony, considerably the foraging.

Beside the significance of bees' pollen, nectar, resin, propolis and water collection for their colonies, plants pollination is of great importance. The reason of being strong and successful pollinator is that they have great contribution in seeds production, seeds quality and quantity of the various crops such as broccoli, sunflower, apple and peers, almonds, etc. Another important advantage beside pollination is the nitrogen deposition (around 2.27 to 2.69 g per month) on plants during foraging time and having competency to distribute the bio-control agents such as in *Erwinia herbicola* of fire blight. A technique called Triwaks is being used in order to obtain maximum benefit from their distributing behavior of bio-control agents. The foragers also work as excellent bio-indicators (direct or indirect) for detecting the contamination through different supply resources such as heavy metals, drugs, pesticides, etc. Foragers are trained for this intention to detect TNT, drugs, pests, species monitoring and other stressors through magnifying their proboscis.

The bees foraging pattern is also followed by the computer science experts for applying in research. The honeybees never collect the nectar, pollen, water and other floral resources found on their way but they choose the most favorable foraging way. Therefore, the computer experts use to follow the foraging skills, pattern and behavior to resolve their optimization related issues. The swarming phenomenon in the honeybees is much important for further innovations and being recently followed by the folks working on the artificial intelligence.

Followings are the important factors that influence the foraging behavior of the honeybees;

- Foraging tasks
- Foraging time
- Foraging distance
- Foraging preference
- Foraging behavior of honeybee sub species
- Factors impacting foraging activities
- Monitoring of foraging activities
- Controlling of foraging activities

10.2.2 Foraging Tasks

The worker bees are usually categorized into the scout and reticent bees. The scout perform their duties such as collecting food from field while the reticent are those remain in the hive and wait to listen from foragers regarding food through dance orientation. The reticent bees are maximum strength of the colony population. The general food resources are pollen, nectar and water, but forager bees sometime collect the wax from scale insects (*Ceroplastes* spp). The foraging activities are the exclusive decision of the foragers keeping in-view the colony condition and foraging type is depend upon the genetic elements and exaggerated by the bee genetic strain. Another most essential foraging assignment category is sucrose response threshold and a significant difference exists among their thresholds response and further these differences are associated with labor workers. So far the water foragers showed lowest response as compared to the other foragers. The affected sucrose response is due to the environmental circumstances and the use of different chemical compounds such as Bovine insulin treatment in the apiaries. There insulin receptor substrates (IRS) manipulate and impinge on the foraging preference among the pollen and nectar foraging.

10.2.3 Foraging Time

Foraging time have a great importance regarding foraging of honeybees. Generally, they perform foraging in the early day time and returns to hive before evening, however, it depends on the area where the colonies are kept such as in deserts, they start foraging at 8:00 am and finish at 10:00 am. The fluctuating foraging schedule is a general phenomenon. Sometimes bees find high-quality and high quantity pollen in the morning and low quality in the evening and some contrary to that. The foraging bees have efficiency of naturally memorizing the foraging timings and simultaneously when and where the higher floral resources are present in the field. The maximum time a bee spent in an area during foraging is five minutes because she prefers to visit maximum number of sites and flowers to find best pollen and nectar quality.

10.2.4 Foraging Distance

The foraging distance is of great importance in foraging attributes. Various hypothesis had drawn for accurate distance estimation such as the initial energy used by foragers for flight was standard for foraging distance estimation but later on replaced by optical flow–speed motion of image of the ground which received by retina and certainly both are purposeful for measuring. Various sound statement calculations are also recorded such as the distance covered by the pollen collector bees in simple ecosystem was 1743 m while in complex it was 1543 m likewise in July, in small hives the average distance was 670 m while in large it reduced up to 620 m. These differences fluctuate (45 m-5983 m) with respect to monthly weather conditions and rise up to 1430 m for small and 2850 m for large hives. In deserts, bees fly up to 2 km for searching water, so there are difference foraging choices and distances according to the environmental conditions. The foraging behavior varies by race, colony strength, days and months timings and floral resources in similar and various locations.

10.2.5 Foraging Preferences

Generally, honeybees do not collect the flora present on their way but they go for better possible resources such as if they have to collect water, they go to forage to unusual source where water flow is clean, fresh and continuous as compared to stable standing water. And they collect it from the nearby shady place and do not fly long away. For nectar, forager visit variety of blooming flowers as per their specific locations and normally many flowers are visited by them. Bees collect nectar from upper part because some bees collect the nectar from top part of flower and some like to collect from base. Similarly, the pollen collector showed different behavior and visit many locations and use to collect pollen from middle of tree and sometime from branches near base and further also varies from crop to crop. However, their preferences could be different in the dearth periods comparatively.

10.2.6 Foraging Behavior of Subspecies

Like different honeybees, their sub-species of honeybees present different foraging behavior. So far, the difference is observed among three species; Yemeni, Italian and Carniolan. In the deserts, the Yemeni spp showed greater foraging activities following the Italian and Carniolan species and particularly in June and August, Yemeni spp constantly show higher foraging rate than Carniolan species in different day timings. Generally, the foraging efficiency of the bee species are measured studying their morphological characters, for instance, if they have complete and strong wings, they could have strong foraging capability as compared to those with small and fractured wings. The strong-winged bees intended to fly far away from the hive than the ones having weak or damaged wings and with low foraging exposure. Besides all that, the temperature fluctuations, humidity level, other biotic and abiotic factors and environmental resources (discussed below) greatly impinge on the sub-species of honeybee' foraging efficiency.

10.2.7 Factors Influence the Foraging

Various factors including inside the colony and outside the colony impact on the colony. The colonies containing queen are productive with higher pollen and nectar while the queenless colony is less productive and low foraging activities. The bees' higher foraging activity might be influenced pollen needs, brood rearing and colony strength. The foraging colony is subjective by the hive type and diseases and parasites, and possibly faces failure to homecoming after foraging. They generally do not permit the foragers to go so long away due to their weakness. Another reason is the genotype of the bee strain that strongly influence the foraging behavior of the bees. As long as the out-colony factors concerned, major factor is food source availability that impinge on foraging. Foraging got fluctuate sue to the environmental conditions and temperature, however, the best foraging done in the ambient temperature. It is known that the foraging below 10°C and up to 43°C can badly affect the foragers and floral resources and present very slow/ low foraging. Therefore, high and low temperature affects the pollen foragers particularly, however, the humidity stress does not affect the foraging flight, etc.

The bees' pests are also regarded as the foraging influencer such as pollen beetle *Meligethes aeneus* (Nitidulidae) often present on oilseed rape flower and cause hindrance in flowers opening and create hurdle for foragers in visiting the place and finding pollen. The other predators such as hornets and bee-eaters do affect the forgers indirectly. The chemical compounds such as imidacloprid, neonicotinoid, etc. being sprayed in the agricultural settings are also affecting the foraging strength. Bees expose to insecticides during foraging and after feeding, return to colony. The insecticides infiltrate into the foragers' body and influence their foraging and also transferred to the colony and make it weak.

10.2.8 Monitoring of Foraging

In order to measure monitor foraging, the parameters precisely and accurately use are; commencement time, foraging ending time, number of bees leaving or entering the hive, foraging time fluctuation, foraging speed, foraging distance and its estimation through waggle dance. Some other factors associated to monitor foraging includes; plants quantity visited by forager, quantity of forager per plant, time spent on each plant, nectar, pollen, water collection method, foragers sitting position on plant, plant position, pollen and nectar percentage from total foraging, pollen and nectar role and type, foraging type, and contention with other pollinators.

10.2.9 Controlling the Foraging

The treatment of honeybees with different chemicals, such as suspended glass tubes enclosing synthetic brood pheromones with isopropanol, might enhance the foraging activities of the honeybees after 1 hr treating the colony.

10.3 ARTIFICIAL FEEDING

10.3.1 Carbohydrates

10.3.1.1 Colony nutrition

The nectar is the original source of carbohydrate that is generally collected during foraging and further stored in the cells as honey in the colony. The conversion of nectar in to honey is and instigate through the recurring flight to the hive (Nicolson and Human, 2008). Further, the water content are condensed up to 16–20%, enzymes are inserted, which describe the sugar content in the honey (averagely fructose 38%, glucose 31% and saccharides) (Doner, 1977).

The yearly honey production differs greatly because of environmental conditions, beekeeping practices, and the floral accessibility and calculated as 19.2 kg (McLellan, 1977), 24.3–31.3 kg (Avni et al., 2009) or over 50 kg for the colonies fed on carbohydrate-supplemented diet (Severson and Erickson, 1984). The enduring continuation of food availability facilitates the colonies to continue existing for a long period of time exclusive of food incoming form the environment. The vigorous cost of overwintering is elevated. Seeley and Visscher (1985) calculated the weight loss of undersized colonies – generally because of honey eating. There is less weight and more energetic cost of brood rearing in winter season (0.84 kg per week) than the thermoregulation cost obligatory for the growth, development and colony survival exclusive of brood (0.42 kg per week).

10.3.1.2 Adult nutrition

The honeybees necessitate carbohydrates to fulfill their energy requirements. The fully grown workers are completely reliant on the food stored in the colony and cannot endure for a long time devoid of feeding since they don't have substantial carbohydrates, proteins or lipids stored in the bodies (Kunert and Crailsheim, 1988). Comparatively the adult bees contain low stores of glycogen (0.05–0.47 mg / worker, (Hrassnigg and Crailsheim, 2005). When the need energy for foraging tasks, the worker bees utilize the sugar present in the honey store through trophallactic interaction.

The adult worker bees require 4 mg sugar to use for their endurance (Barker and Lehner, 1974). The bees in captivity survive successfully provided with exclusively carbohydrates; sucrose (LT50 = 56.3 days) than honey (31.3 d) or high-fructose corn syrup (Barker and Lehner, 1978). The forager bees when aimed to fly, need sugar mostly polysaccharides. A number of sugars such as galactose, stachyose, etc. are poisonous for bees and are present in the pollen and nectar of Tiliaceae or tulips (Barker, 1990). Around 40% of the sugars present in the pollen supplemental diets are poisonous for bees until this toxicity abridged through the abundance of nectar or can be diluted through 4% with 50% sucrose solution testing (Barker, 1977).

As a consequence of heat treatment or during storage of honey, a material is formed known as hydroxymethylfurfural (HMF), is also detrimental for bees. Its high level is a threat to feed. The sugar solution holding 150 ppm HMF cause 58.7% death rate in 20 days in captivity, while the solution holding 30 ppm, grounds only 15.0%, and found non-significant comparative to the controls (12.5%) treatment and hence can be observed as secure for bees. Recently, the HMF concentration in commercially obtainable HFCS was found as 3.1 and 28.7 ppm (LeBlanc et al., 2009). They further revealed that the HFCS stored at 40°C for a long period of time like 69 days might reach 250 ppm which significantly causes the mortality.

10.3.1.3 Larval nutrition

Nurse bees generally observed the larvae and fed them if required, and always adequately supplied with food. The sugar content in their food are different as per their age such as 18% up to 3 days of larvae growth and 45% up to 5 days. Rortais et al. (2005) nurture the larvae on total of 59.4 mg of carbohydrates to a single worker through its growth period. The deficiency of carbohydrates decreased the strength of larvae in the colony raised in spring season, when there are deprived nectar resources and the winter food stores previously worn-out, or subsequent to extracting the honey devoid of sufficient carbohydrates' substitution.

The larvae do not eat on the voluntarily present vegetal diet; rather nosh by the adult bees. The adult sister bees in the colony nourish workers on the worker jelly (royal jelly and worker jelly are different after three days with respect to composition according to their age group. Such as the Pantothenic acid is detected in higher amount in worker jelly, however, here were no difference found in the vitamin content. A larval development is regularly examined and increasingly fed 135-143 times on worker jelly (produced by mandibular glands and hypopharyngeal glands) throughout. According to a toothed judgment, one worker larva consumes around 59.4 mg carbohydrates and 25-37.5 mg protein in their diet throughout the duration (Hrassnigg and Crailsheim, 2005). Up to five days, the worker larvae feed additionally unprocessed pollen untreated pollen, however, after six days, larvae close down the eating and larval cells are closed and preserved by their adult worker bees (pupation period). The larvae in the cells utilize the food material for anabolism and catabolism through the pupation duration.

10.3.1.4 Feeding carbohydrates

In order to fulfill the nutritional requirement of bees after honey extraction or through dearth period, the carbohydrates are provided. So the sucrose solution, sugar/fruit syrups, etc., are provided inside the colony (Neupane and Thapa, 2005). The grapes syrup is not recommended since it creates dysentery and decrease the longevity as well (Barker and Lehner, 1978). After feeding the colonies in the autumn season, in the temperate areas apiaries, bees consume around 20–25 kg of sugar and fruit syrups between November to April (Severson and Erickson, 1984). The syrups are fed by the bees in the hive, and the supplemental carbohydrates are gathered in the food flow of colony and further transferred toward the combs contain food stores (DeGrandi-Hoffmann and Hagler, 2000). The bees feeding on the sugar solutions through the feeders in the hive are observed young by (Free, 1965) and later on through same method were found of same age as nectar collector (Brodschneider et al., 2007).

In the temperate region, there are no detrimental effects of feeding HFCS to beehives in a temperate climate because they observed the weight gain, average seasonal honey production and the quality of brood (head, thorax, dry weight). The harmful possessions are mostly experiential subjected to the colony receiving exclusively carbohydrates from HFCS, as is the situation when beehives are supplied food prior to winter season. Colonies fed on supplemental diet with HFCS in autumn season had less capped brood in spring season than in the hives fed on sucrose, however, this hindrance in spring upsurge did not impinge on the honey production later. The findings of the survey carried out among beekeepers (Brodschneider et al., 2010) also recommend that the category of carbohydrates provided to the beehives (sucrose, starch syrup, etc.) may not distress the overwintering death rate of hives when alternative supplements were applied from July to October.

10.3.2 Protein

10.3.2.1 Colony nutrition

The merely protein resource for bees is the pollen and the colonies accumulate almost 10–26 kg of pollen yearly on average (Wille et al., 1985) and (Crailsheim et al., 1992) described that pollen necessity of two colonies containing 10-frame is 13.4 and 17.8 kg yearly, respectively. Contrary to the honey, just a little pollen quantity is accumulated in the colony, and food stores speedily reduced during the non-foraging periods or non-floral availability (Schmickl and Crailsheim, 2001, 2002).

In beehive, bees combine the pollen with reiterated nectar, honey and glandular discharge in order to generate bee bread, which greatly diverge from recently composed pollen, in containing a low pH and low starch level as well (Ellis and Hayes, 2009). The nourishing worth of bee bread to bees is greater as compared to the fresh one, stored in the laboratory, or freeze with little exclusion (Pernal and Currie, 2000). The modifications in the stored pollen quality (bee bread), is ascribed to microorganisms connected with honeybee (Gilliam, 1997). The lactic acid bacteria present in bee stomach belong to *Lactobacillus* and *Bifidobacterium*, and both have concerned with bee bread fermentation and might be accountable for increasing the nutritious value through vitamins (Vásquez and Olofsson, 2009). The protein concentration in pollen of varying species and areas fluctuate extensively about 2.5-61%, (Roulston et al., 2000). Honeybees require ingesting almost ten amino acids regarded as indispensable to their food (De Groot, 1953). The amino acid are so much obligatory in the diet even they can hinder the colony growth and development. The pollen contains different tastes and nutritional ingredients from different plants and is essential in accomplishing the bioassays of varying pollen or protein foods and their influence on the brood manufacturing (Dietz and Stevenson, 1980), longevity (Schmidt et al., 1987) or other various physiological characters.

The colonies holding no pollen cannot survive for a long period of time and because firstly they end up the bee bread and then depleting the body reserve. In order to react towards the changes in pollen ratio and brood demand, the bees have a strange mechanism; they cannibalize the offspring and thus obtain protein which they utilize for other larvae feeding. Comparatively, the young larvae are cannibalized than the older ones which are maintained in the colony (Schmickl and Crailsheim, 2001, 2002). And if the shortage of pollen continues, no more brood production took place in the colony.

10.3.2.2 Adult nutrition

Proteins compose around 66-74% of dry matter of full grown worker bees (Hrassnigg and Crailsheim, 2005). Protein concentration enhanced through the first few days because of protein anabolism (Crailsheim, 1986), and then somewhat reduced in the aged bees. An adult worker bee devours averagely 3.4-4.3 mg pollen/day (Crailsheim et al., 1992). The protein quantity in the hemolymph also enhanced over the initial some days subsequent to emergence and is calculated as 11.4-27.6 μg/μL (Cremonez et al., 1998) and on sixth day being fed on the supplemental diets observed as 6.0-9.4 μg/μL (De Jong et al., 2009).

The protein content fluctuates according to the weather conditions and seasons. In late summer, bees' last generation demonstrate distinctive physiological uniqueness (Kunert and Crailsheim, 1988) and a distinctly improved longevity as compared to the short-lived bees in the summer season. This variation is supported by the bees' competency of strengthening big stage of hemolymph proteins, which permit them for endurance of about numerous months on carbohydrates merely. Vitellogenin, the foremost protein stored in the hemolymph and forerunner for numerous other proteins (Amdam et al., 2003), accomplish concentration up to 80 μg/μL in 20 days older bees in winter season. The lipoprotein' significance regarding the foraging commencement, endurance and overwintering has been discussed by (Amdam and Omholt, 2002, 2003; Amdam et al., 2004).

The adult bees spent a prolonged longevity on the carbohydrates, which they require regarding power metabolism, however, the bees permitted to consume the pollen as well that demonstrate larger endurance (Manning et al., 2007). The demand of bees for age-specific protein is extraordinary: the newly adults individuals go through the physiological modifications for instance the distress growing of flight muscles and substandard pollen food cause hindrance the time duration that workers take to accomplish their highest thorax masses (Hagedorn and Moeller, 1968). Pollen intake is compulsory as well to build up the hypopharyngeal glands (Alqarni, 2006) and ovaries (Hoover et al., 2006) and their maturity is straightforwardly associated with the protein utilization. A protein-less food decreases the metabolic doings of hemocytes and enhances the hemocytes' quantity (Szymas and Jedruszuk, 2003). Alaux et al. (2010) authenticated this enhancement in the hemocyte concentration and found glucose oxidase movement (social immunity indicator) and fat body mass (immune-competence indicator of bees) to reliant on the protein diet.

10.3.2.3 Larval nutrition

A larvae is consume 25–37.5 mg protein to be reared healthy through pollen—only diet straight fed to the larvae in small quantity; the maximum protein attained by a larva is offspring' food prepared by the adult bees. The protein originate from the pollen is around 5% (Babendreier et al., 2004). Further nurse bees use their hypopharyngeal glands and the enzyme activities to convert the protein into first-class food for larvae for their rapid growth and development (Moritz and Crailsheim, 1987). The food concentration and feeding quantity is adjusted according to the larval age, however, it is observed that the young larvae feed less as compared to older larvae.

The colonies finish to rear brood rather to generate emaciated pupae, and this supports them in upholding the pupal quality (Imdorf et al., 1998). The underfed larvae can be raised, however, they demonstrate the many impairments such as longevity and weight fluctuation (Eischen et al., 1982), lacking protein substance (Kunert and Crailsheim, 1988), morphometrics (Herbert et al., 1988), suffering through wing and reduced body volume, and the possessions of larval underfeeding on lifetime flying functions or foraging commencement cannot be debarred.

The workers lacking pollen are regarded having less developed ovaries comparative to those having sufficient pollen. When both groups were provided with same combination of royal jelly and honey, the bee' ovaries raised over deprived conditions fails to develop as compared to the others (Hoover et al., 2006). Larval nourishment possibly affects the functioning and behavior worker bees when applied protein-supplemented diet in spring season, observed higher longevity in one year and short longevity of worker bees in second year. Hence, it depicts that protein nutrition is not only accountable for the consequences but other factors as well.

10.3.2.4 Feeding proteins

In order to enhance colony development and growth in spring season, the protein-rich food can be provided to the colonies in the dearth period, when deprived pollen quality (Somerville and

Nicol, 2006) is available, and in time of single source pollen access (Schmidt et al., 1995). Application of supplemental diets in fall season widens brood nurturing prior to winter season, however, it does not assist the wintering process. Therefore, fall reared colonies did not achieve superior in subsequent spring. There is a clear difference among the pollen supplement (synthetic diet containing pollen) and pollen substitutes (synthetic food with no pollen but are protein-rich) (Standifer et al., 1977). The pollen substitute could be more costly alternative as compared to pollen supplements.

The protein utilization is greater in young caged honeybees, in an short time experiment as compared to long duration experiment (Pernal and Currie, 2000). The honeybees consume proteins with pollen very proficiently. So the protein mixed diets results in great numbers of offspring when the provided food is utilized in a significantly considerable amount. The first proper symptom of accepting artificial food is the consumption. Pollen are utilized more enthusiastically from the supplemental diet as compared to others because of the presence of phagostimulants (Schmidt and Hanna, 2006), and fabricating pollen an excellent additive for bees. Fermented foods are highly utilized eagerly as compared to the unfermented feed. The pollen utilization by bees is also reliant on the application method such as soy flour mixed enlarges surface patties (Avni et al., 2009) consumed more eagerly. However, there is a great variation in the pollen feeding and brood production by the bees over a variety of supplemental diets.

The protein-mixed diets enhance the protein level in hemolymph in the captive bees (De Jong et al., 2009). The captive bees' longevity and development of hypopharyngeal gland depends on the feeding protein-mix food (particularly pollen-mixed) to be alike, but not better, than those fed on bee bread. The brood rearing by the colony is only possible when there are sufficient food resources available. Thus, the colony must be furnished with sufficient pollen and protein-enriched supplemental diets for better consumption and high-quality brood strength. When the bees are fed with the supplemental diet enriched with pollen in the spring, they start rearing brood in the early season and at least one pollen substitute is required by the colony for brood rearing (DeGrandi-Hoffmann et al., 2008). The addition of pheromones to colonies nourish on the protein supplements for the higher brood area, adult strength and food consumption as compared to the colonies nourish on only protein (Pankiw et al., 2008). It is documented that addition of carbohydrates with the supplemental diets to colonies gives greater yields and that also leads towards higher longevity (van der Steen, 2007).

10.3.3 Other Nutrients

Beside the macronutrients (carbohydrates and proteins), lipids, minerals and vitamins are also obligatory to ingest for the honeybees. The supplemental feeding transport colonies with extra vitamins, fats, etc., however their tangible needs and most favorable levels linger comparatively unfamiliar. Bees attain the lipids absolutely from pollen, and the lipid content of different species lies between 0.8-18.9% (Roulston and Cane, 2000). The addition of lipid extract around 2–4% in the bees' diet can help them to rear more brood.

Lipids are mostly utilized at brood stage and considered imperative energy supply, and as forerunner for the advance biosynthesis (Cantrill et al., 1981). The presence of sterols in the pollen is also important that is transformed by worker bees into 24-methylenecholesterol–major sterol for bees, sitosterol and isofucosterol (Svoboda et al., 1982). It is documented that food enclosed with 0.1% cholesterol or 24-methylenecholesterol facilitate the bee survival and high brood production. Apart from the presence of sterols in the diet, the workers are naturally uniquely capable of converting sterols into larvae and uphold the 24-methylenecholesterol level (Svoboda et al., 1986).

The water soluble vitamins are frequently present in the pollen as compared to fat soluble vitamins, (Roulston and Cane, 2000), however the substance of vitamin C, such as, differ all over the duration along with various floral resource. The bees are generally assumed to be capable of synthesizing the vitamin C. The pyridoxine presence in the diet is also indispensable for the larval growth and development. Anderson and Dietz (1976) confirmed that the nuclei are simply capable of rearing the brood when provided with 4 mg pyridoxine per 500 g, and 5.4 µg of pyridoxine is compulsory to raise one larva to capping stage. Though the soluble vitamins (A, K, E, D) are not considered macro and essential but their presence in feed significantly enhances the brood quantity.

Bees attain the inorganic ingredients generally from pollen, and are raised in the pollen dearth enclosed equal amount of the majority minerals comparative to the ones reared under encouraging foraging situations. The other minerals such as potassium, calcium, magnesium, sodium, zinc, manganese, iron and copper must be investigated keeping in-view the honeybee' health requirement. Water is very important to continued existence, and the bees may pay great cost bringing it into the hive.

10.3.4 Nutrition-related Threat for Bees

10.3.4.1 Starvation

After honey harvest, a long dearth period and carbohydrates insufficiency, leads towards starvation – the general symptom of bees mortality in winter season (vanEngelsdorp et al., 2010). The starvation manipulates the behavioral growth and development in bees: food scarcity encourage the premature foraging commencement in young worker bees, which impinge on the foragers' longevity and hive's demography (Schulz et al., 1998). The parasites exerting nourishing strain on bees strengthen the starvation upshot. The worker bees contaminated with *Nosema ceranae* demonstrate higher hunger level and the modification in the trophallactic behavior (Naug and Gibbs, 2009). The parasites affect the bees through making the bees in the colonies sluggish and less energetic and sometimes diminish the forager survival even (Mayack and Naug, 2009). The vitellogenin stroing capability of the contaminated workers is because of another parasite, *Varroa destructor*, that further negatively affects the overwintering capacities of bees.

10.3.4.2 Monocultures

Honeybees typically collect lots of pollens through visiting many plant species (Dimou and Thrasyvoulou, 2009) that facilitate a reasonable and diverse food. The colonies placed in the agricultural settings for pollination, faces more or less same floral pattern with less diversification that not covers all the indispcnsible food ingredients. Only a small number of homogeneous pollen feed such as sweet clover or mustard (Singh and Singh, 1996) are regarded better compared to the mixed (even poor) pollens food, and usually the food containing single pollen are substandard than mixed food (Alaux et al., 2010). The bees with the exposure of only seesam and sunflower are considered as under stress and must be nourished with additional pollen and or protein-enriched food. The different pollen mixed diet might be capable of compensating with essential nutrients, i.e., arginine present in the dandelion. If the arginine element is available in the diet, the bees cannot devour more low-quality pollen for the reparation and its deficiency in the colony and related nutritional stress can cause high mortality in the colony (Naug, 2009).

10.3.4.3 Transgenic products

The effect of genetically modified crops on bees is inadequately stated. Researchers (Malone and Pham-Delègue, 2001) explored the consequence of nutritional transgene goods on bees, due to potential intimidation originating from plant species enclosed with Bacillus thuringiensis (Bt) contaminants. Malone et al. (2004) demonstrated that the exposure of insecticidal transgenic plant proteins to the captive honeybees did not manipulate the survival and glands development. Extremely high doses of Cry1Ab proteins possibly will impinge on the learning or nourishing activities (Ramirez-Romero et al., 2008), however the longevity is not abridged.

10.3.4.4 Pesticide residues and plant poisoning

When the bees visit field for foraging in the vicinity of extreme agricultural settings, they came into contact with the chemical compounds applied for the insect pest control and inescapably bring them with the collected and contaminated flora in to the colony (Johnson et al., 2010). The maximum exposure of the chemicals becomes lethal for bees. Recently, leaf drops from corn, develops from neonicotinoid-coated seeds, documented to containing chemical compounds at the doses that are poisonous and deadly to the honeybees (Girolami et al., 2009). There are some plants species even regarded as honey plants produce the noxious chemicals that are detrimental for the honeybees. For instance, toxins from protoanemonin exist in flowers in Ranunculaceae and suppose to cause the May disease. In the flora of *Aesculus* species (Hippocastanaceae), a poison saponin is present.

10.4 CONCLUSION

Honeybees are essential agricultural pollinators that are kept for their honey and other hive products. The honeybees' foraging behaviors are incredibly specific and link the bee colonies to their current environmental conditions. The bees' specific behavior helps with reproduction and other health-related issues. The queen, worker, and drone bees are the three castes of these social insects, and the workers are only thought of as foragers. The foraging habits of bees are likewise distinctive through which they perform various food searching tasks. Bees never collect the nectar, pollen, water, or other floral resources they come across; instead, they select the best foraging route.

Natural foraging is how honeybees obtain their food, and a sufficient and healthy diet is a crucial colony strength metric. Alternatively, colonies are fed with supplemental foods during times of scarcity to meet their nutritional needs. A crucial factor in colony strength is a proper food. For honeybee immunity, overwinter survival, and brood production, abundant floral nutrients are essential. Sugars, proteins, lipids, vitamins, minerals, and water are essential for the growth, maintenance, and reproduction of honeybees. The main sources of carbohydrates for bees in their diet are nectar and honeydew, while pollen provides all the other essential nutrients. While water and carbohydrates (honey or sucrose) can keep adult bees' alive, the growth and development of young bees as well as the rearing of larvae require proteins, lipids or fats, minerals, and vitamins. Pollen, nectar, and water are the main dietary sources, however forager bees occasionally gather wax from scale insects.

While pollen is the only supply of proteins, lipids, and minerals, nectar provides carbohydrates. For many reasons, pollen nutrition is very important. First, pollen replaces bees that may be dying quickly, promoting colony expansion and enhancing resilience to environmental shocks. Second, bad weather or poor foraging conditions can cause a colony's pollen reserves to run out soon. In order to maintain population development, pollen must be accessible in adequate amounts throughout the brood-producing season. The nutritional needs of bees must also be met by a variety of flower sources because pollen's composition differs significantly with respect to plant species.

Artificial diets are often given to bee colonies keeping in mind the colony population and bee physiology. Young worker bees are given food within the hive so they can eat it and store it as energy for raising brood. For large-scale beekeepers, supplemental feeding is a management approach with high labor and financial expenses. Supplemental nutrition may help the colony survive or sustain brood rearing and colony development in the early spring, before pollen and nectar are available, or at other times of the year when these materials are not available for bees in the field or in the hive. Several brewer's yeast products, wheat, and soybean flour can be fed separately or in combination to improve the nutrition of colonies when natural pollen is limited, but none of the protein-supplement diets offered to honeybees can completely replace natural pollen. However, artificial diet can also enhance the health and performance of commercially maintained honeybee colonies.

REFERENCES

- Alghamdi AA (2002) The effect of pollen supplementary feeding on some activities of honeybee colonies during summer season in Riyadh, Saudi Arabia. Saudi J Biol Sci 9: 85–93
- Alaux C, Ducloz F, Crauser D, Le Conte Y (2010) Diet effects on honeybee immunocompetence, Biol Lett DOI: 10.1098/rsbl.2009.0986
- Alqarni AS (2006) Influence of some protein diets on the longevity and some physiological conditions of honeybee *Apis mellifera* L. workers. J Biol Sci 6: 734–737
- Amdam GV, Omholt SW (2002) The regulatory anatomy of honeybee lifespan. J Theor Biol 216: 209–228
- Amdam GV, Omholt SW (2003) The hive bee to forager transition in honeybee colonies: the double repressor hypothesis. J Theor Biol 223: 45–464
- Amdam GV, Hartfelder K, Norberg K, Hagen A, Omholt SW (2004) Altered physiology in worker honeybees (Hymenoptera: Apidae) infested with the mite *Varroa destructor* (Acari: Varroidae): a factor in colony loss during overwintering? J Econ Entomol 97: 741–747

- Avni D, Dag A, Shafir S (2009) The effect of surface area of pollen patties fed to honeybee (*Apis mellifera*) colonies on their consumption, brood production and honey yields. J Apic Res 48: 23–28
- Anderson LM, Dietz A (1976) Pyridoxine requirement of the honeybee (*Apis mellifera*) for brood rearing. Apidologie 7: 67–84
- Barker RJ (1977) Some carbohydrates found in pollen and pollen substitutes are toxic to honeybees. J Nutr 107: 1859–1862
- Barker RJ (1990) Poisoning by plants, in: Morse R.A., Nowogrodzki R. (eds), honeybee pests, predators, and diseases, Cornell University Press, Ithaca NY and London. 306–328.
- Barker RJ, Lehner Y (1974) Acceptance and sustenance value of naturally occurring sugars fed to newly emerged adult workers of honeybees (*Apis mellifera* L.). J Exp Zool 187: 277–285
- Barker RJ, Lehner Y (1976) Galactose a sugar toxic to honeybees found in exudate of tulip flowers. Apidologie 7: 109–112
- Brouwers EVM, Ebert R, Beetsma J (1987) Behavioral and physiological aspects of nurse bees in relation to the composition of larval food during caste differentiation in the honeybee. J Apic Res 23: 94–101
- Brodschneider R, Moosbeckhofer R, Crailsheim K (2010) Surveys as a tool to record winter losses of honeybee colonies – a 2-year case study in Austria and South Tyrol, Tyrol. J Apic Res 49: 23–30
- Brodschneider R, Hrassnigg N, Vollmann J, Petz M, Riessberger-Gallé U, Crailsheim K (2007) Liquid nutrition within a honeybee colony – who feeds? Apidologie 38: 492
- Cantrill RC, Hepburn HR, Warner SJ (1981) Changes in lipid composition during sealed brood development of African worker honeybees. Comp Biochem Physiol B 68: 351–353
- Crailsheim K, Schneider LHW, Hrassnigg N, Bühlmann G, Brosch U, Gmeinbauer R, Schöffmann B (1992) Pollen consumption and utilization in worker honeybees (*Apis mellifera carnica*): dependence on individual age and function. J Insect Physiol 38: 409–419
- Crailsheim K (1986) Dependence of protein metabolism on age and season in the honeybee (*Apis mellifica carnica* Pollm). J Insect Physiol 32: 629–634
- Crailsheim K, Schneider LHW, Hrassnigg N, Bühlmann G, Brosch U, Gmeinbauer R, Schöffmann B (1992) Pollen consumption and utilization in worker honeybees (*Apis mellifera carnica*): dependence on individual age and function. J Insect Physiol 38: 409–419
- Cremonez TM, de Jong D, Bitondi MMG (1998) Quantification of hemolymph proteins as a fast method for testing protein diets for honeybees (Hymenoptera: Apidae). J Econ Entomol 91: 1284–1289
- Dimou M, Thrasyvoulou A (2009) Pollen analysis of honeybee rectum as a method to record the bee pollen flora of an area. Apidologie 40: 124–133
- Doner LW (1977) The sugars of honey: A review. J Sci Food Agric 28: 443–456
- DeGrandi-Hoffman G, Hagler J (2000) The flow of incoming nectar through a honeybee (*Apis mellifera* L.) colony as revealed by a protein marker. Insectes Soc 47: 302–306

- De Jong D, da Silva EJ, Kevan PG, Atkinson JL (2009) Pollen substitutes increase honeybee hemolymph protein levels as much as or more than does pollen. J Apic Res 48: 34–37
- Dietz A, Stevenson HR (1980) Influence of long term storage on the nutritional value of frozen pollen for brood rearing of honeybees. Apidologie 11: 143–151
- De Groot AP (1953) Protein and amino acid requirements of the honeybee (*Apis mellifica* L.). Physiol Comp Oecol 3: 197–285
- Ellis AM, Hayes GW, JR (2009) An evaluation of fresh versus fermented diets for honeybees (*Apis mellifera*). J Apic Res 48: 215–216
- Eischen FA, Rothenbuhler WC, Kulincevic JM (1982) Length of life and dry weight of worker honeybees reared in colonies with different worker-larva ratios. J Apic Res 21: 19–25
- Free JB (1965) The behavior of honeybee foragers when their colonies are fed sugar syrup. J Apic Res 4: 85–88
- Gilliam M (1997) Identification and roles of nonpathogenic microflora associated with honeybees. FEMS Microbiol Lett 155: 1–10
- Girolami V, Mazzon L, Squartini A, Mori N, Marzaro M, Di Bernardo A, Greatti M, Giorio C, Tapparo A (2009) Translocation of neonicotinoid insecticides from coated seeds to seedling guttation drops: a novel way of intoxication for bees. J Econ Entomol 102: 1808–1815
- Hrassnigg N, Crailsheim K (2005) Differences in drone and worker physiology in honeybees (*Apis mellifera* L.). Apidologie 36: 255–277 http://dx.doi.org/10.1051/apido:2005015
- Hagedorn HH, Moeller FE (1968) Effect of the age of pollen used in pollen supplements on their nutritive value for the honeybee. I. Effect on thoracic weight, development of hypopharyngeal glands, and brood rearing. J Apic Res 7: 89–95
- Hoover SE, Higo HA, Winston ML (2006) Worker honeybee ovary development: seasonal variation and the influence of larval and adult nutrition. J Comp Physiol B 176: 55–63
- Hrassnigg N, Crailsheim K (2005) Differences in drone and worker physiology in honeybees (*Apis mellifera* L.). Apidolocgie 36: 255–277
- Herbert EW, Sylvester HA, Vandenberg JD, Shimanuki H (1988) Influence of nutritional stress and the age of adults on the morphometrics of honeybees (*Apis mellif*era L.), Apidologie 19: 221–230
- Imdorf A, Rickli M, Kilchenmann V, Bogdanov S, Wille H (1998) Nitrogen and mineral constituents of honeybee worker brood during pollen shortage. Apidologie 29: 315–325
- Johnson RM, Ellis MD, Mullin CA, Frazier M (2010) Pesticides and honeybee toxicity USA. Apidologie 41: 312–331
- Jung-Hoffmann I (1966) Die Determination von Königin und Arbeiterin der Honigbiene. Zeitschrift für Bienenforschung 8: 296–322

- Kunert K, Crailsheim K (1988) Seasonal changes in carbohydrate, lipid and protein content in emerging worker honeybees and their mortality. J Apic Res 27: 13–21
- Lindauer M (1952) Ein Beitrag zur Frage der Arbeitsteilung im Bienenstaat. Zeitschrift für Vergleichende Physiologie 34: 299–345
- LeBlanc BW, Eggleston G, Sammataro D, Cornett C, Dufault R, Deeby T, Cyr EST (2009) Formation of hydroxymethylfurfural in domestic high-fructose corn syrup and its toxicity to the honeybee (*Apis mellifera*). J Agric Food Chem 57: 7369–7376
- Mattila HR, Otis GW (2006) Influence of pollen diet in spring on development of honeybee (Hymenoptera: Apidae) colonies. J Econ Entomol 99(3): 604–613
- Mishima S, Miyata T, Suzuki KM, Araki Y, Akao Y, Isohama Y (2005) Estrogenic effects of royal jelly. J Trad Med 22(1): 171–175
- Manning R, Rutkay A, Eaton L, Dell B (2007) Lipid-enhanced pollen and lipid-reduced flour diets and their effect on the longevity of honeybees (*Apis mellifera* L.). Aust J Entomol 46: 251–257
- McLellan AR (1977) honeybee colony weight as an index of honey production and nectar flow: A critical evaluation. J Appl Ecol 14: 401–408
- Moritz B, Crailsheim K (1987) Physiology of protein digestion in the midgut of the honeybee (*Apis mellifera* L.). J Insect Physiol 33: 923–931
- Mayack C, Naug D (2009) Energetic stress in the honeybee *Apis mellifera* from *Nosema ceranae* infection. J Invertebr Pathol 100: 185–188
- Malone LA, Pham-Delègue MH. (2001) Effects of transgene products on honeybees (*Apis mellifera*) and bumblebees (*Bombus* sp.). Apidologie 32: 287–304
- Malone LA, Todd JH, Burgess EPJ, Christeller JT (2004) Development of hypopharyngeal glands in adult honeybees fed with a Bt toxin, a biotin-binding protein and a protease inhibitor. Apidologie 35: 655–664
- Naug D, Gibbs A (2009) Behavioral changes mediated by hunger in honeybees infected with *Nosema ceranae*. Apidologie 40: 595–599
- Naug D (2009) Nutritional stress due to habitat loss may explain recent honeybee colony collapses. Biol Conserv 142: 2369–2372
- Nicolson SW, Human H (2008) Bees get a head start on honey production. Biol Lett 4: 299–301.
- Neupane KR, Thapa RB (2005) Alternative to offseason sugar supplement feeding of honeybees. J Inst Agric Anim Sci 26: 77–81
- Nabors R (2000) The effects of spring feeding pollen substitute to colonies of *Apis mellifera*. Am Bee J 140: 322–323.
- Pernal SF, Currie RW (2000) Pollen quality of fresh and 1-year-old single pollen diets for worker honeybees (*Apis mellifera* L.). Apidologie 31: 387–409
- Pankiw T, Sagili RR, Metz BN (2008) Brood pheromone effects on colony protein supplement consumption and growth in the honeybee (Hymenoptera: Apidae) in a subtropical winter climate. J Econ Entomol 101: 1749–1755

- Rortais A, Arnold G, Halm MP, Touffet-Briens F (2005) Modes of honeybees exposure to systemic insecticides: estimated amounts of contaminated pollen and nectar consumed by different categories of bees. Apidologie 36: 71–83
- Roulston TH, Cane JH, Buchmann SL (2000) What governs protein content of pollen: pollinator preferences, pollen–pistil interactions, or phylogeny? Ecol Monogr 70: 617–643
- Ramirez-Romero R, Desneux N, Decourtye A, Chaffiol A, Pham-Delègue MH (2008) Does Cry1Ab protein affect learning performances of the honeybee *Apis mellifera* L. (Hymenoptera, Apidae)? Ecotoxicol Environ Saf 70: 327–333
- Rembold H, Dietz A (1965) Biologically active substances in royal jelly. Vitamins and Hormones 23: 359–383.
- Rhein WV (1933) Über die Entstehung des weiblichen Dimorphismus im Bienenstaate. Wilhelm Roux Archiv für Entwicklungsmechanik der Organismen 129: 601–665.
- Saffari AM, Kevan PG, Atkinson JL (2004) A promising pollen substitute for honeybees. Am. Bee J.144: 230–231.
- Seeley TD (1985) Honeybee Ecology: A Study of Adaptation in Social Life. Princeton University Press, Princeton.
- Somerville D (2000) Honeybee nutrition and supplementary feeding. AGNOTE, NSW Agriculture. DAI/178: 1–8
- Schmidt JO, Thoenes SC, Levin MD (1987) Survival of honeybees, *Apis mellifera* (Hymenoptera: Apidae), fed various pollen sources. J Econ Entomol 80: 176–183
- Schmickl T, Crailsheim K (2001) Cannibalism and early capping: strategies of honeybee colonies in times of experimental pollen shortages. J Comp Physiol 187: 541–547
- Schmickl T, Crailsheim K (2002) How honeybees (*Apis mellifera* L.) change their broodcare behavior in response to non-foraging conditions and poor pollen conditions. Behav Ecol Sociobiol 51: 415–425
- Seeley TD, Visscher PK (1985) Survival of honeybees in cold climates: the critical timing of colony growth and reproduction, Ecol. Entomol 10: 81–88
- Severson DW, Erickson EH (1984) Honeybee (Hymenoptera: Apidae) colony performance in relation to supplemental carbohydrates. J Econ Entomol 77: 1473–1478
- Szymas B, Jedruszuk A (2003) The influence of different diets on hemocytes of adult worker honeybees, *Apis mellifera*. Apidologie 34: 97–102
- Somerville DC, Nicol HI (2006) Crude protein and amino acid composition of honeybee collected pollen pellets from south-east Australia and a note on laboratory disparity. Aust J Exp Agr 46: 141–149
- Standifer LN, Moeller FE, Kauffeld NM, Herbert EW, Shimanuki H (1977) Supplemental feeding of honeybee colonies. USDA Agr Inform Bull 413: 8
- Svoboda JA, Herbert, EW, Thompson MJ, Feldlaufer MF (1986) Selective sterol transfer in the honeybee: Its significance and relationship to other hymenoptera. Lipids 21: 97–101
- Svoboda JA, Thompson MJ, Herbert EW, Shortino TJ, Szczepanik-Vanleeuwen PA (1982) Utilization and metabolism of dietary sterols in the honeybee and the yellow fever mosquito. Lipids 17: 220–225

- Schulz DJ, Huang ZY, Robinson GE (1998) Effect of colony food shortage on the behavioral development of the honeybee, *Apis mellifera*. Behav Ecol Sociobiol 42: 295–303
- Singh RP, Singh PN (1996) Amino acid and lipid spectra of larvae of honeybee (*Apis cerana* Fabr) feeding on mustard pollen. Apidologie 27: 21–28
- Vásquez A, Olofsson TC (2009) The lactic acid bacteria involved in the production of bee pollen and bee bread. J Apic Res 48: 189–195
- VanEngelsdorp D, Hayes J, Underwood RM, Pettis JS (2010) A survey of honeybee colony losses in the United States, fall 2008 to spring 2009. J Apic Res 49: 7–14
- Van der Steen J. (2007) Effect of a home-made pollen substitute on honeybee colony development. J Apic Res 46: 114–119
- Wille H, Wille M, Kilchenmann V, Imdorf A, Bühlmann G (1985) Pollenernte und Massenwechsel von drei *Apis mellifera*-Völkernauf demselben Bienenstand in zwei aufeinanderfolgenden Jahren. Rev Suisse Zool 92: 897–914
- Zahra A, Talal M (2008) Impact of pollen supplements and vitamins on the development of hypopharyngeal glands and brood area in honeybees. J. Apic. Sci. 52(2): 5–12

11

Plant Biochemicals in Insect Pollination for Crop Improvement

Swagatika Sahoo[1], Prabhu Prasanna Pradhan[2] and Rudra Narayan Borkakati[3]

ABSTRACT

Pollination in plants is essential for the existence of many species, including the plant, the pollinators, and ultimately us – the human beings. Pollination contributes to more than one-third of crop yield worldwide. Pollination by animals is a key ecosystem service and the list is long and diverse, bees (insects) top the list. They need some cues or rewards from plants for pollinating them. Plants produce a wide range of specialized metabolites that serve a variety of functions throughout their life cycle and some of these metabolites are essential components of the aroma and flavor of flowers and fruits. The emitted biochemical compounds by plants or their parts mediate plant-pollinator interactions and this helps in establishing both positive and negative relationships. So, for optimal pollination, a better understanding of the role of the metabolic biochemicals in plants is needed and should be considered for exploring the strategies for maintaining the plant-pollinator interaction, and crop improvement. In this chapter, we have highlighted the various biochemicals responsible for pollinators attraction to plants and some ways to exploit it for increasing pollination thereby its productivity.

Keywords • *Pollination* • *Biochemicals* • *Crop improvement* • *Metabolites*

[1]Department of Entomology, Odisha University of Agriculture and Technology, Bhubaneswar, Odisha, India

e-mail: swagatikasahoo81@gmail.com

[2]Department of Entomology, Odisha University of Agriculture and Technology, Bhubaneswar, Odisha, India

e-mail: prabhuprasannapradhan@gmail.com

[3]AICRP on Biological Control, AAU, Jorhat Centre

Department of Entomology, Assam Agricultural University, Jorhat-785013

E-mail: rnbk.agri@gmail.com

11.1 INTRODUCTION

The result of crop pollination is the source of all the food we consume today. It is crucial to the process of plant reproduction. And this pollination can be defined as the transfer of pollen (male) from the anther of a flower to a stigma (female), where it germinates and fertilizes an ovule to produce seed (Inouye, 2013). The stability of agricultural food production appears to be largely dependent on pollination, which accounts for more than one-third of crop yield globally (Klatt et al., 2013). The evolution of flowers has largely been associated with the behavior of pollinators that are rewarded with nectar and pollen when they help plants to reproduce (Lucas-Barbosa et al., 2016). The self-incompatible plants, cross-pollinated ones, and even the plants in which the transfer of pollen from one flower to another flower occurs in the same crop (geitonogamy), require external agents for pollination. That is accomplished either through abiotic agents like wind and water or biotic ones like animals, birds, insects, bats, etc. Most angiosperms are self-incompatible and require pollen from one flower to be carried by an animal to the style of a second for successful fertilization (Harborne, 2001). The relationship between the plants and pollinators is commonly assumed to be mutualistic, with the plants benefiting from pollen transfer and the pollinators receiving a reward.

11.2 WHY INSECT POLLINATORS

Pollination by animals is a key ecosystem service. The value per hectare of crops that depend on insect pollinators for their production was on an average much higher than that of crops not pollinated by insects (Palmer et al., 2009). The list of species of animals that serve as pollinators is long and diverse. The pollinators are insects (the largest group), both flying and nonflying mammals, birds, and at least one reptile have been recorded as pollinators. Although flower visitation has been observed by species from at least 16 orders of insects, only four species, out of them regularly pollinate flowers that seem to have been involved in coevolutionary interactions with plants. These are the beetles (Coleoptera), flies (Diptera), butterflies and moths (Lepidoptera), and ants, bees, and wasps (Hymenoptera). Also, the thrips (Thysanoptera) which are pollen eating species that may be pollinators, some stoneflies (Plecoptera) and true bugs (Hemiptera) visit flowers and eat pollen and nectar, and some lacewing (Neuroptera), scorpionfly (Mecoptera), and caddisfly (Trichoptera) species eat nectar. Additional work is needed to confirm whether most of these lesser-known flower visitors are indeed pollinators (Inouye, 2013).

The world's most staple nourishments, including wheat, corn, and rice (comprising 65% of global food production) reproduce without insect pollination, still leaving as much as 35% depending on insects for pollination (Klein et al., 2007). Besides this most meat and dairy too rely on bees for pollination which play a significant role in the pollination of clover and Lucerne (Dias et al., 1999). As many as 2,00,000 different animal species are thought to act as pollinators, though some estimates put this figure as high as 3,50,000. Around 1000 are vertebrates, including small mammals (such as bats) and birds. Insects are by far the largest group and bees are the most important group of pollinating insects. Insects are the dominant pollinators in agricultural systems and bees top this list, providing pollination services for many of our crops.

Bees are a highly diverse group. Over 20,000 species have been described globally. 80 percent of crop pollination is provided by only two percent of bee species and threatened or rare species are rarely observed in crop fields (Kleijn et al., 2015). Recent estimates of the importance of pollinators in agriculture, Melathopoulos et al. (2015) noted that pollinator dependence of any single crop is confounded by effects of variety (genotype), environment, and management practices. Nevertheless, without wild and managed bees, various fruit and nut crops would be unavailable and other crops would be less abundant or more costly. Honeybees, mainly *Apis mellifera*, is an economically valuable pollinator of crop monocultures worldwide and the maximum yield of some fruit, seed, and nut crops depends without these pollinators (Klein et al., 2007). Over 50 million hectares under crops are benefited by the bee pollination services in India and due to this yields have been enhanced in oilseeds, pulse crops, vegetables, and fruits (Tanda, 2021). These pollinators (bees) not only encourage the yield of the crop but also induce a longer shelf-life by maintaining their firmness as seen in strawberries (Klatt et al., 2014), oriental melons (Shin et al., 2007), cucumber (Gajc-Wolska et al., 2011), blueberries (Isaacs and Kirk, 2010) and coffee (Klein et al., 2003). Land-use changes like urbanization, the intensification of agriculture, and other factors have negatively affected local pollinator populations in different parts of the world (Garibaldi et al., 2011). 70% of the world's crops benefit from animal pollination (Klein et al., 2007), but due to agricultural intensification there is decline of wild pollinators in many parts of the world (Kremen et al., 2002; Potts et al., 2010). More than a decade of active pollination research has led to greatly improved general understanding of animal pollination benefits to crop yields worldwide (Bartomeus et al., 2014).

11.3 PLANT BIOCHEMICALS

Plants produce both primary metabolites (amino acids, organic acids, or nucleosides) which are essential for their metabolism, and secondary metabolites (phenols, terpenoids, alkaloids, etc.) which are responsible for defense, flavor in fruits, pollination, or sometimes act as an attractant to herbivores too. Every plant tends to produce different secondary metabolites–pigments, scents, essential oils, and volatiles to attract their favored pollinator. Pollination in angiosperms requires pollinator services and, in return, they are rewarded (Raguso and Willis, 2005). Along with the production of floral fragrances to attract pollinators, these chemicals are responsible for providing defense against herbivores and pathogens (Tan and Nishida, 2012). Floral scent or "floral bouquet" which is the combination of volatile compounds of the different chemical groups is considered a key attribute in the evolution of the plant-pollinator interactions (Wright and Schiestl, 2009). It is the biochemistry of these chemical attractants, which determine the fidelity of a pollinating animal to visit a particular plant species (Harborne, 2001). So far, more than 1700 floral volatile compounds have been identified in approximately 900 species of 90 angiosperms (Knudsen et al., 2006) and they are grouped as terpenoids, benzenoids / phenylpropanoids, and fatty acids derivatives (Muhlemann et al., 2014; Ramya et al., 2017). Most pollinator species orient themselves by exploiting visual and odor cues simultaneously, nevertheless colors and odors of flowers can be equally important in determining pollinator preferences (Burger et al., 2010). Attraction and fidelity to a particular plant species, however, are enhanced by the co-occurrence of nectar and floral traits (odors) (Wright and Schiestl,

2009). So, it is conceivable that secondary metabolites might also act as attractants and cues for pollinators, thus acting as chemical filters and enhancing pollination behaviors (Couvillon et al., 2015).

It is still unclear about the presence of secondary metabolites nectar (Heil, 2011). They could be transported from phloem or xylem through nectary cells in a similar way like carbohydrates. In buckwheat, trichomes in nectary glands secrete sugars into nectaries from phloem via nectar parenchyma (Cawoy et al., 2008). Furthermore, the nectar, anthers, corollas, stems, and pollen of *different plants* contain some chemicals differing in their concentration.

The term secondary metabolite describes natural chemicals produced by plants, fungi, and other organisms, which are not used in primary metabolic pathways (Pichersky and Gang, 2000). It is possible that their occurrence in nectar is regulated for ecological functions: to enhance pollination service or to protect the flower and/or pollinator rewards (Adler, 2001; Irwin et al., 2014; Manson et al., 2012). Pollinators impose selection pressure on plants, especially on floral traits, so it is reasonable to expect that pollination and chemical defense may not have evolved independently (Campbell et al., 2015).

Aromatic volatile compounds such as benzenoids have been widely associated with pollinator attraction (Knudsen & Gershenzon, 2006). Knudsen et al. (2006), listed 12 common compounds, namely, limonene, (E)-ocimene, myrcene, linalool α and β-pinene, benzaldehyde, methyl 2- hydroxybenzoate, benzyl alcohol, 2-phenyl ethanol, caryophyllene, and 6-methyl-5-hepten-2-one which are detected in floral scent from over 50% of seed plant families. Changes in the biosynthesis of one of these groups of compounds may cause changes in the production of the others and alter the color-odor combination in flowers. Specificity in color-odor combinations may be influenced by phenotypic changes induced upon pollinator visitation (Dormont et al., 2014). The odor or scent of a flower plays a major role as an attractant for pollinators. Odor is of special importance in night-flying insects, where the visual stimulus is lacking: bat-pollinated and moth-pollinated plants are generally strong smelling. Bees are especially responsive to fragrant scents, where the active volatile is the sesquiterpene. Insects are sensitive to even small concentrations of volatile chemicals present in flower odors. Maximum scent production is coordinated with the time when the pollen is ripe and the flower is ready for pollination (Harborne, 2001). Floral scent can be an attractant for either food (nectar and pollen) or a suitable mating site for certain moth species.

Not only the floral scent is influenced by plant volatiles and causes pollinators attraction but also the presence of pigments (found in flower tissues or chromoplasts) in flower colors attracts the pollinators by eliciting their visual cues (Abrol, 2012). The most important group of flower pigments are the flavonoids and carotenoids imparting blue color and yellow with some red color to flowers respectively. Flavonoids contribute to a high coloration range and are widely distributed in plants (Borghi et al., 2017). However, other parts of the flowers also have pigments. For instance, pollen grains, the external layer of the pollen grains that can accumulate pigments such as flavonoids, anthocyanins, and carotenoids (Lunau, 1995; Koski and Galloway, 2018). The most common color of pollen is yellow, but white, orange, blue, red, and black has also been reported (Miller et al., 2011). It is believed that the pollens (anthers) can create a strong contrast over the flowers which could increase the attraction of pollinators (Lunau, 1995).

Furthermore, fruits also have aroma (volatile compounds) which is prominent mainly at the ripening stage and this is, of course, a signal for different seed dispersers. Immature fruits also produce volatiles, but in this case, their role is mainly to protect the growing organ against pathogens and predators (Araguez and Valpuesta, 2013). Still, studies are very few regarding the role of fruits aroma in attracting pollinators or seed dispersers.

11.4 GENERAL PLANT BIOCHEMICALS FOR POLLINATOR'S ATTRACTION

The presence of plant biochemicals as floral scent plays a major role in pollination. The floral scent is an important source of communication between pollinators and flowering plants for their evolution, particularly during long-range communication (Dudareva and Pichersky, 2000; Farré-Armengol et al., 2013). The information sent by the floral volatiles depends on eliciting a distinct behavioral response and composition of their emission towards the respective pollinators. Floral biochemicals can be in the form of secondary metabolites, volatile organic compounds (VOCs), essential oils that attracts pollinators. The information sent by the floral volatiles depends on eliciting a distinct behavioral response and composition of their emission towards the respective pollinators. Some most studied examples are noted below.

11.4.1 Secondary Metabolites

Plants produce several low molecular weight organic compounds, based on their function they are divided as primary metabolites, secondary metabolites, and hormones (Erb and Kliebenstein et al., 2020). Plant secondary metabolites (PSMs) include different nitrogen and sulfur-containing metabolites, namely, alkaloids, amines, cyanogenic glucosides, glucosinolates, non-protein amino acids, organic acids, terpenoids, phenolics, quinones, polyacetylenes and peptides (War et al., 2020) and are present in the form of floral volatile organic compounds (VOCs), essential oils, nectar scents, flower color, etc. These compounds can be either defensive to insect herbivores or attractive to pollinators (Stevenson et al., 2017). Pollinators attract by the smell and color of the flowers. The color is mostly due to the production of flavonoids, anthocyanins, or carotenoids, whereas the smell is due to the terpenoids, amines and phenylpropanoids which exhibit distinctive odors that are recognized by pollinators (Wink, 2018; Afroz et al., 2021).

Presence of secondary metabolites in nectar benefits pollinators through antimicrobial activity. Pyridine alkaloids in *Nicotiana* nectar affect the richness and composition of its bacterial communities, having nicotine with the strongest antimicrobial effect (Aizenberg-Gershtein et al., 2015). There is some evidence that these secondary metabolites in nectar could increase pollinators resistance to parasite and pathogen infection. As studied by Manson et al. (2010), the presence of gelsemine from *G. sempervirens*, given in artificial nectar to *Bombus impatiens* reduced pathogen loads after infection. Responses to secondary metabolites in nectar are highly dependent on the concentrations and the pollinator species (Stevenson et al., 2017).

Pollinators know to find flowers with high quality nectar, and are also aware of visiting the flowers with toxic nectar (secondary compounds) (Wright et al., 2010).

11.4.1.1 Volatile organic compounds (VOCs)

Volatile organic compounds (VOCs) play vital roles in pollinator attraction, defense, and interaction with the environment (Ramya et al., 2020). Synthesis of VOCs occur in all plant organs, including roots, stems, leaves, seeds, fruits, as well as flowers, which are reported to emit the largest amounts and diversity of VOCs (Knudsen, 2006). Floral volatiles play a crucial role in reproductive process of plants by attracting the pollinators (Piechulla and Pott, 2013). Floral scent bouquets may contain from one to 100 volatiles, but most species emit between 20 and 60 different compounds (Knudsen and Gershenzon, 2006). Volatile compounds are synthesized via biochemical pathways, and various forms of enzymatic modifications such as hydroxylations, acetylations, and methylations (Dudareva et al., 2004). VOCs released from fruits and seeds attract various animal dispersers and are essential for the spatial dynamics of plant populations within ecosystems (Dudareva et al., 2013). Volatiles emitted from flowers not only attract insects for food availability, but also provide mating and oviposition opportunities. Certain orchid species employ flower scent to imitate pheromone blends of female pollinators, thereby triggers the copulation attempts of male pollinators with flowers and cause pollination (Schiestl et al., 1999). Pollinators use their multisensory inputs like thermal, olfactory, electromagnetic, visual senses for locating food and the site of breeding. Potential pollinators quickly identify and locate the scented flowers, promoting association between pollinators and plants via individual ratios of general compounds or special compounds (Wright and Schiestl, 2009). Floral scents are grouped into three major clusters: terpenoids, phenylpropanoids, and fatty acid derivatives based on origin, biosynthesis, and function (Ramya et al., 2020).

Terpenoids are the most dominant group of floral VOCs including 556 scent compounds (Muhlemann et al., 2014). Terpenoids as floral scent are a major cue for attracting pollinators (animals, insects, mammals, birds, and bats) and serve as agents for transferring pollens (Abrol, 2012; Farré-Armengol et al., 2015). Geraniol, linalool and their derivatives are some major components found in scented *Phalaenopsis bellina* (orchid) flowers (Ramya et al., 2020). Floral scent emission contributes in guiding pollinators to flowers, especially the night-emitting plants, which produce a high intensity of scent to prevail over the low conspicuousness of flowers in illumination. For example, *Petunia axillaris* emits larger amounts of volatile terpenoids than day-emitting bee-pollinated plants of the same genus *Petunia integrifolia* (Waelti et al., 2008; Dudareva et al., 2013). Terpenoid volatiles are not only emitted by petals but also by pollens as pollen odors, such as β-ocimene, a universal attractant emitted by plants that attract long-range pollinators (Knudsen et al., 2006), such as bees, moths, butterflies, and beetles (Okamoto et al., 2007), thus benefitting pollinator foraging. They also have been found to attract bumblebees and honeybees (Granero et al., 2005). Volatile terpenoids attract certain pollinators by releasing pheromones; for example, orchids employ floral scent that mimics a blend of pheromones from female insect pollinators, triggering copulation attempts by male pollinators with flowers (Bohman et al., 2014).

Phenylpropanoids and benzenoids are the second most dominant group of volatile compounds (Knudsen et al., 2006). By undergoing various biosynthetic pathways, they release floral scents like hydroxycinnamic acids, aldehydes, alcohols, eugenol and metyl eugenol (Dudareva et al., 2006). Moth pollinated plants emit high amounts of benzenoids and, while, terpenoids and nitrogen containing to a lesser extent (Dobson, 2006). Petunia flowers specifically emit deterrent compounds such as isoeugenol and benzyl benzoate, to avoid florivory (Dudareva et al., 2103). Presence of phenylacetaldehyde in closely related species of *Silene dioica* and *S. latifolia*, emits blend of similar constituents but in different relative amounts which influenced reproductive isolation between two plant species (Waelti et al., 2008). Floral volatiles such as phenylpropanoids and benzenoids are also emitted from scentless flowers of *Phalaenopsis equestris* (Huber et al., 2005). In flowers, the diversity of phenylpropanoid/ benzenoid compounds is further increased by modifications such as by methylation, hydroxylation, and acetylation of direct scent precursors and enhancing the volatility of scent compounds (Gupta et al., 2012; Muhlemann et al., 2014).

Fatty acid derivatives constitute the third class of flower VOCs, which derive from the unsaturated C18 fatty acids, linolenic, and linoleic acid (Muhlemann et al., 2014). Aldehydes and alcohols are important constituents in the floral volatile bouquet of plant species such as carnation and wild snapdragon (Suchet et al., 2011). Information regarding the synthesis/ pathways of fatty acids and/or their derivatives in flowers is limited (Ramya et al., 2020). Flowers of *Antirrhinum majus* have 20 fatty acid derivatives (Suchet et al., 2011). Furthermore, biosynthesis of fatty acids yields methyl jasmonate and jasmonic acid as derivatives (Muhlemann et al., 2014), in the floral scent pathway of *C. ensifolium* and *C. faberi*. Various volatile fatty acids were also found in the orchid genus *Ophrys* in which alkenes have an important role as mediators in attracting pollinators (Borg-Karlson et al., 1993; Muhlemann et al., 2014).

More about the complex chemistry between floral volatiles and pollinators is yet to be unraveled. Biosynthetic pathways and regulatory mechanisms that are responsible the production of floral scents in different plant species, is essential for better understanding the relevant genes involved and its utilisation for improving pollination technology through molecular and breeding approaches.

11.4.1.2 Essential oils

Essential oil is any concentrated, hydrophobic liquid containing volatile aroma compounds, produced by plants. Essential oils are synthesized in various parts of plants, including leaves and stems (e.g., fennel, parsley, tarragon, rosemary, basil, mints, sage, wintergreen, spicebush, eucalyptus, pine, lemongrass, bay, oregano), seeds (e.g., almond, anise, celery, cumin), berries (e.g., juniper, allspice), bark (e.g., cinnamon, sassafras), fruit peel or rind (e.g., grapefruit, lemon, citron, orange, lime), roots (e.g., valerian), rhizomes (e.g., ginger), and flowers (e.g., chamomile, clove, geranium, jasmine, lavender, orange, and rose). The flowers of many plant species attract pollinators by producing different complex mixtures of essential oil compounds within the various floral organs (i.e., stigma, style, ovary, filaments, petals, sepals, and/or nectaries) or in special scent gland tissues (called osmophores) most commonly located on the epidermal cells of the petals. It is the combination of the constituents of this scent mixture that gives each flowering plant species a unique fragrance (Galen, 1985; Dodson, 1995).

More than 1700 individual aroma compounds have been identified so far from over 990 taxa belonging to 90 families and 38 orders. The two largest groups are the terpenoid compounds and the aliphatic compounds (Cseke et al., 2007). Indole and 30 other nitrogenous compounds were identified, to which over 30 insect species, were of Coleoptera and Diptera, (Skubatz et al., 1996). The amount of floral compounds produced varies from low picograms to more than 30 micrograms per hour (Knudsen et al., 2004). The flowers of many beetle and moth pollinated plants produce the highest quantities of scent compounds. The quality and quantity of floral scent composition vary within and between plant species, and such variation allows the sensory mechanisms of potential pollinators to perceive differences between species, even from a great distance.

There are some chemicals of floral scents linked to the primary pollinator group, such as methoxylated benzenoid compounds attracting tropical scarab beetles, various terpenoids for bees and wasps, benzenoids (phenylacetaldehyde, 2-phenyl ethanol, benzaldehyde, benzyl alcohol), terpenoids (linalool, *trans*-β-ocimene) linked with butterflies and benzenoids (phenylacetaldehyde, benzaldehyde, esters), terpenoids (linalool, β-ocimene, lilac compounds attracting nocturnal moths (as reviewed by Cseke et al., 2007).

Some of the plants release floral volatiles which have distasteful odors (fecal-smelling indole) and foul-smelling methyl derivative (skatole) and both of these have been reported in the nauseating odor of the voodoo lily, *Sauromatum guttatum* (Borg-Karlson et al., 1994). ß-Ionone (terpenoid) is a natural plant volatile compound. It is found in the essential oil of *Osmanthus fragrans*, Petunia hybrid, and *Rosa bourboniana*, and fruits and vegetables, such as carrot, tomato, melon, raspberry, apricot, plum, apple, and fig. It provides flavor, aroma, pigment, defensive compound or as a cue for insect attractant. β-Ionone was found to be a significant volatile organic compound (VOC) in melon and watermelon, as a flavouring agent of tomato, stone fruits, plum (*Prunus domestica*). It is an essential contributor to fragrance in the flowers of *Boronia megastigma*, in petunia (*Petunia hybrida*), and in the yellow and red *Lawsonia inermis* L. (henna) flowers (as reviewed by Paparella et al., 2021).

There is a generalist pollination syndrome, where the flowers are pollinated by a diversity of insects (beetles, flies, bees, butterflies) that feed on the exposed nectar and pollen (Proctor, 1996). The plant families displaying this pollination syndrome include Apiaceae, Arecaceae, Rosaceae, and Ranunculaceae. Coleopterans (beetles) often feed on pollen, floral tissues, and other floral exudates (Endress, 1994). Diptera (flies) is also an important order of flower pollinators, contributing to a major portion of the pollinators at higher elevations and latitudes, where they replace the small bees that are most prevalent at lower altitudes (Eberling and Olesen, 1999). Thrip-pollinated flowers tend to be of medium size, white to yellow, have floral structures that provide shelter, and are sweetly scented (Mound, 2005). Perhaps the best-known insect pollinators are bees and wasps, pollination by bees melittophily, gets attracted to immensely varying floral morphology and color, as well as fragrance, with no obvious trends emerging in scent chemistry (Endress, 1994; Proctor, 1996). Flowers pollinated by nocturnal moths are usually characterized as having nocturnal anthesis, nectar in floral tubes or spurs, light color to be seen at night, and a generally pleasant and often very strong scent containing acyclic terpene alcohols (e.g., linalool), benzenoid compounds, and some nitrogen-containing compounds. Adult butterflies visit the flowers in which the floral scents are weak, fresh, and sweet (Cseke, 2007).

Insects have evolved considerably more advanced mechanisms with which to distinguish between the different constituents of the floral scent mixtures coming from diverse floral species. The coevolution between essential oil production in the flowers of plants and the highly specific sensing/detection systems in insects for these scent compounds has resulted in highly successful and highly specific pollination syndromes (Cseke et al., 2007). All those above specific chemicals attracting specific insects can be utilized in a way either by natural extraction or by artificial production for attracting the pollinators to those plants which does not have specific floral scents.

Different varieties of crops differ in the emission of flower volatile compounds, which causes differences in bee visitation rates and can be expected to affect pollination success and thereby, yield and quality. Hence, breeding of crop varieties should focus more closely on flower volatiles, to increase fruit set and thus the market value of fruits via the attraction of a wider range of pollinator species (Dotterl et al., 2010; Klatt et al., 2013).

A current initiative of plant breeders is to restore and/or alter floral scents, especially because of public demand, commercial potential, and the need to restore the attraction of diverse kinds of pollinators to improve the productivity of various crops. One relatively new field devoted to controlling how flowers smell is called "scent engineering" (Pichersky and Dudareva, 2007). Many groups of investigators are now beginning to focus on "scent genes" to understand how the expression of these genes can be used to manipulate floral scent and essential oil production. Some examples are, introduction of the *LIS* gene into transgenic Petunia (*Petunia hybrida*) (Lücker et al., 2001) and carnation (*Dianthis caryophyllus*) (Lavy et al., 2002) to produce linalool from leaves or flowers, introduction of three lemons (*Citrus* × *Limon*) terpenoid synthases in tobacco (*Nicotiana tabacum*) flowers and leaves, using the constitutive 35S promoter, for the emission of native terpenoids (β-pinene, limonene, and γ-terpinene) and production of (+)-*trans*-isotranspipiterinol from (+)-limonene by introducing mint (*Mentha* spp.) limonene-3-hydroxylase genes into those transgenic tobacco plants (Lücker et al., 2004). The metabolic pathways and the genes that regulate the synthesis of the enzymes that produce terpenes, phenylpropanoids, or fatty acid derivatives, are the largest and best understood of the scent compound categories (Cseke et al., 2007).

11.5 EFFECT OF PLANT SECONDARY METABOLITES ON POLLINATORS

Attraction and fidelity of pollinators to a particular plant species is enhanced by the co-occurrence of nectar and floral traits (odors) (Wright & Schiestl., 2009; Kessler et al., 2015a). One of the major functions of floral volatiles is the attraction of pollinators, either through triggering innate behavioral responses or by providing a stimulus that can be associated with the presence of rewards (Schiestl and Johnson, 2013). Thus, plant secondary metabolites might also act as attractants and cues for pollinators, which would enhance pollination behaviors (Couvillon et al., 2015). Pollinator attraction is often specific and specificity increases the efficiency of pollen transfer between conspecific stigmas. Specificity in pollination can be achieved with floral filters (Raguso, 2008), which means through specialized morphology that allows access only to a few potential pollen vectors and provide rewards. Floral scent is an important component of the floral phenotype and represents a decisive communication channel between plants and animals. It facilitates the attraction of pollinators and promotes pollinator specificity by the intensity of the signal, the presence of unique VOCs, and exclusive multicomponent blends of ubiquitous compounds (Raguso, 2008; Schiestl and Dotterl, 2012; Farre-Armengol et al., 2013). Therefore, floral scent is typically more important in encoding highly specific signals, because of its chemical complexity, and the high number of olfactory neurons in insects (Haverkamp et al., 2018).

Flowers attract pollinators through highly regulated visual and olfactory stimuli. The role of floral scent volatiles in attracting as well as eliciting landing, feeding, and in some cases mating behaviors on flower varies with each flower-animal interaction (Raguso, 2001; Weiss, 2001). The floral scent is mostly detected by insects with neurons located on the antennae, but also by the olfactory neurons present on the tip of the proboscis (Haverkamp et al., 2016). Each sensillum of antennae normally houses two to five olfactory receptor neurons (ORNs), but rarely more than 100 (Keil, 1999). The ORNs are bipolar cells connected directly to the brain (Steinbrechet, 1997). Visual and olfactory signals help guiding insects to the flowers and favor pollination, which are then consequently expected to play an important role in the ecology and evolutionary diversification of plant-pollinator interactions (Peakall et al., 2010; Schaffler et al., 2012).

The floral scent is a signal which pollinators can use to discriminate a particular flower whose nectar and/or pollen is the reward. Flowers could be identical in their color or shape but no two floral scents could be the same, as there is a large diversity of volatile compounds, their relative abundances, and interactions within the scent bouquet. Thus, in addition to attracting insects and guiding them to food resources within the flower, floral volatiles are also essential in differentiating different plant species and even individual flowers of a single species (Dudareva et al., 2006). Floral scent bouquets may contain from one to 100 volatiles, but most species emit between 20 and 60 different compounds (Knudsen and Gershenzon, 2006). Floral volatile profiles are species specific and tend to reflect the type of pollinator. For example, plant species pollinated by moths emit high amounts of benzenoids and, to a lesser extent, terpenoids and nitrogen containing compounds (Dobson, 2006). The floral scent is usually released at the right time of day for the particular pollinator, e.g., during the day for bee-pollinated flowers. For moth-pollinated species, it may be at dusk or even later in the night. Odor is of special importance in night-flying insects and other animals, where the visual stimulus is practically absent; bat-pollinated and moth-pollinated flowers are generally strong-smelling (Abrol, 2012).

The plant species which are closely related, depending on different types of insects for pollination, produce different odors, reflecting the olfactory sensitivities or preferences of the pollinators (Henderson, 1986; Raguso and Pichersky, 1995). By providing species-specific signals, flower fragrances facilitate an insect's ability to learn particular food sources, thereby increasing its foraging efficiency. Within a species, the level of scent emission changes in response to endogenous diurnal rhythms, flower age, pollination status, and environmental conditions such as light, temperature, and moisture status (Dudareva et al., 2004).

Pollinators learn to find flowers with high quality or copious amounts of nectar, and they can also learn to avoid visiting flowers with nectar containing toxic secondary compounds. This is possible as the pollinators taste the secondary metabolites in nectar or learn from after ingestion effects of toxic nectar (Wright et al., 2010). Insect pollinators detect chemicals using contact chemosensilla which respond to sugars, salts, acids, water, and non-nutrient compounds (Inoue et al., 2009; Wright et al., 2010). When neurons in this location detect toxic or bitter substances such as quinine, this leads bees to reject food (Wright et al., 2010). In situations where insect pollinators initially face difficulty in detecting the toxic secondary metabolites in nectar, they may learn to associate floral cues such as odors with the after-effects of ingesting toxins (Wright et al., 2010). Depending on the spatial distribution of the chemical compounds within a flower petal, changes induced by pollination may result in changes in the visual cues used by pollinators such as nectar guides or the overall flower color (Miosic et al., 2013).

The ability of honeybees to distinguish between cultivars expands with the increasing intensity of the floral scent, as in four different snapdragon cultivars emitting the same volatile compounds but at different levels (Wright et al., 2005). In contrast to bees, moths use odor cues over longer distances (Dobson, 1994) and therefore odor quantity has to be higher to be detected. Moth-pollinated *Clarkia breweri* flowers emit a strong sweet floral scent rich in linalool and aromatic esters in contrast to closely related bee-pollinated *Clarkia concinna* flowers which emit fewer compounds and at lower levels (Raguso and Pichersky, 1995). Similarly, hawk moth pollinated *Petunia axillaries* flowers release high levels of several compounds and elicit significantly higher responses from the *Manduca sexta* antenna than flowers of the closely related bee-pollinated *Petunia integrifolia*, which emit almost exclusively benzaldehyde (Hoballah et al., 2005).

Another, often highly specific and scent-driven, pollination system is nursery pollination, in which pollinators oviposit into flowers and the pollinators' larvae consume floral tissue of developing seeds, for example, it is seen in jackfruit (*Artocarpus heterophyllus*), which is a monoecious crop pollinated by gall midges. The insect oviposits only onto male fungus-infected inflorescences, so when the highly specific scents are released during nursery stages, it attracts the insect to oviposit in both the inflorescence (infected and un-infected) and cause pollination (Bouwmeester et al., 2019).

The presence of secondary metabolites in nectar helps in maintaining the nectar quality by preventing its spoilage because of microbes. Nectar is a rich medium for microbial growth which can be contaminated with yeasts and bacteria and transferred to the bodies of pollinators (Herrera et al., 2009; Fridman et al., 2012). Microbial enzymes metabolize glucose or fructose, leading to an imbalance in the carbohydrate ratio and hence a reduced carbohydrate reward for the pollinators (Vannette et al., 2013). Microbial contamination also alters the amino acid composition of nectar (Peay et al., 2012). These changes in nectar chemistry may have an impact on pollinator attractiveness and consequently on plant fitness (Schaeffer & Irwin, 2014). However, due to the presence of secondary metabolites in nectar, it makes a less conducible environment for pathogens. Volatiles involved in antimicrobial defense are often produced in pistils and/or nectaries, as linalool and linalool oxide in flowers of *Clarkia* species and for sesquiterpene and monoterpene formation in Arabidopsis flowers (Dudareva et al., 2006). That shows that interactions with microbes add a new level of complexity to the potential ecological functions of secondary metabolites in nectar. Aizenberg-Gershtein et al. (2015) studied that pyridine alkaloids in Nicotiana nectar has a negative effect (antimicrobial) on the richness and composition of its bacterial communities. These secondary metabolites provide some resistance to the pollinators against microbes or parasites, as investigated by Manson et al. (2010), the infection by a trypanosomatid gut parasite on bumblebee has been reduced after it fed on the nectar having gelsemine from *Gelsemium sempervirense* (yellow jasmine). Pollinators achieve some level of protection by selecting a mixed pollen diet that includes both the correct blend of nutrients and also secondary compounds that confer resistance to disease (Stevenson et al., 2017).

In nature, plants are exposed to both herbivores and pollinators at the same time, and in response to herbivores, plant metabolites release defensive chemicals which induce important phenotypic changes in flowers. Plant responses to pollination are essential in optimizing reproductive success and the herbivore-induced changes in plants may affect the attraction of pollinators because such changes can interfere with local responses to pollination, influencing the cues exploited by flower visiting insects and are needed to be studied in detail (Lucas – Barbosa et al., 2016).

Traditionally, ants were considered nectar thieves or sometimes get repelled by the plants as they emit volatile (allomones), but Clara et al. (2014) studied ants and *Cytinus* flower interaction and found that there is a mutualistic relationship between them. Floral volatiles alone are sufficient to elicit the ants to pollinate the flowers effectively and, in return, provide floral rewards. (E)-cinnamyl alcohol and (E)-cinnamaldehyde, play an important role in the attraction of pollinators to *Cytinus* flowers. Some of the volatile compounds released by *Cytinus* flowers are known to attract bees and butterfly pollinators but were not found to be potential pollinators of *Cytinus* flowers.

There are examples where pheromones and flower scents have become interwoven in insect behavior (Abrol, 2012). In the flowers of the orchids (genus *Ophrys*), one distinctive pollination by male bees of the genus *Andrena* was observed. Here, the floral volatiles have a pheromonal role, as well as being a pollen attractant. The major constituents are short-chain aliphatic compounds, monoterpenes, and some bicyclic sesquiterpenes of the cadinene series (Abrol, 2012). As the orchid flower is having the same shape, same color, and same scent as the female bee, so there happens pseudocopulation by the male bee (exclusively). In other words, there is a degree of matching between some of the floral constituents and the pheromonal volatiles of the female bee (Bergstorm, 1978; Borg-Karlson and Tengo, 1986) and that is 'Dufour's gland' in female bees secretions contain open-chain mono- and sesquiterpene esters. Release of methyl eugenol from the blossoms of golden flower tree *Cassia fistulosa* (Leguminosae) and *Zieria srnithii* (Rutaceae) for the attraction of oriental fruit fly (*Dacus orientalis*), which uses the same volatile as sex pheromone (Abrol, 2012).

Sometimes it also happens that the insects get 'hooked' by the floral scents of some specific flowers. As it is noted in flowers of *Brugmansia × candida* releasing some attractive compounds present in the nectar affect the hawkmoth pollinator. The moth becomes erratic in flight after visiting the flowers and shows all the signs of being 'hooked' by the plant (Harborne, 2001).

The perceptual properties of any mixture of volatile compounds rely on the perceptual properties of their constituents, their relative concentrations, the intensity of the scent that arises from all the compounds within a bouquet, and also on the previous scent experiences of the pollinator (Wright and Smith, 2004a, 2004b). There is scarce information about how insects respond to individual components found in floral scents, even though it is known that they are able to distinguish between complex floral scent mixtures. Also, it is still unclear whether insect pollinators use only a few compounds present in a scent for floral identification, or whether they use information from all the scent compounds (Dudareva et al., 2006).

Talking about the flower color induced from the pigments of the plants, it was observed that a pollinator in daylight approaches the flower by discriminating between plant species based on flower color and floral structure. Evidence for visual discrimination by insects can be seen in flowers that change color after pollination. As seen in *Lantana camara*, the color changes from yellow (carotenoid) to red (anthocyanin). The pollinator may be aware of the pollinated red flowers, from a distance but as it approaches it selectively visits only the nectar-rich yellow flowers (Weiss, 1991). In the larkspur, *Delphinium nelsonii*, the white mutants have inferior nectar guides and it takes the bees longer to locate for the nectar. The pollinators in this case, experience lower net rates of energy intake and concentrate their attention on blue flowers (Waser and Price, 1983). These white mutants are unable to maintain themselves, their seed set and viability become poor as they got discriminated or not preferred against the colored variants.

Most of the insects including bees have a trichromatic color vision; flies and beetles are dichromatic; and other insects such as butterflies can be trichromatic, tetrachromatic or pentachromatic (Shrestha et al., 2013; Briscoe and Chittka, 2001). Some plant species produce more than one flower color to attract pollinators. In *Pedicularis* (Scrophulariaceae), the scarlet corollas attract hummingbirds, while the magenta calyces and bracts, with high ultraviolet and reflective hairs, attract bumblebees. In the Scarlet gilia, *Ipomopsis aggregata*, there are both red flowers to attract hummingbirds and white flowers to attract hawkmoths. Color shifting in these plants occurs seasonally (Paige and Whitham, 1985). Some examples of color preferences by pollinators are: bees to yellow and blue intense color; beetles to dull; cream or greenish color; butterflies to vivid red and purple color; flies to dull brown, purple, or green; moths to red, purple, white, or pale pink and wasps to brown color (Harborne, 2001). This is only a general guide otherwise there are exceptions. Floral scent and floral color attracting insect pollinators are summarised in Table 11.1.

The color of the nectar is also due to some secondary metabolites of the plant (Sosenski and Parra-Tabla, 2019). It is also stated by some researchers that colored nectar can be an honest reward signal for the most effective pollinators, which could increase the efficiency of pollen transfer and the fitness of plants (Olesen et al., 1998; Zhang et al., 2012). It has been proposed that colored nectar could repel some other floral visitors. For example, while the dark nectar of *Aloe vryheidensis* seems to visually attract birds that pollinate its flowers but also repels nectar robbers due to its bitter taste (Johnson et al., 2006). It has been observed that bees learn quicker the color of a flower whose nectar contains nicotine than the color of another flower that lacks nicotine even though it contains the same composition and concentration (Wright et al., 2013; Baracchi et al., 2017).

Table 11.1 Proposed chemical profiles of floral scents and flower color attracting primary insect pollinators

Insect type		Chemical profile of floral scents	Color of flower
General pollinators	Diverse insects	Fatty acid derivatives, terpenoids and benzenoids (terpenoids are dominant) (Dobson, 2006)	–
Bees and wasps	Foraging bees	Mainly terpenoids, low amounts of fatty acid derivatives and benzenoids (Dobson, 2006)	Bees-white and yellow to blue (Dobson, 2006)
	Male bees	Benzenoid and monoterpenoids (*p*-cymene, limonene, camphene and myrcene) (Dobson, 2006)	–
	Fig wasps	Limonene, β-caryophyllene, alkanes (Dobson, 2006) Benzenoids and Terpenoids (Okamota and Su, 2021)	Wasps-dull color (green, reddish brown) (Dobson, 2006)
	Nectar-seeking wasps	Eugenol, monoterpenoids, benzyl alcohol, benzaldehyde (Dobson, 2006)	
Thrips	Generalist flower inhabiting thrips - *Frankliniella occidentalis*	Phenylpropanoids, monoterpenes, sesquiterpenes, few fatty-acid derivatives and amines. (Koschier, 2008)	White to yellow (Mound, 2005)
Flies	Food-seeking flies	Fatty-acid derivatices, alcohols and N-containing compounds	In general-pale and dull to dark brown or purple (Dobson, 2006 and USDA forest service)
	Male fruit flies	Methyl eugenol	Hover fly-white and yellow flowers (Kleca et al., 2018)
	Carrion flies	Oligosulphides (Johnson, 2016), alochols, ketones (Cseke et al., 2007)	
Moths and butterflies	Micropterygid moths	Fatty-acid-derived esters (Cseke et al., 2007)	
	Yucca moth	Fatty-acid derivative hydrocarbons, alcohols and sesquiterpenes (Cseke et al., 2007)	
	Nocturnal moths	Benzenoids (phenylacetaldehyde, benzaldehy-date,esters), terpenoids (linalool, lilac compounds), sometimes fatty-acid derivatives and N-compounds	Moths-Light colored flowers (Dobson, 2006)
	Butterflies	Benzenoids (benzaldehyde, benzyl alcohol, phenylacetaldehyde, methyl eugenol), terpenoids (linalool, trans-β-ocimene) (Dobson, 2006)	Bright colors like red, yellow, orange, pink and purple (Andersson, 2003, nwf.org)
Coleoptera	Pollinating beetle like scarabaeids	fatty-acid derived esters, benzoid esters Methoxylated benzoid compounds, terpenoids	Dull color- cream, green or brownish (Dobson, 2006)

11.6 NEGATIVE CONSEQUENCES

The chemical compounds in nectar and pollen can have a range of effects on pollinator preference and performance, from negative to neutral to positive. The toxic consequence of post-ingestion of nectar and their repellence by a plant before feeding can be considered a negative reaction by the pollinator. Secondary compounds, can have deleterious consequences such as reduced mobility, ovary development, and survivorship (Irwin, 2014). Sometimes, it depends upon the concentration of the chemical compounds according to which it will affect the pollinator. Honeybees and bumblebees show preference for sucrose solutions containing concentrations of nicotine <15 μ but repel if present in more concentration (Singaravelan et al., 2005). Responses to secondary metabolites are highly dependent on the concentrations of secondary metabolites in the nectar (Singaravelan et al., 2005; Manson et al., 2013) and the pollinator species (Tiedeken et al., 2016). For example, caffeine occurs in nectar and improves honeybee memory for odors associated with food rewards which enhances pollen transfer at naturally occurring concentrations but is repellent to honeybees at higher concentrations (Stevenson et al., 2017). Looking into the reproduction-defense trade-off, the production of herbivore-induced plant volatiles attract carnivorous insects which can conflict with the attraction of pollinators that are essential for the plant reproductive success (Lucas-Barbosa et al., 2016).

Recent studies have demonstrated that leaf herbivory alters floral VOC emissions, usually with negative consequences for pollinator attraction. In wild tomato, this effect is jasmonate-mediated, and therefore floral VOCs triggering resistance responses creates challenges for the attraction of pollinators (Kessler et al., 2011). It was observed in glasshouse tomato flowers primarily emit four monoterpenes (β-phellandrene, α-pinene, p-cymene, (+)-2- carene), all of which are known as herbivore-induced leaf volatiles, used in defense and possess toxic properties. Emission of these compounds was negatively correlated with bumblebee visitation, suggesting that this largely repellent-dominated floral VOC blend caused pollination failure in glasshouse tomatoes (Morse et al., 2012). However, there might be very few breeding programs that exist to select floral fragrances that are 'pollinator-friendly' (Dudareva et al., 2013).

Further, some ecological complications are also a hindrance to plant fitness. As in an experiment on the wild gourd, *Cucurbita pepo* var. Texana, the floral emission of 1,4-dimethoxybenzene was elevated and resulted as the most attractive compound to specialist squash bee pollinators but decreased plant reproduction as it concomitantly attracted florivorous beetles (Theis & Adler, 2012).

11.7 METABOLIC RESISTANCE TO THE TOXIC METABOLITES OF NECTAR

There are several biochemical and physiological mechanisms in floral visitors (pollinators) for dealing with xenobiotics after their ingestion (Despres et al., 2007; Irwin et al., 2014). The most important mechanisms involve the metabolism and elimination of these toxic compounds. As reviewed by Berenbaum & Johnson (2015) detoxification pathways in insects are of three phases, phase I (oxidation to render the toxin more soluble), phase II (conjugation to a carrier molecule), and phase III (excretion). The most common enzymes responsible for this, are cytochrome P450 monooxygenases (P450s), glutathione transferases (GSTs), and carboxylesterases, as reported in honeybees by Claudianos et al. (2006). Snyder et al. (1994), reported honeybees convert nicotine into several phase I metabolites with 20C-oxidation of nicotine being the main catabolic pathway. Pollinators may also have enzymes in saliva, the crop, or the gut that breaks down secondary metabolites. Phenolics in the nectar of *Aloe littoralis* are lost during processing into honey (Liu et al., 2007).

'Social detoxification' is seen in honeybees which refers to the behavior of detoxifying the nectar. It is done by mixing the pollen and nectar, collected from various sources, and colony food processing by microbial fermentation that causes dilution and thus reduces the quantity of some toxins needed for detoxification, as explained by Berenbaum & Johnson (2015).

11.8 FLORAL METABOLITE CUES OF SOME COMMON CROPS FOR POLLINATORS

Caffeine present in nectars of coffee and citrus plants, improves honeybee memory of a reward (nectar) and its associated cue (odor), suggesting caffeine encourages bees to make repeat visits to flowers of both plant genera (Wright et al., 2013) and as a result, will help in increasing the crop yield. Carter et al. (2006), suggested that a wild relative of soybean (*Glycine* spp.) influences bee foraging due to the presence of high levels of proline in nectar. In wild *Brassica rapa*, the amount of the floral volatile phenylacetaldehyde was correlated with floral reward (sugar and pollen per flower), and bumblebees learned a positive response to the volatile after foraging on plants (Knauer and Schiestl, 2015). Preference for higher levels of floral volatiles in strawberries attracted wild bees (*Osmia* spp.), but the correlation of volatiles with reward was not tested (Holzschuh et al., 2012; Klatt et al., 2013). The floral pigments also influence pollinators in foraging, as in apple cultivars appearance of flowers is important for pollination; when white-flowered apples were planted with several crabapples as pollen donors, honeybees showed a strong preference for white crabapples (Mayer et al., 1989). Wild *Brassica rapa* and oilseed rape vary for the presence of nectar guides, an ultraviolet floral pattern visible to bees, which increase pollinator visits to plants (Brock et al., 2016). Honeybees appear to evaluate alfalfa at a distance, as the floral display size of individual plants positively influenced honeybee visitation (Bauer et al., 2017). Production of methyl eugenol from the flowers of papaya and *Colocasia* sp. was reported to attract *Bactrocera* sp. fruit flies (as reviewed by Tan and Nishida, 2012). Commercial sunflower hybrids may be capable of self-pollination, yields are generally improved by bees DeGrandi-Hoffman and Chambers (2006); Prasifka et al. (2018), also reported that there is a positive association between bee foragers as the nectars of sunflowers are rich in sucrose content. Though some hybrids receive no benefit from pollinators still open pollination by insects increased yields by 26% (Mallinger and Prasifka, 2017). *Annona coriacea* (custard apple) attracts its beetle pollinator *Cyclocephala atricapilla* (Coleoptera: Dynastidae) by producing a flower with the thermogenetic function that warms to 34°C, may be as much as 15°C above ambient air temperature. The purpose of this thermogenesis is to volatilize a characteristic spicy odor that attracts the beetle for pollination (Gottsberger, 1990).

11.9 METABOLIC ENGINEERING OF PLANT VOLATILE ORGANIC COMPOUNDS

In nature, volatiles contribute to seed dispersion by increasing fruit attractiveness. Volatiles released from vegetative parts of plants may also be attractive to some insects as foodstuff, even though they are unpalatable to most other herbivores. The presence of volatiles in fruits, vegetables and herbs has important influence on the cultivation of many plant species (Dudareva and Pichersky, 2008). As extensive breeding programs are undertaken to maximize certain attributes of crops – overall yield (i.e., size), total solids, sugar content-less attention is devoted for enhancing or even maintaining volatile production. As a result, many current cultivars of domesticated plant species produce less volatiles than their wild relatives or earlier cultivars (Gutterson, 1993). Floral VOCs have been inadvertently selected to retain some baseline number of pollinator-attracting traits, without which the crop would simply fail to yield fruit (Dudareva et al., 2013). For example, self-incompatible pummelo (*Citrus grandis*) blossoms release higher amounts of total VOCs than species (e.g., orange, grapefruit, lemon) with flowers containing both male and female organs (Jabalpurwala et al., 2009). Pollination services can be improved by producing more fragrant flowers as in natural ecosystems, as pollinators select the one emitting more VOCs (Parachnowitsch et al., 2012). Many modern crop cultivars are deficient in VOC production as more focus is on increased growth or yield, breeding techniques negatively affect secondary metabolism (VOCs) (Dudareva et al., 2013). Flowers produced by horticultural cultivars of cacao (*Theobroma cacao*) are small, barely fragrant, and only attract opportunistic flies that are inefficient pollinators, whereas wild *Theobroma* sp. possess large aromatic flowers that are predominantly bee-pollinated (Young & Severson, 1994).

In recent years, metabolic engineering emerged as a practical tool to boost or change the production of volatile metabolites in the whole plant or specific organs (fruits/ flowers). Successful manipulation of the plant's volatile metabolic profile requires an understanding of the relevant biosynthetic pathways and identification of the involved intermediates and products (Peled-Zehavi et al., 2015). Floral scent bouquet has been modified in several plant species, including petunia (via downregulation of scent biosynthetic genes or upregulation of Myb transcription factor PAP1). The first investigation of the consequences of metabolic engineering of floral VOC on pollinator visitation was performed in transgenic tobacco *Nicotiana attenuata* plants with diminished concentrations of benzyl acetone (Kessler et al., 2008). Monitoring the activity of floral visitors in their native habitat revealed that plants lacking benzyl acetone attracted fewer hawkmoths and hummingbirds than flowers emitting this volatility. It was also shown that changes in more than one volatile compound can influence pollinator's behavior.

Only a few studies have experimentally evaluated the impact of floral odor on crop pollination. Bee visitation (number of visits per unit time) was compared between wild-type and *Bacillus thuringiensis* (Bt)-expressing eggplant. Bt genetic transformation (cry3Bb gene) has elevated emission of five (methyl salicylate, Z-jasmone, α-pinene, α-methyl styrene, d-2-carene) out of 13 total floral compounds from transgenic flowers. All of those volatiles elicited positive electro-antennogram responses from bee antennae. When both plant types were subjected to commercial bumblebees in a glasshouse, bees strongly preferred Bt-eggplant despite significantly smaller flowers and overall plant size, illustrating the dominance of floral scent in pollinator behavior (Arpaia et al., 2011).

Interestingly, the same VOCs that lure pollinators sometimes serve as signals for floral antagonists or florivores. In the case of *Cirsium arvense*, emission of VOCs correlates with the activity of pollinators and is low when florivores are active (Theis et al., 2007). To resolve the function of single volatile compounds in deterrence and attraction, five transgenic *Petunia hybrida* RNAi lines silencing individual scent biosynthetic genes were used in field and glasshouse experiments (Kessler et al., 2012).

Volatile organic compounds released from fruits and seed/spores attract various animal dispersers and are essential for the spatial dynamics of plant populations within ecosystems. In contrast to floral VOCs, direct involvement of chemical cues in seed disperser interaction has only been seen in a few instances. A blend of VOCs emitted by seeds of the epiphyte *Peperomia macrostachyam* attracts ant garden *Camponotus femoratus* in the Amazonian rainforest and triggers their seed collecting behavior (Youngsteadt et al., 2008). Then these ants store the seeds in ant gardens where they later germinate. Thus, these VOCs, if produced artificially or by these seeds directly with other crop seeds in attract pollinators for seed dispersal. Likely, the aroma of the fruits has also been enhanced by knowing the biosynthesis of the volatiles and identifying the genes responsible for expressing them. Few examples are there: The expression of a *geraniol synthase* (*GES*) from basil (*O. basilicum*) in tomato, through metabolic engineering, under the control of a ripening-specific promoter. It is responsible for the biosynthesis of the monoterpenes geraniol, nerol, citronellol, citronellic acid, geranic and neric acid, which were absent in control fruits (reviewd by Araguez and Valpuesta, 2013). Production of anthocyanin pigment 1 (PAP1) transcription factor into rose flower, which resulted in increased levels of phenylpropanoid-derived color and scent compounds, as well as higher levels of terpenoid scent compounds. This, in turn, helps in the attraction of bees by providing an intensified olfactory cue (Zvi et al., 2012). Identification of a number of QTLs associated to the production of several volatiles in tomato (*S. lycopersicum*) through classical breeding in an intraspecific cross between a cherry tomato line with good overall aroma and an inbred line with bigger fruit and medium flavour (Saliba-Columbani et al., 2001).

Application of omics approaches, such as transcriptomics and metabolomics, to both plants and insects, will undoubtedly promote our understanding of the mechanisms underlying plant-insect interactions and help to achieve desired effects such as maximizing pollination while minimizing pest attraction (Dudareva et al., 2013).

11.10 CONCLUSION

Crop pollination is a valuable insect service and the science of pollination ecology has not been adequately advanced. First, it is very important to understand the crop-pollinator relationship and then use that relationship as a cue for developing strategies for enhancing the interaction which could increase the productivity of crops. In this chapter, different biochemical aspects of plant pollination have been considered: the role of flower color, flower scent, nectar, and pollen. One or other biochemical factor may dominate in a particular interaction, whereas in others, may all the characters be required to attract a particular pollinator. It is acknowledged that crops with floral attractiveness for insects will increase the crop yield and also pollinator conservation.

So, more functional studies on floral VOCs are needed to highlight how individual organisms respond to different signals and also the genes encoding biosynthesis of those metabolites. There is a range of factors that probably influence the evolution of floral signaling strategies in the context of defense and reproduction which are needed to be studied. The development of crop varieties with specific nectar or nectar related traits to attract and retain pollinating insects is an appealing strategy to help address the needs of agriculture and pollinators for several reasons. Thus, there is both scope and requirement of more research and studies for utilizing the plant biochemical traits through, biotechnology and breeding, for sustainable crop production.

REFERENCES

- Abrol DP (2012) Biochemical basis of plant-pollination interaction. In: Pollination biology. Springer 413–458
- Afroz M, Rahman M, Amin R. (2021) Insect plant interaction with reference to secondary metabolites: A review. Agricultural Reviews, 42(4): 427–433
- Aizenberg-Gershtein Y, Izhaki I, Santhanam R, Kumar P, Baldwin IT, Halpern M (2015) Pyridine-type alkaloid composition affects bacterial community composition of floral nectar. Scientific Reports 5: 11536
- Andersson S (2003) Foraging responses in the butterflies *Inachis io*, *Aglais urticae* (Nymphalidae), and *Gonepteryx rhamni* (Pieridae) to floral scents. *Chemoecology* 13(1): 1–11
- Aragüez I, Valpuesta V (2013) Metabolic engineering of aroma components in fruits. Biotechnology Journal 8(10): 1144–1158
- Arpaia S, De Cristofaro A, Guerrieri E, Bossi S, Cellini F, Di Leo GM, Germinara GS, Iodice L, Maffei ME, Petrozza A, Vitagliano S (2011) Foraging activity of bumblebees (*Bombus terrestris* L.) on Bt-expressing eggplants. Arthropod Plant Interactions 5: 255–261
- Baracchi D, Marples A, Jenkins AJ, Leitch AR, Chittka L (2017) Nicotine in floral nectar pharmacologically influences bumblebee learning of floral features. Scientific Reports 7(1): 1–8
- Bartomeus I, Potts SG, Steffan-Dewenter I, Vaissiere BE, Woyciechowski M, Krewenka KM, Tscheulin T, Roberts SPM, Szentgyörgyi H, Westphal C, Bommarco R (2014) Contribution of insect pollinators to crop yield and quality varies with agricultural intensification. Peer J 2: e328
- Bauer AA, Clayton MK, Brunet J (2017) Floral traits influencing plant attractiveness to three bee species: consequences for plant reproductive success. Am. J. Bot. 104, 772–781.
- Berenbaum MR, Johnson RM (2015) Xenobiotic detoxification pathways in honeybees. Current Opinion in Insect Science 10: 51–58
- Bergström G (1978) Role of volatile chemicals in ophrys-pollinator interactions. In: Harborne G (ed.) Biochemical Aspects of Plant and Animal Coevolution. Academic, New York, pp. 207–230
- Bohman B, Phillips RD, Menz MH, Berntsson BW, Flematti GR, Barrow RA, Dixon KW, Peakall R (2014) Discovery of pyrazines as pollinator sex pheromones and orchid

semiochemicals: implications for the evolution of sexual deception. New Phytol 203: 939–952

- Borghi M, Fernie AR, Schiestl FP, Bouwmeester HJ (2017) The sexual advantage of looking, smelling, and tasting good: the metabolic network that produces signals for pollinators. Trends in Plant Science 22(4): 338–350
- Borg-Karlson AK, Groth I, Agren L, Kullenberg B (1993) Form–specific fragrances from *Ophrys insectifera* L. (Orchidaceae) attract species of different pollinator genera: Evidence of sympatric speciation? Chemoecology 4: 39–45
- Borg-Karlson AK, England FB, Unelius CR (1994) Dimethyl oligosulphides, major volatiles released from *Sauromatum guttatum* and *Phallus impudicus*. Phytochemistry 35: 321–323
- Borg-Karlson AK, Tengo J (1986) Odour mimetism? Key substances in *Ophrys lutea* Andrena pollination relationship (Orchidaceae: Andrenidae). Journal of Chemical Ecology 12: 1927–1941
- Bouwmeester H, Schuurink RC, Bleeker M, Schiestl F (2019) The role of volatiles in plant communication. The Plant Journal 100(5): 892–907.
- Briscoe AD, Chittka L (2001) The evolution of color vision in insects. Annu Rev Entomol 46: 471–510
- Brock MT, Lucas LK, Anderson NA, Rubin MJ, Cody-Markelz, RJ, Covington MF, Devisetty UK, Chapple C, Maloof JN, Weinig C (2016) Genetic architecture, biochemical underpinnings and ecological impact of floral UV patterning. Mol. Ecol. 25: 1122–1140
- Burger H, Dotterl S, Ayasse M (2010) Host-plant finding and recognition by visual and olfactory floral cues in an oligolectic bee. Functional Ecology 24: 1234–1240
- Carter C, Shafir S, Yehonatan L, Palmer RG, Thornburg R (2006) A novel role for proline in plant floral nectars. Naturwissenschaften 93: 72–79
- Clara CD, Carlos MH, Dotterl S (2014) Floral volatiles play a key role in specialized ant pollination. Perspectives in Plant Ecology Evolution and Systematics 16: 32–42
- Claudianos C, Ranson H, Johnson RM, Biswas S, Schuler MA, Berenbaum MR, Feyereisen R, Oakeshott JG (2006) A deficit of detoxification enzymes: pesticide sensitivity and environmental response in the honeybee. Insect Molecular Biology 15(5): 615–636
- Couvillon MJ, Al Toufailia H, Butterfield TM, Schrell F, Ratnieks FLW, Schurch R (2015) Caffeinated forage tricks honeybees into increasing foraging and recruitment behaviors. Current Biology 25: 1–4
- Cseke LJ, Kaufman PB, Kirakosyanb A (2007) The biology of essential oils in the pollination of flowers. Natural Products Communications 2(12): 1318–1336
- DeGrandi-Hoffman G, Chambers M (2006) Effects of honeybee (Hymenoptera: Apidae) foraging on seed set in self-fertile sunflowers (*Helianthus annuus* L.). Environ. Entomol. 35: 1103–1108
- Despres L, David JP, Gallet C (2007) The evolutionary ecology of insect resistance to plant chemicals. Trends in Ecology and Evolution 22: 298–307

- Dötterl S, Vereecken NJ (2010) The chemical ecology and evolution of bee flower interactions: a review and perspectives. Can J Zool 88: 668–697
- Dobson HE (2006) Relationship between floral fragrance composition and type of pollinator. In Biology of floral scent 147–198. CRC Press
- Dobson HEM (1994) Floral volatiles in insect biology. In: Insect-Plant interactions 5: 47–81. Bernays, E. (ed.), CRC Press, Boca Raton, FL
- Dobson HEM (2006) Relationship between floral fragrance composition and type of pollinator. In Biology of floral scent. Dudareva NA, Pichersky E (eds.). CRC Press. Boca Raton, FL. 147–198
- Dodson HEM (1994) Floral volatiles in insect biology. In Insect–Plant interactions, Vol. 5. Bernays EA (ed.). CRC Press, Boca Raton, FL
- Dormont L, Delle–Vedove R, Bessi_ere JM, Schatz B (2014) Floral scent emitted by white and coloured morphs in orchids. Phytochemistry 100: 51–59
- Dudareva N, Klempien A, Muhlemann JK, Kaplan I (2013) Biosynthesis, function and metabolic engineering of plant volatile organic compounds. New Phytologist 198(1): 16–32
- Dudareva N, Negre F, Nagegowda DA, Orlova I (2006) Plant volatiles: recent advances and future perspectives. Critical Reviews in Plant Sciences 25(5): 417–440
- Dudareva N, Pichersky E (2000) Biochemical and molecular genetic aspects of floral scents. Plant Physiol 122: 627–634
- Dudareva N, Pichersky E, Gershenzon J (2004) Biochemistry of plant volatiles. Plant Physiol 135: 1893–1902
- Dudareva N, Pichersky E (2008) Metabolic engineering of plant volatiles. Current Opinion in Biotechnology 19: 1–9
- Eberling H, Olesen JM (1999) The structure of a high latitude plant–flower visitor system: the dominance of flies. Ecography 22: 314–323
- Endress PK (1994) Diversity and Evolutionary Biology of Tropical Flowers. Cambridge University Press
- Erb M., Kliebenstein D.J. (2020) Plant secondary metabolites as defenses, regulators, and primary metabolites: the blurred functional trichotomy. Plant Physiology, 184(1): 39–52
- Farré-Armengol G, Filella I, Llusia J, Peñuelas J (2013) Floral volatile organic compounds: between attraction and deterrence of visitors under global change. Persp Plant Ecol Evol Syst 15(1): 56–67
- Farré-Armengol G, Filella I, Llusia J, Peñuelas J (2015) Relationships among floral VOC emissions, floral rewards and visits of pollinators in five plant species of a Mediterranean shrubland. Plant Ecol Evol 148: 90–99
- Fridman S, Izhaki I, Gerchman Y, Halpern M (2012) Bacterial communities in floral nectar. Environmental Microbiology Reports 4: 97–104
- Galen C (1985) Regulation of seed set in *Polemonium viscosum*: floral scents, pollination and resources. Ecology 66: 792–797

- Garibaldi LA, Steffan-Dewenter I, Kremen C, Morales JM, Bommarco R, Cunningham SA, Carvalheiro LG, Chacoff NP, Dudenhöffer JH, Greenleaf SS, Holzschuh A, Isaacs R, Krewenka K, Mandelik Y, Mayfield M, Morandin LA, Potts SG, Ricketts TH, Szentgyöorgyi H, Viana BF, Westphal C, Winfree R, Klein AM (2011) Stability of pollination services decreases with isolation from natural areas despite honeybee visits. Ecology Letters 14(10): 1062–1072
- Gottsberger G (1990) Flowers and beetles in the South–American tropics. Bot Acta 103(4): 360–365
- Granero AM, Sanz JMG, Gonzalez FJE, Vidal JLM, Dornhaus A, Ghani J, Serrano AR, Chittka L (2005) Chemical compounds of the foraging recruitment pheromone in bumblebees. Naturwissenschaften 92: 371–374
- Gupta AK, Akhtar TA, Widmer A, Pichersky E, Schiestl FP (2012) Identification of white campion (*Silene latifolia*) guaiacol O-methyltransferase involved in the biosynthesis of veratrole, a key volatile for pollinator attraction. BMC Plant Biology 12(1): 1–13
- Gutterson NC (1993) Molecular breeding for color, flavor and fragrance. Sci Hort 55: 141–160
- Harborne BJ, (2001) Secondary metabolites: attracting pollinators. Encyclopedia of Life Sciences pp: 1–5
- Haverkamp A, Hansson BS, Knaden M (2018) Combinatorial codes and labeled lines: how insects use olfactory cues to find and judge food, mates, and oviposition sites in complex environments. Front. Physiol. 9: 49
- Haverkamp A, Yon F, Keesey IW, Missbach C, Koenig C, Hansson BS, Baldwin IT, Knaden M, Kessler D (2016b) Hawkmoths evaluate scenting flowers with the tip of their proboscis. eLife 5: e15039
- Herrera CM, De Vega C, Canto A, Pozo MI (2009) Yeasts in floral nectar: a quantitative survey. Annals of Botany 103: 1415–1423
- Hoballah ME, Stuurman J, Turlings TCJ, Guerin PM, Connetable S, Kuhlemeier, C (2005) The composition and timing of flower odour emission by wild *Petunia axillaris* coincide with the antennal perception and nocturnal activity of the pollinator *Manduca sexta*. Planta 222: 141–150
- Holzschuh A, Dudenhoffer J-H, Tscharntke T (2012) Landscapes with wild bee habitats enhance pollination, fruit set and yield of sweet cherry. Biol Conservat 153: 101–107
- Huber F, Kaiser R, SauterW, Schiestl F (2005) Floral scent emission and pollinator attraction in two species of Gymnadenia (Orchidaceae). Oecologia 142: 564–575
- Inoue TA, Asaoka K, Seta K, Imaeda D, Ozaki M (2009) Sugar receptor response of the food-canal taste sensilla in a nectar-feeding swallowtail butterfly, *Papilio xuthus*. Naturwissenschaften 96: 355–363
- Inouye WD (2013) Role of pollinators. Encyclopaedia of Biodiversity 6: 140–146
- Irwin RE, Cook D, Richardson LL, Gardner DL (2014) Secondary compounds in floral rewards of toxic rangeland plants: impacts on pollinators. Journal of Agricultural and Food Chemistry 62: 7335–7344

- Isaacs R, Kirk AK (2010) Pollination services provided to small and large highbush blueberry fields by managed bees. J. App. Ecol. 47: 841–84
- Jabalpurwala FA, Smoot JM, Rouseff RL (2009) A comparison of citrus blossom volatiles. Phytochemistry 70: 1428–1434
- Johnson SD (2016) Carrion flowers. Current Biology 26(13): R556–R558.
- Kamal AHM, Takashina T, Egashira H, Satoh H, Imanishi S (2001) Introduction of aromatic fragrance into cultivated tomato from the 'peruvianum complex'. Plant Breed 120: 179–181
- Keil TA (1999) Morphology and development of the peripheral olfactory organs. Insect Olfaction. Springer, Berlin. pp: 5–47.
- Kessler A, Halitschke R, Poveda K (2011) Herbivory-mediated pollinator limitation: negative impacts of induced volatiles on plant-pollinator interactions. Ecology 92: 1769–1780
- Kessler D, Diezel C, Clark DG, Colquhoun TA, Baldwin IT (2012) Petunia flowers solve the defence/apparency dilemma of pollinator attraction by deploying complex floral blends. Ecology Letters 16(3): 299–306
- Kessler D, Gase K, Baldwin IT (2008) Field experiments with transformed plants reveal the sense of floral scents. Science 321: 1200–1202
- Kessler D, Kallenbach M, Diezel C, Rothe E, Murdock M, Baldwin IT (2015a) How scent and nectar influence floral antagonists and mutualists. eLife 5: e15039
- Klatt BK, Burmeister C, Westphal C, Tscharntke T, von Fragstein M (2013) Flower volatiles, crop varieties and bee responses. PLoS One 8(8): e72724
- Klatt BK, Holzschuh A, Westphal C, Clough Y, Smit I, Pawelzik E, Tscharntke T (2014) Bee pollination improves crop quality, shelf life and commercial value. Proc. R. Soc. B 281: 20132440
- Klecka J, Hadrava J, Biella P, Akter A (2018) Flower visitation by hoverflies (Diptera: Syrphidae) in a temperate plant-pollinator network. PeerJ 6: e6025
- Kleijn D, Winfree R, Bartomeus I, Carvalheiro LG, Henry M, Isaacs R, Potts SG (2015) Delivery of crop pollination services is an insufficient argument for wild pollinator conservation. Nat. Commun. 6(1): 1–9
- Klein AM, Steffan-Dewenter I, Tscharntke T (2003) Bee pollination and fruit set of *Coffea arabica* and *C. canephora* (Rubiaceae). Am J Bot 90: 153–15
- Klein AM, Vaissière BE, Cane J, Steffan-Dewenter I, Cunningham SA, Kremen C, Tscharntke T (2007) Importance of pollinators in changing landscape for world crops. Proc. R. Soc. B. Biological Sciences 274(1608): 303–313
- Knauer AC, Schiestl FP (2015) Bees use honest floral signals as indicators of reward when visiting flowers. Ecol. Lett. 18, 135–143
- Knudsen JT, Eriksson R, Gershenzon J, Stahl B (2006) Diversity and distribution of floral scent. Botanical Reviews 72: 1–120.
- Knudsen JT, Eriksson R, Stahl B (2006) Diversity and distribution of floral scent. Botanical Review 72: 1–120

- Knudsen JT, Gershenzon J (2006) The chemistry diversity of floral scent. Biology of Floral Scent (eds. N. Dudareva & E. Pichersky), pp. 27–52. CRC Press, Boca Raton, Florida
- Knudsen JT, Tollsten L, Groth I, Bergstrom G, Raguso RA (2004) Trends in floral scent chemistry in pollination syndromes: floral scent composition in hummingbird-pollinated taxa. Botanical Journal of the Linnean Society 146: 191–199
- Koschier EH (2008) Essential Oil Compounds for Thrips Control – A Review. Natural Product Communications 3(7)
- Koski MH, Galloway LF (2018) Geographic variation in pollen color is associated with temperature stress. New Phytologist 218: 370–379
- Kremen C, Williams NM, Thorp, RW (2002) Crop pollination from native bees at risk from agricultural intensification. PNAS, 99(26): 16812–16816
- Lavy M, Zuker A, Lewinsohn, E, Larkov O, Ravid U, Vainstein, A, Weiss D (2002) Linalool and linalool oxide production in transgenic carnation flowers expressing the *Clarkia breweri* linalool synthase gene. Molecular Breeding 9: 103–111
- Liu F, Chen J, Chai J, Zhang X, Bai X, He D, Roubik D (2007) Adaptive functions of defensive plant phenolics and a non-linear bee response to nectar components. Functional Ecology 21(1): 96–100
- Lucas–barbosa D, Sun P, Hakman A, van Beek TA, van Loon JJ, Dicke M (2016) Visual and odour cues: plant responses to pollination and herbivory affect the behaviour of flower visitors. Functional Ecology 30(3): 431–441
- Lücker J, Bouwmeester HJ, Schwab W, Blaas J, van der Plas LH, Verhoeven HA (2001) Expression of *Clarkia* S-linalool synthase in transgenic petunia plants results in the accumulation of S-linalyl-beta-D-glucopyranoside. Plant Journal 27: 315–324
- Lücker J, Schwab W, van Hautum B, Blaas J, van der Plas LH, Bouwmeester HJ, Verhoeven HA (2004) Increased and altered fragrance of tobacco plants after metabolic engineering using three monoterpene synthases from lemon. Plant Physiology 134: 510–519
- Lunau K (1995) Notes on the color of pollen. Plant Systematics and Evolution 198 (3): 235–252
- Mallinger RE, Prasifka JR (2017) Benefits of insect pollination to confection sunflowers differ across plant genotypes. Crop Sci. 57, 3264–3272. doi:10.2135/cropsci2017.03.0148
- Manson, J.S., Otterstatter, M.C., Thomson, J.D. (2010) Consumption of a nectar alkaloid reduces pathogen load in bumble bees. Oecologia, 162: 81–89
- Mayer DF, Johansen, CA, Lunden JD (1989) Honeybee foraging behavior on ornamental crabapple pollenizers and commercial apple cultivars. HortScience 24, 510–512
- Melathopoulos AP, Cutler G, Tyedmers P (2015) Where is the value in valuing pollination ecosystem services to agriculture? Ecol Econ 10959–70
- Miller R, Owens SJ and Rørslett B (2011) Plants and color: flowers and pollination. Optics & Laser Technology 43: 282–294

- Miosic S, Knop K, Hoslscher D, Greiner J, Gosch C, Thill J, Kai M, Shrestha BM, Schneider B, Crecelius AC, Schubert US, Svatoš A, Stich K, Halbwirth H (2013) 4-Deoxyaurone formation in *Bidens ferulifolia* (Jacq.) DC. PLoS One 8: e61766
- Morse A, Kevan P, Shipp L, Khosla S, McGarvey B (2012) The impact of greenhouse tomato (Solanales: Solanaceae) floral volatiles on bumblebee (Hymenoptera: Apidae) pollination. Environmental Entomology 41: 855–864
- Mound LA (2005) Thysanoptera: diversity and interactions. Annu Rev Entomol 50: 247–269
- Muhlemann JK, Klempien A and Dudareva N (2014) Floral volatiles: from biosynthesis to function. Plant, Cell & Environment 37(8): 1936–1949
- Nwf.org – https://www.nwf.org/Garden-for-Wildlife/Wildlife/Attracting–Butterflies
- Okamoto T, Kawakita A, Kato M (2007) Interspecific variation of floral scent composition in Glochidion and its association with host-specific pollinating seed parasite (Epicephala). J Chem Ecol 33: 1065–1081
- Okamoto T, Su, ZH (2021) Chemical analysis of floral scents in sympatric *Ficus* species: highlighting different compositions of floral scents in morphologically and phylogenetically close species. Plant Systematics and Evolution 307(4): 1–12
- Olesen JM, Rønsted N, Tolderlund U, Cornett C, Molgaard P, Madsen J, Olsen CE (1998) Mauritian red nectar remains a mystery. Nature 393: 529–529
- Paige KN, Whitham TG (1985) Individual and population shifts in flower color by scarlet Gilia. Science 227: 315–317
- Palmer RG, Perez PT, Ortiz-Perez E, Maalouf F, Suso MJ (2009). The role of crop-pollinator relationships in breeding for pollinator-friendly legumes: from a breeding perspective. Euphytica 170(1): 35–52
- Paparella A, Shaltiel–Harpaza L, Ibdah M (2021) β-Ionone: Its Occurrence and Biological Function and Metabolic Engineering. Plants (Basel) 10(4): 754
- Parachnowitsch AL, Raguso RA, Kessler A (2012) Phenotypic selection to increase floral scent emission, but not flower size or color in bee-pollinated *Penstemon digitalis*. New Phytologist 195: 667–675
- Peakall R, Ebert D, Poldy J, Barrow RA, Francke W, Bower CC, Schiestl FP (2010) Pollinator specificity, floral odour chemistry and the phylogeny of Australian sexually deceptive Chiloglottis orchids, implications for pollinator–driven speciation. New Phytol. 188: 437–450
- Peay KG, Belisle M, Fukami T (2012) Phylogenetic relatedness predicts priority effects in nectar yeast communities. Proceedings of the Royal Society of London Soc B 279: 749–758
- Peled-Zehavi H, Oliva M, Xie Q, Tzin V, Oren-Shamir M, Aharoni A, Galili G (2015) Metabolic engineering of the phenylpropanoid and its primary, precursor pathway to enhance the flavor of fruits and the aroma of flowers. Bioengineering, 2(4): 204–212.
- Pichersky E, Dudareva N (2007) Scent engineering: toward the goal of controlling how flowers smell. Trends in Biotechnology 25: 105–110

- Piechulla, B.; Pott, M.B. (2003) Plant scents-mediators of inter-and intraorganismic communication. Planta, 217: 687–689
- Pichersky E, Gang DR (2000) Genetics and biochemistry of secondary metabolites in plants: an evolutionary perspective. Trends in Plant Science 5: 439–445
- Potts SG, Biesmeijer JC, Kremen C, Neumann P, Schweiger O, Kunin WE (2010) Global pollinator declines: trends, impacts and drivers. Trends in Ecology and Evolution, 25(6): 345–353
- Prasifka JR, Mallinger RE, Portlas ZM, Hulke BS, Fugate KK, Paradis T, Hampton ME and Carter CJ (2018) Using nectar-related traits to enhance crop-pollinator interactions. Front. Plant Sci. 9: 812. doi: 10.3389/fpls.2018.00812
- Proctor M, Yeo P, Lack A (1996) The Natural History of Pollination. Timber Press, HarperCollins Publisher.
- Raguso RA (2001) Floral scent, olfaction, and scent-driven foraging behavior. In Cognitive ecology of pollination. Chittka L, Thomson JD (eds). Cambridge University Press. 83
- Raguso RA (2008) Wake up and smell the roses: the ecology and evolution of floral scent. Annu. Rev. Ecol. Evol. Syst. 39: 549–569
- Raguso RA, Pichersky E (1995) Floral volatiles from *Clarkia breweri* and *C. concinna* (Onagraceae): Recent evolution of floral scent and moth pollination. Plant Sys. Evol. 194: 55–67
- Ramya M, Kwon OK, An HR, Park PM, Baek YS, Park PH (2017) Floral scent: regulation and role of MYB transcription factors. Phytochemistry Letters 19: 114–120
- Ramya M, Jang S, An HR., Lee SY, Park PM, Park PH (2020) Volatile organic compounds from orchids: From synthesis and function to gene regulation. Int J Mol Sci, 21(3), 1160; doi: 10.3390/ijms21031160
- Saliba-Colombani V, Causse M, Langlois D, Philouze J, Buret M (2001) Genetic analysis of organoleptic quality in fresh market tomato. 1. Mapping QTLs for physical and chemical traits. Theor Appl Genet 102(2): 259–272
- Schaeffer RN, Irwin RE (2014) Yeasts in nectar enhance male fitness in a montane perennial herb. Ecology 95: 1792–1798
- Schaffler I, Balao F, Dotterl S (2012) Floral and vegetative cues in oil-secreting and non-oil-secreting *Lysimachia* species. Ann. Bot. 110: 125–138
- Schiestl FP, Ayasse M, Paulus HF, Löfstedt C, Hansson BS, Ibarra F, Francke W (1999) Orchid pollination by sexual swindle. Nature 399(6735): 421–422
- Schiestl FP, Dotterl S (2012) The evolution of floral scent and olfactory preferences in pollinators: coevolution or pre-existing bias? Evolution 66: 2042–2055
- Schiestl FP, Johnson SD (2013) Pollinator-mediated evolution of floral signals. Trends Ecol. Evol. 28(5): 307–315
- Schulz S, Francke W, Edgar J (1988) Volatile compounds from androconial organs of danaine and ithomiine butterflies. Z. Naturforsch. C 43: 99–104

- Shin YS, Park SD, Kim JH (2007) Influence of pollination methods on fruit development and sugar contents of oriental melon (*Cucumis melo* L. cv. Sagyejeol–Ggul). Sci Horticult 112, 88–392
- Shrestha M, Dyer AG and Burd M (2013) Evaluating the spectral discrimination capabilities of different pollinators and their effect on the evolution of flower colours. Communicative & Integrative Biology 6: 301–310
- Skubatz H, Kunkel DD, Howald WN, Trenkle R, Mookhergee B (1996) The *Sauromatum guttatum* appendix as an osmophore: excretory pathways, composition of volatiles and attractiveness to insects. New Phytol 134(4): 631–640
- Sosenski P, Parra-Tabla V (2019) Secondary Metabolites: Attracting Pollinators. Encyclopedia of Life Sciences 1–9
- Steinbrecht RA (1997) Pore structures in insect olfactory sensilla: a review of data and concepts. International Journal of Insect Morphology and Embryology 26: 229–245
- Stevenson CP, Nicolson SW, Wright GA (2017) Plant secondary metabolites in nectar: impacts on pollinators and ecological functions. Functional Ecology 31: 65–75
- Suchet C, Dormont L, Schatz B, Giurfa M, Simon V, Raynaud C, Chave J (2011) Floral scent variation in two *Antirrhinum majus* subspecies influences the choice of naïve bumblebees. Behavioural Ecology and Sociobiology 65, 1015–1027
- Tan KH, Nishida R (2012) Methyl eugenol: Its occurrence, distribution, and role in nature, especially in relation to insect behavior and pollination. J Insect Sci 12: 56
- Tanda AS (2021) Why insect pollinators important in crop improvement? Indian J Entomol pp: 1–14
- Theis N, Adler LS (2012) Advertising to the enemy: enhanced floral fragrance increases beetle attraction and reduces plant reproduction. Ecology 93: 430–435
- Theis N, Lerdau M, Raguso RA (2007) The challenge of attracting pollinators while evading floral herbivores: patterns of fragrance emission in *Cirsium arvense* and *Cirsium repandum* (asteraceae). International Journal of Plant Sciences 168: 587–601
- USDA Forest service- https://www.fs.fed.us/wildflowers/pollinators/animals/flies.shtml
- Vannette RL, Gauthier MPL, Fukami T (2013) Nectar bacteria, but not yeast, weaken a plant-Pollinator mutualism. Proceedings of the Royal Society of London Soc B 280: 20122601
- Waelti M, Muhlemann J, Widmer A, Schiestl F (2008) Floral odour and reproductive isolation in two species of *Silene*. J Evol Biol 21: 111–121
- War AR., Buhroo AA, Hussain B, Ahmad T, Nair RM, Sharma, HC (2020). Plant defense and insect adaptation with reference to secondary metabolites. Co-Evolution of Secondary Metabolites. 795–822
- Waser NM, Price MV (1983) Pollination behaviour and natural selection for flower color in *Delphinium nelsonii*. Nature 302: 422–424
- Weiss MR (1991) Floral color changes as cues for pollinators. Nature 354: 227–229
- Weiss MR (2001) Vision and learning in some neglected pollinators: beetles, flies, moths, and butterflies. In Cognitive Ecology of Pollination. Chittka L, Thomson JD (eds.). Cambridge University Press. 171

- Wright GA, Baker DD, Palmer M J, Stabler D, Mustard JA, Power EF, Borland, AM, Stevenson PC (2013). Caffeine in floral nectar enhances a pollinator's memory of reward. Science 339(6124): 1202–1203. doi: 10.1126/science.1228806
- Wright GA, Lutmerding A, Dudareva N, Smith BH (2005) Intensity and the ratios of compounds in the scent of snapdragon flowers affect scent discrimination by honeybees (*Apis mellifera*). J. Comp. Physiol. A Neuroethol. Sens. Neural Behav. Physiol. 191: 105–114
- Wright GA, Mustard JA, Simcock NK, Ross-Taylor AAR, McNicholas LD, Popescu A, Marion-Poll F (2010) Parallel reinforcement pathways for conditioned food aversions in the honeybee. Current Biology 20: 2234–2240
- Wright GA, Schiestl FP (2009) The evolution of floral scent: the influence of olfactory learning by insect pollinators on the honest signalling of floral rewards. Funct Ecol 23: 841–851
- Wright GA, Smith BH (2004a) Variation in complex olfactory stimuli and its influence on odour recognition. Proc. R. Soc. Lond. B. Biol. Sci. 271: 147–152
- Wright GA, Smith BH (2004b) Different thresholds for detection and discrimination of odors in the honeybee (*Apis mellifera*). Chem. Senses 29: 127–135
- Young AM, Severson DW (1994) Comparative analysis of steam distilled floral oils of cacao cultivars (*Theobroma cacao* L., Sterculiaceae) and attraction of flying insects - implications for a Theobroma pollination syndrome. Journal of Chemical Ecology 20: 2687–2703
- Youngsteadt E, Nojima S, Haberlein C, Schulz S, Schal C (2008) Seed odor mediates an obligate ant-plant mutualism in Amazonian rainforests. Proceedings of the National Academy of Sciences 105: 4571–4575
- Zhang FP, Cai XH, Wang H, Ren ZX, Larson-Rabin Z, Li DZ (2012) Dark purple nectar as a foraging signal in a bird–pollinated Himalayan plant. New Phytologist 193: 188–195.
- Zvi MM, Shklarman E, Masci T, Kalev H, Debener T, Shafir S, Ovadis M, Vainstein A (2012) *PAP1* transcription factor enhances production of phenylpropanoid and terpenoid scent compounds in rose flowers. New Phytol 195(2): 335–345

12

Urbanization and Its Impact on Native Pollinators

Amarjit S. Tanda*

ABSTRACT

Urbanization is a principal anthropogenic operator of decline for ecologically and economically important insect pollinators including native bees. With their adverse influence on pollinators, urbanization can exhibit an amazing degree of biodiversity in comparison to other landscapes. The bee populations observed in these habitats are inclined to interact with local elements and landscape features that comprise the daily expanding city structures. Within and across taxa, the landscape covers and local environment interact with native pollinators which ultimately affects bee community composition in different ways that contribute to functional trait homogenization and reduced phylogenetic diversity. Bees and other pollinators show functional trait characteristics, for example, polylectic food, nesting behavior in cavities, emergence, and behaving in the urban habitats differently. Green covers such as parks and urban gardens have been observed as potential refuges to protect especially endangered specialists bee species, which have become diminished in urban habitats. These bee environments are dispersed in the city matrix and differ in the availability of pollinators' food resources. For pollinators assessing such habitats are crucial. This chapter describes how city features affect pollinators in addition to ways these green spaces can be utilized to encourage maximum pollinator populations and sustainable urban bee diverse ecosystem.

Keywords *Urbanization • Native pollinators • Conservation • Bees • Urban heat island effect • Phylogenetic diversity • Habitat • Fragmentation • Microclimate • Green spaces*

12.1 INTRODUCTION

In addition to maximum human population and size, urbanization is a harmful, anthropogenic change of the bee habitat described by an enhanced proportion of impervious and roads, sidewalks, parking lots, etc. [Wu, 2014]. The development of cities and intensified farming, have been found as primary factors for the destruction of biodiversity, with declines of many

* School of Horticulture, University of Western Sydney, Richmond, Australia and CSIRO North Ryde NSW, Australia.

E-mail: drtanda101@gmail.com

coleopterans, lepidopterans, dipterans, and hymenopterans insect pollinators [Cardoso and Gonçalves, 2018; Fenoglio et al., 2020; Kurylo et al., 2020; Piano et al., 2019; Sattler et al., 2010; Tanda, 2019, 2020 a-b, 2021a-i]. The depletion in biodiversity is mostly due to different interactive factors of the city's habitat. These elements vary from fragmentation and reduction of habitat, the alien plants and insect pollinators, the urban heat island (UHI) effect, and the loss of suitable environment (Fig. 12.1) [Harrison and Winfree, 2015].

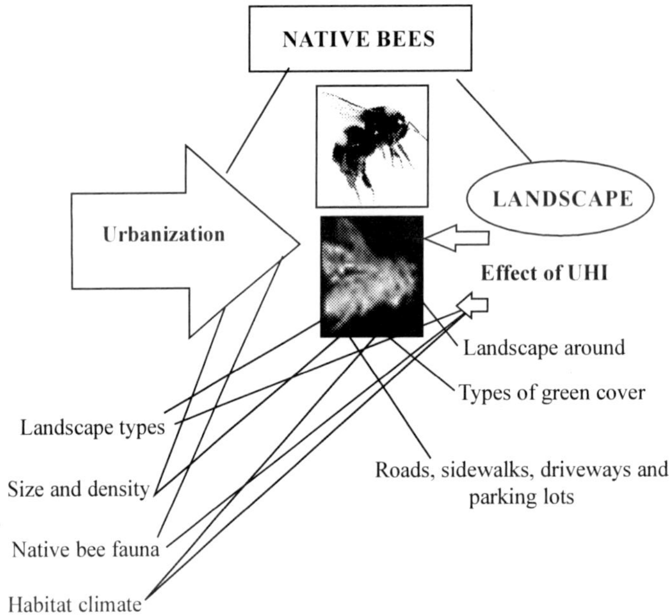

Fig. 12.1 Schematic representation of interactions between landscape and local elements and the urban heat island (UHI) effect, on the deterioration of bee habitat quality (*Schematic model designed by Dr Amarjit S. Tanda, 2021*)

Urban habitats are complex contributing to an overall loss in biodiversity, which can help animal communities and bee pollinators. Both ecologically and economically bees are significant, and conserving them is a great challenge due to urbanization, and the loss of neighboring landscapes. Few measures to support bees in urban areas have been taken to establish and encourage city green covers. How bees and other pollinators react to these green places in addition to other city landscapes and local elements should be examined. This chapter elaborates on the current knowledge and recent approaches in native bee pollinator's response to the urban landscape and local changed habitats. Few functional characteristics of bees including dietary span and nesting site plans are known and supported, and socially considered less generalized. Insufficiency of reports on bee reactions to urban elements is in part due to the individual variation shown in various bee species. To encourage maximum biodiversity in city landscape, research should aim on specific reactions to urban local and landscape features and how these green covers can be assessed for bee conservation and sustainable biodiversity (Tanda, 2020 a-b, 2021a-i).

All these landscapes or local components of city environment can affect insect pollinators [Ballare et al., 2019; Beninde et al., 2015; Egerer et al., 2017; Geslin et al., 2016; Quistberg et al., 2016; Tanda, 2020 a-b, 2021a-i]. Thus, expanding cities have been considered as filtering system that influence the urban bee groups by functional characteristics that affect health, and life cycle plans [Dale and Frank, 2018; Williams et al., 2009; Wenzel et al., 2020]. City filtering environment may homogenize bee communities according to functional characteristics (Fig. 12.2) [Deguines et al., 2016; Banaszak-Cibicka and Zmihorski, 2012].

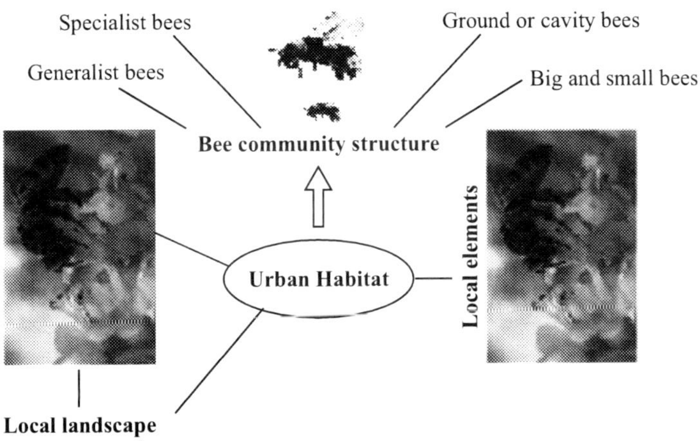

Fig. 12.2 Schematic representation of depicting the urban habitat, local elements, landscape and influence on bee community structure. (Schematic model designed by Dr Amarjit S. Tanda, 2021)

Negative impact of the city matrix exerts on native bees, city areas can represent biodiversity that work as refuge for several ecologically significant groups especially insect pollinators [Egerer et al., 2017; Banaszak-Cibicka and Zmihorski, 2012; Hamblin et al., 2017; Nagy et al., 2018]. Cities may shelter maximum bee diversity and populations than nearby farm and semi-natural landscapes [Baldock et al., 2015; Banaszak-Cibicka et al., 2018; Kaluza et al., 2016; Sirohi et al., 2015]. The biodiversity and population found within urban areas are in part due to the heterogeneity of cities, and the landscapes in which different environments or green covers survive [Tommasi et al., 2004]. Parks, gardens, and home backyards are noteworthy habitats that can offer foraging and nesting sites for bees and other pollinators [Baldock et al., 2019; Dylewski et al., 2020; Tanda, 2020 a-b, 2021a-i]. City green covers have also been utilized to enhance habitat association by developing green corridors or stepping-stone environments [Braaker et al., 2017]. Green spaces to encourage biodiversity and alleviate urban influences has been an area of interest in these days. How these urban spaces can be improved to encourage bee strength and population is important to understand, especially when accepting the development of city areas required to adjust the increasing world population. About 55% of the world population lived in cities according to the United Nations, 2018 report, which will be 68% by 2050 [United Nations, 2020]. For example, Canada and the US showing maximum rates of urbanization, where >80% of the population can be observed living in cities [Government of Canada, 2020; Center for Sustainable Systems, 2020].

Insect pollinators are a crucial research group due to their fast response to the habitat changes [Jones and Leather, 2012]. Insect pollinator reaction to the single element of urbanization is important not only because declining of bee pollinators [Potts et al., 2010], but also because they offer crucial pollination services to the bio-environment. To reproduce successfully, about 90% of flowering plants rely on animal pollination and a major role of which is performed by bee pollinators [Ollerton et al., 2011; Kearns et al., 1998]. This mechanism not only influence plant reproduction, but it also affect the pollinators that depends on the fruits, seeds, and other food resources for such biological jobs. Considering the scale of farming and food production, humans are no exception to this. As there is increasing urbanization, farm practices have even started to be absorbed into city contexts, which also greatly rely on insect pollinators [Lovell, 2010; Siegner et al., 2018]. Native pollinators foraging has been shown to enhance the seed set and fruit in agricultural and urban contexts even more so than domesticated Honeybees [Garibaldi et al., 2013; Lowenstein et al., 2015; Tanda, 2021a-i]. So, wild pollinators can contribute in the economy of urban agriculture, by sustaining food security in urban contexts. Butterflies and bees are among the potential pollinators can also be used as a means of facilitating community participation in and for their protection. Southon et al., [2017] determined community response to city meadows and observed that not only did participants show a preference for meadows over managed, mown areas, but tolerance for urban meadows enhanced when people were informed of the benefits such habitats can offer for wild pollinators. Pawelek et al., [2009] also described the participation of community in pollinator protection as many gardeners showed a desire to learn more about the native pollinators foraging their gardens and growing flowers that can allure pollinators. The purpose of this chapter is:

To explain the efficiency of green covers and management strategies that have been implemented in protecting the pollinators.

To present the urbanization aspects affecting pollinators, their strength, population, size of body, nesting environment, sociality, generality, and behavior mainly targeting wild bees.

To know the subject of future research so that a holistic comprehension of urbanization impacts on pollinators can be set to develop more efficient policies and conservation measures (Tanda, 2021a-i).

12.2 NATIVE POLLINATORS AND URBAN GREEN HABITATS

Various city green covers variably affect native pollinators, the impacts of which rely on the city landscapes that constitute important habitats. Landscape features concern habitat traits that surround environment gaps, whereas local features refer to the features of such habitats [Pardee et al., 2014; Kennedy et al., 2013]. Impervious surfaces can be grouped in to city landscapes and local features [Ballare et al., 2019; McKinney, 2008]. Impervious surfaces may be more impressive from a landscape outlook and local features are assumed to affect pollinators diversity the maximum [Matteson and Langellotto, 2010; Williams and Winfree, 2013]. The development of surrounding landscapes together with local features can encourage native pollinator biodiversity [Viana, 2012; Williams and Kremen, 2007; Tanda, 2021a-i].

12.3 URBAN LANDSCAPES

12.3.1 Segmentation, Habitat and Diverseness

Segmentation and loss of living environment have been suggested as the most important operators contributing to pollinator's strength and decline in population [Winfree, 2009]. Urbanization, new roads, and buildings in city landscapes makes a fragmented city matrix having environments of small size and quality [Stenhouse, 2004]. Destruction of habitat and fragmentation also affect landscape structures and geography in such a way that changes biological and bee densities including their movement and bee behavior [Hadley and Betts, 2011]. Landscape segmentation can complicate the access to isolated covers, for tiny bees with less mobility [Greenleaf et al., 2007; Hennig and Ghazoul, 2012]. Cities can be quite diverse, due to the various management practices implemented in each fragmented regions. So, the landscape heterogeneity can develop idiosyncratic environments with limited food resources.

12.3.2 Urbanization and Warming

Cities local climates becoming warmer and drier than surrounding areas due to the extensive impervious surfaces all over the city landscapes [Oke, 1973]. This phenomenon of warming, is known as the urban heat island (UHI) effect, can raise urban temperatures 2–4 °C greater than rural areas, even up to 12 °C [Kłysik and Fortuniak, 1999]. This considerable change on urban climate can affect plant densities and advance their floral phenology along with affect the physiology and abundance of insect pollinators. The UHI effect has eased the movement of exotic wasp species *Sceliphron curvatum* into warmer city areas [Polidori et al., 2021]. Similar movements also occur in exotic bees, however, some reports have reported that city UHI effect has no effect on alien bee species [Fitch et al., 2019]. This report may make sense as several alien exotic bees observed are living in cavities and availability of nesting sites may be influencing them more [Wilson and Jamieson, 2019]. The other changes related to UHI effect influence differently to the pollinators [McIntyre et al., 2001]. Within the same taxa, interspecific influences can happen as found in butterflies showing desiccation and heat resistance [Kaiser et al., 2021]. New studies have utilized thermal maximum (CTmax) and critical water content (CWC) values as proxies for assessing bee tolerance to city UHI effect [Burdine and McCluney, 2019; Hamblin et al., 2018]. These metrics has showed morpho-group specific reactions to city habitats comprising enhanced bumblebee thermal tolerance yet low desiccation tolerance in comparison to sweat bees [Kaiser et al., 2021]. Eventually, urban landscapes may be discriminatory suitable to bees and other pollinators having higher heat tolerances than heat susceptible pollinators [Kaiser et al., 2021; Burdine and McCluney, 2019; Hamblin et al., 2018]. Depending on urban area, the UHI effect may help bee growth and development in areas that are cooler, as reported by Burdine et al., [2019]. However, in hot urban landscapes may push bee species to thermal tolerances and ultimately influence their populations [Kaiser et al., 2021]. Research in the US demonstrated that pollinator's abundance reduced by 41% per 1 °C increase [Hamblin et al., 2018]. So, the green space development process may play an important role in urban cooling island (UCI) effects in fragmented green covers [Bowler et al., 2010; Chang et al., 2007]. In the morning buildings develop UCI effects however, green covers can absorb additional heat during noon time depending on the density of impervious surfaces and green spaces [Chang et al., 2007; Gonçalves et al., 2018].

12.3.3 Density of Impervious Surfaces and Landscape Types

The landscape type of a city region can influence the existence of pollinators in a region, especially within agroecosystems where pollinator can enhance fruit set [Kennedy et al., 2013; Klein et al., 2012; Ricketts et al., 2008]. Great pollinator's strength, richness and abundance should be in close proximity to natural resources regions those are away from city centers [Jones and Leather, 2012]. In Poland and France have substantiated this hypothesis while examining urbanization influences on wild bees along city development gradients. By adding more green landscape can enhance the bee richness and abundance [Tonietto et al., 20121]. The presence of semi-natural areas and green spaces within city areas can be helpful for pollinator foraging as these features increase the total number of food resources in or near landscapes [Phillips and Gardiner, 2015]. Semi-natural areas near orchards, have gastropod shells that can be used by bees for nesting sites [Bogusch et al., 2020]. Many reports have studied the influence of impervious on the abundance of animal pollinators, their richness, and structure using the gradients of urbanization [Egerer et al., 2017; Geslin et al., 2016; Senapathi et al., 2017]. In spite of criticisms to use impervious as a metric for urbanization, its utilization has become standardized within city bee pollinator bio-ecology [Wenzel et al., 2020]. In the methods of classification inconsistency may affect as over or underestimation of the actual city impact on native pollinators. Nevertheless, pollinator richness and abundance mostly work in association with enhanced proportions of imperviousness [Kuussaari et al., 2020]. Impervious surface intensity has also been related with movement in pollinator structure where maximum proportions of impervious surface has a higher population of cavity-nesting bees in comparison to ground-dwelling bees [Kuussaari et al., 2020; Cane et al., 2006]. So imperviousness reduce the available nesting space that can be used by ground-nesting bees; but, such approaches are not generalized as urban make-up may differ to influence the bee abundance. Research should be undertaken to prove the filtering out of soil nesting bees.

12.4 ATTRIBUTES OF URBAN STRUCTURES

12.4.1 Local Habitat Climatic Conditions

Local environment or microclimates, within habitats, can differ in cities and even within habitat patches based on the biotic and abiotic components of a landscape [Pincebourde et al., 2016; Fisogni et al., 2020]. Habitat heating and local microclimate can be influenced by the type of vegetation and density of green cover [Crum and Jenerette, 2017]. Mostly surrounding vegetation can decrease the habitat temperature [Chang et al., 2007]. Daffodils develop an independent microclimates which, work as warming pads for bee pollinators at a finer scale [Herrera, 1995]. Shady vegetation can also affect the local microclimates for nesting by cavity-nesting bees [Richards et al., 2020]. Within the nest these microclimates can affect the bee larvae and their metabolic and development rate [Richards et al., 2020; Radmacher and Strohm, 2009]. Nitrogen deposition, which is wide spread in city environment as an abiotic factor can interact synergistically with environmental conditions and may affect micro habitat climate [WallisDeVries and Van Swaay, 2006]. These components can expedite plant growth in spring which has been previously shown to influence the development of thermophilic butterflies by lowering their population [WallisDeVries and Van Swaay, 2006] and similar pattern may also be found in bees and other native pollinators. Only a little work has been carried out in microclimates in spite of the information that landscapes can show microclimate heterogenity [Burdine and McCluney, 2019; Hamblin et al., 2018]. It is important to comprehend how bee species react to a specific habitat features as shifts in microclimate may influence pollinator abundance either through extinction or dispersion.

12.4.2 City Landscapes

In city areas within home yards and gardens, alien and ornamental species can be observed considerably [Threlfall, 2016]. Plant biodiversity adds to the enhanced species strength and abundance which may increase the richness of insect pollinators [Wilson and Jamieson, 2019; Leston and Koper, 2017]. This correlation rely on enhanced plants attractive to bees and lepidopterans rather than just increased plant richness [Martins et al., 2017]. Flower diversity, abundance and consistency, but, may be better impactful than encouraging richness itself [Lanner et al., 2019; Stewart et al., 2018]. Enhanced floral resources have also been shown to affect butterflies and bee population [Matteson and Langellotto, 2010; Werrell et al., 2009; Tanda, 2021a-i]. Great prevalence of alien plants within urban habitats, determining non-native effect on the local landscape and native pollinators is significant. Studies have reported that exotic plants can negatively influence pollinators by affecting foraging and the successful reproduction of native plants [Morales and Traveset, 2009]. Recent studies showed that non-native plants are not approved over native plants among insect pollinators. This preference lacks in part and attributed to potential structure flaws in the reports [Harrison and Winfree, 2015]. Preferences may not differ drastically in generalists, oligolectic species related native species are susceptible to changes in plant groups [Hostetler and McIntyre, 2001]. So, conserving native plant species is one important aspect of sustaining specialist and wild bees. Exotic species may have little influence pollinator presence reports suggest [Mach and Potter, 2018; Giovanetti et al., 2020; Rollings and Goulson, 2019]. In fact, exotic plants may benefit areas by extending floral resource availability throughout the season and contributing to the overall abundance of floral resources [Salisbury et al., 2015]. More attention should be given towards understanding the influence of alien plants having on bee pollinators more conclusively. Native and perennial species show maximum bee population as compared to exotic and annual plants [Lowenstein et al., 2019].

12.4.3 Size of Green Habitat

The size of green cover is crucial in affecting the pollinator present by limiting the food resources and to share the green habitat area. To adequately accommodate sufficient floral resources and nesting sites needed by native pollinators, landscape should be of ample size. The total area required to support pollinators remains poorly appreciated and is perhaps rely on bee movements [Aronson et al., 2017; Lepczyk et al., 2017]. Recent reports on habitat spaces mainly study components of green space area like the proportion of floral resources and, to a lesser extent, sites for nesting availabile [Bogusch et al., 2020]. Beninde et al., [2015] studied green cover needs for pollinators and reported that areas more than 50 ha should be developed to reduce the losses in the richness of species. Too much green cover may not offer any advantages for some taxa as bees have established away from 100-m area [Hennig and Ghazoul, 2012]. Bigger habitats have been observed to positively influence the richness of butterflies and bees, population, and biodiversity [Stewart et al., 2018 ; Konvicka and Kadlec, 2011; Tzortzakaki et al., 2011; Jauker et al., 2012]. These differences in size may be having more effect on the fecundity and population of small bees than larger one, for example *Bombus.*

12.4.4 Landscape Composition

Habitat types or landscape composition can vary in assisting pollinators due to the differences in food resources and its maintenance. Vegetation remnants, city parks, home neighborhood, and golf courses display plant variations in composition and green covers which can affect their resources to bee pollinators [Tonietto et al., 2011; Threlfall et al., 2016]. The food resources and nesting in green spaces depends on the functions of those green covers. Home yards are established for esthetic values and, follow frequent mowing that affect the bee abundance and diversity by weed cutting and useful flowers [Lerman et al., 2018]. City grasslands, need less frequent management, have been reported to live bigger butterflies, Hover flies, and bees in comparison to parks and housing estates which needs minimum management preserves floral resources and nesting sites for pollinators [Dylewski et al., 2019]. Wildflower patches can also be crucial as they can host ground nesting bees such as Andrenidae spp. in spite of occasional mowing [Sirohi et al., 2015]. City gardens can also be hotspots for bees and trait diversity due to the presence of a number of plants, many of which are dependent upon native pollinators [Lanner et al., 2019; Normandin et al., 2017]. Normandin et al., [2017] reported that gardens are less effective in attracting pollinators as compared to city parks being.

Small in size. Various types of parks can also differ in their capability to assist bee pollinators. Development of more recreational parks can assist in encouraging floral resources and food quality [Daniels et al., 2020] and growing maximum gardens within these spaces may be one means of achieving this target [Middle et al., 2014]. Green covers including interspersed in residential regions may be helpful to bees if they display variation in floral resources, food and management [Fischer et al., 2016]. Brownfields, which represent urbanized industrial sites previously, may also be used in many ways that encourage biodiversity in bee abundance [Hunter, 2014] when sites are not isolated and have sufficient foraging flowers [Twerd et al., 2019; Turo et al., 2021]. Due to absence of use, derelict and other post-industrial locations may also be of interest as they can be unmanaged and using minimum sprays and/or chemical exposure, unless located at previous chemical plants [Twerd et al., 2019; Baldock, 2020]. Reports have reported the quality of many green covers, aiming on few specific types mostly [Normandin et al., 2017; Buchholz and Egerer, 2020]. Within an urban environment, assessing the significance of green spaces is crucial to understand which spaces are especially useful for pollinator's conservation and which should be studied further to improve their environment quality.

12.5 EFFECT OF URBAN DEVELOPMENT

City development has been previously reported as a filter that selectively helps the pollinators having different functional attributes which render these taxa less susceptible to densely disturbed habitats. In spite of a city's capacity to help large biodiversity, the cities may develop more homogenized groups depending upon functional characteristics or phylogenetic relations [Martins et al., 2017]. Regarding urban bee taxa and functional traits, however, reports may not be generalized due to specific differences in response [Buchholz and Egerer, 2020; Cariveau and Winfree, 2015]. Common functional traits of pollinators have been found in city areas due to variations in diet, nesting plan, size of bees, their behavior, and phylogenetic assortment among pollinators.

12.5.1 Nesting Sites

Plan for nesting sites applies to bees as the nesting of butterflies and syrphids is dependent on the nearby landscapes. Bees are either subterranean soil-nesters or above-ground nest in cavities. More than 20,400 bee species found globally are ground-nesting; but, cavity-nesting bees are disproportionately distributed in urban studies as a result of the changing of bare ground into impervious surfaces [Buchholz and Egerer 2020; Michener 2007]. City spaces possess novel, above-ground nesting sites, for example, cracks and holes in buildings and, to a lesser extent, bee hotels that can be used by above-ground nesters [Verboven et al., 2014]. *Xylocopa virginica* is observed in proximity of anthropogenic disturbance now using milled lumber as a nesting site [Hurd, 1978]. City gardens and urban parks may encourage soil nesting bees, as these green covers often have bare soil and stem-nesting sites [Lanner et al., 2019].

12.5.2 Pollinator Lecty and Food

Species composition throughout the city have found general trends regarding pollinator dietary breadth and lecty. There is sufficient data to suggest that urbanization help more populations of generalist bee species showing wide foraging needs compared to specialist species having restricted dietary linking's of plant hosts [Lanner et al., 2019; Normandin et al., 2017]. Though specialists are under-represented and even rare in urban [Hernandez et al., 2009], their presence is not entirely lacking, particularly if species-specific needs are available in the green spaces [Lanner et al., 2019; Da Rocha-Filho et al., 2018]. City habitats may help generalist species as floral landscapes, are largely dominated by alien, ornamental and other species which are non-hosts to specialists [Garbuzov et al., 2017; Frankie et al., 2005]. Those flowers having deep corollas may restrict the resources for bees and other pollinators with short tongues [Geslin et al., 2013]. Insufficient host plants in an area may thus efficiently turn city spaces into resource deserts for specialist bees. Knowing the relations between pollinators and the food plants they utilize is important and efficient protection measures are to effectively conserve specialist's species. Recent report in the field on pollinators fail to record species-specific interactions using morpho-group classifications. While the use of morpho-groups offer relevant knowledge about plant–pollinator relationships, interactions identified to the species will be most useful for the diversity and specificity of relationships present in between crop plants and insect pollinators.

12.5.3 Functional Traits and Phylogenetic Diversity

Though many studies target on functional trait homogenization however reports have found closeness in evolutionary outlook [Harrison et al., 2018; Odanaka and Rehan, 2019]. Phylogenetic diversity may concur with functional traits, however, phylogenetic diversity changes may be more useful than functional traits alterations showing taxonomic changes in a group [Harrison et al., 2018; Xu et al., 2019]. In city environments, phylogenetic with functional trait studies may also be more useful in knowing vulnerable bees [Baiser and Lockwood, 2010].

Despite the potential advantage of phylogenetically based studies, some have occurred in city habitats. These approaches executed in urban areas, however, were able to assertion family level phylogenetic homogenization, or predominance of few families, or homogenization was not evident [Harrison et al., 2018]. Harrison et al. (2018) reported that urban areas within the New Jersey region have phylogenetic homogenization represented by bees belonging to Halictidae. Changes in phylogenetic diversity from urbanization is imperative, in view of the reduction in such diversity equal to a bee evolutionary history loss and a depletion in environmental services [Grab et al., 2019]. Phylogenetic reports have shown that *Lasioglossum* (*Dialictus*) and *Bombus* are more resilient than others, such as *Andrena* and *Nomada* who may be vulnerable to urban changes [Grab et al., 2019; Matteson et al., 2008]. These clades may be more susceptible to urbanization due to their special diets, small flight seasons, and other traits that are phylogenetically associated [Harrison et al., 2018]. Few vulnerable cleptoparasites, may be beneficial in working as bio-indicators of community health [Sheffield et al., 2013].

12.5.4 Urbanization and Pollinator Body Size

On the pollinator body size urbanization has shown that cities may encourage smaller body size pollinators. Few reports found higher proportions of large-bodied bees and butterflies, however others observed an opposite trend across urban landscapes [Fortel et al., 2014; Wray et al., 2014]. The flight distance for bees has been correlated with body size, it has been presumed that large bees, such as *Bombus*, are capable to navigate between patches more efficiently than smaller bees including *Lasioglossum*, which show shorter flight ranges [Gunnarsson and Federsel, 2014]. It is suspected that in semi-natural grasslands, this fast mobility related with the richness of larger bees [Jauker et al., 2012]. This trend may be similar in cities where a large population butterflies has also been found. In spite of more mobility of larger pollinators, however, smaller species may need less resources to live in urban regions [Fortel et al., 2014]. UHI effects and less food availability could be driving smaller body size in traditionally larger pollinators such as *Bombus lapidaries* and *B. pascuorum* [Eggenberger et al., 2019]; however, other sources have found the opposite effect in *B. terrestris* better related to urban fragmentation. It may be due to that various factors not only idiosyncratically affect body size but do so in a way that differs intra and inter-specifically across different taxa also [Merckx et al., 2018; Merckx et al., 2018]. Further observations are required to better extricate which city features are more greatly influencing size of the body.

12.5.5 Urbanization and Pollinator Behavioral Traits

Urbanization influences many behavioral traits of pollinators such as foraging and sites of nesting, anti-predator response, to potentially filtering bees depending on their social behavior. In urban areas, microclimate and nesting sites may change the nest-site decisions of some bees as was found in *Megachile rotundata*. Artificial nest boxes, tubes having lower temperatures were inhabited more frequently than those showing higher temperatures in this study [Wilson et al., 2020]. The degree of urbanization can impact network interactions and foraging rates between wild bees and available landscape plants. Across urban areas, sweep netting has showed the preference for native floral resources and less foraging to alien plant species [Buchholz and Kowarik, 2019]. Whether solitary or social species are most abundant in city landscapes remains inconclusive in reports. However, few studies showed a large population of solitary bees [Martins et al., 2017], others indicate that social species found most frequently in city landscapes [Harrison et al., 2017]. Assessing urban bee functional traits, a systemic review however, found no generalized trend in social behavior of bees [Buchholz and Egerer, 2020]. Wilson and Jamieson [2019] included cleptoparasitic and sub-social bees into their solitary group whereas other reports may assess these categories differently. Research should continue to determine impacts of urbanization on functional traits such as sociality, trends may be specific and not generalized due to the differences of response found in bee species and the study areas [Cariveau and Winfree, 2015]. The urbanization influence on the phenology of pollinators, cities may help later season bees more than early emerging bee species [Banaszak-Cibicka and Zmihorski, 2012]. These may be due to changes in microclimate which can develop mismatches between angiosperms and bees [Goulson et al., 2015]. Pollinators that show more adaptive response to phenological shifts, like social bees with prolonged foraging seasons, may be able to take advantage of resources more efficiently than species with shorter active periods [Harrison and Winfree, 2015].

Impact of floral resources lawn mowing effect (LME), and habitat development for urban pollinators: a case study

City and countryside experiments

To assess the comparative impact on insect pollinators, during 2020-21, City and countryside experimental study sites were selected and marked at Rosehill NSW 2142, Australia (33.82721, 151.01699; 33°49′37.96″S, 151°1′1.16″) in Parramatta, New South Wales. For the comparative studies on the insect pollinators, observations were collected from the marked areas of Richmond, New South Wales Australia (33° 35′ 54″-150°45′ 04″ East). Here we used a paired study design explicitly to test the impact of urbanization on insect abundance, biodiversity, and the ecological natural service of pollinators. To evaluate flying insect diversity, we adopted metabarcoding methodology. To quantify pollination, we used 'pollinometers', potted, yard-raised, insect-pollinator-dependent basil plants (Fig. 12.3).

Plant flowering resources, dish liquid detergent chemical sprays, lawn mowing city study sites.

To assess the impact of chemical applications on pest and pollinator's visitation and density (dish liquid detergent), homemade solution spray was prepared dissolving 12 mL of dish liquid detergent, with a few drops of vegetable oil in 500 mL of water used against the aphid populations on radish and mustard. Plants of okra, Chinese mustard, radish, cucurbits were raised in the home garden which was also having pomegranate, peaches, and basil plants for alluring maximum pollinators and providing extensive floral food resources.

Fig. 12.3 Basil plants were raised for alluring bees and other pollinators for nectar and pollen food resources.

To record the effect of lawn mowing fully or leaving few strips on the pollinator's abundance, foraging, lawn strips in the backyard and in the front yards, were mowed and left with some restrictions. Observations were recorded from the grass strips with flowering weeds such as dandelions and clover which served as food resources to the pollinators.

Measurement of relative abundance of insect pollinator's diversity, in city and countryside study areas

To compare the diversity of flying insects, we observed the main orders of flying insect pollinators: Diptera, Lepidoptera, Coleoptera, and Hymenoptera. Though our experimental plants were primarily visited by bees, these insects helped us to explore their taxonomic diversity in response to urbanization. We employed pan traps to sample insects, which are a standardized and commonly used method for collecting flying insects. The relative abundance of pollinators (number of flowers visited by pollinators) was observed randomly on selected five plants (8.00, 12.00, 13.00, and 16.00 hr) during the peak flowering period. Samples on insect pollinators were collected randomly using hand-nets [Burr et al., 2018; Byrne, 2005]. The observations on foraging behavior, viz., no. of flowers (no. of pollinators/ m^2/min) visited foraging rate (mean number of flowers visited/ min) were observed with an electronic stopwatch.

Weather parameters

Data on weather parameters like temperature, relative humidity (%), wind speed (km/hr), and evaporation (mm) were collected daily from the Rosehill Weather zone Mobile. Detailed 48-hour forecast of weather, wind speeds, temperature, rain, and humidity were updated every hour from the Weather zone.

Statistical Analysis

Data were analyzed and correlation coefficients between flower-visiting insect pollinators and meteorological parameters were computed.

Results

Flower density study in urban versus rural sites

Results showed that the flower resources were higher (35-37%) in flower density at Rosehill residential garden as compared to the rural land site selected at Richmond rural ecosystem, as there were only grasses and other weeds with little blooms almost negligible (1.5%) for any effective foraging or insect visitation (Table 12.1).

Table 12.1 Flower density in residential and Richmond rural ecosystems

Areas/sites under experiment	Vegetable/herb flowers composition (Average %)	Fruit plant flowers (Average %)	Weed flowers (Average %)	Flower density (Average %)
Rosehill urban residential	Okra, Chinese mustard, radish, cucurbits, basil (20%)	Pomegranate, peaches (10%)	Dandelions, Others (5-7%)	35-37%
Rural Richmond	Nil	Nil	Grasses, weeds	1.5%

Pollinator's abundance and density in urban versus rural sites

Our observations revealed that Sydney cities harbor lower insect species richness, especially for Diptera and Lepidoptera than the rural regions. In contrast, we recorded higher Hymenoptera species richness and flower visitation rates by bees in urban areas, positively driven by edge density of green spaces, and associated with enhanced pollination. Pollinator populations were critically recorded in both domestic gardens and rural habitats in varied climatic environments having great variations of flower resources. In the domestic gardens, as the flower constancy and density were greater than the rural land habitat, the pollinators were observed more may be due to the absence of flower resources. About 22-25% higher pollinator's population was recorded in the domestic gardens as compared to the countryside land area. We assessed flying insect diversity at urban and adjacent rural sites, and observed that rural sites supported a higher overall insect species richness and biomass as compared to urban sites. About 15% more pollinator's abundance including honeybees, butterflies, flies, and ants, were recorded in higher floral resource areas with pollinator attractant plants such as basil. Less frequent mowing also attracted more native pollinators like Dandelion flowers and other weeds that acted as sources of nectar and pollen in the domestic gardens. Furthermore, chemical sprays on mustard and radish blossoms reduced the number of pollinators, and 35% more honeybees, butterflies, hovering flies, and ants were observed in the unsprayed vegetable patches (Table 12.2).

Table 12.2 Pollinator's abundance in residential and Richmond rural ecosystems

Insect /Pollinator	Urban domestic gardens abundance (Average %)	Country side area Richmond (Average %)
Apis spp.	23.75	0.00
Butterflies	7.00	0.00
Flies	13.00	0.00
Black ant	14.23	2.00
Small ant (*Solenopsis* sp.)	8.02	1.05

Apis spp. were the most dominant flower visitors with 23.75% followed by black ant with 14.23% and small ant (*Solenopsis* sp.) with 8.02% among total flower-visiting insect pollinators in flowering plant-rich sites.

Effect of climate on pollinator's abundance and foraging

At 45-50% relative humidity, *A. mellifera* bees visit was low (1-2) in numbers. These results agree with those who reported that the flight activity of *Apis* spp. begins at 12°C-14°C and decreased with increasing wind speed in blueberry and cotton. When wind speed was higher, the insect pollinators' almost did not visit the flowers (0.01%). Observations on seasonal abundance revealed that the foraging activity of *A. mellifera* was increased with an increase in temperature from 29°C-34°C. The foraging population of *A. mellifera* was highly significant and positively correlated with maximum temperature, whereas, it was negatively correlated with relative humidity during evening hours.

Effect of dish liquid detergent chemical sprays on pollinator's abundance

To assess the effect of dish liquid detergent chemical, the homemade solution was sprayed against the aphid populations on radish and mustard in the early morning or early evening, as the cooler temperatures slow evaporation. Pollinator activity tends to be low during these hours, still impaired bees and flies foraging activity on mustard and radish blossoms and reduced (4%) the average number of pollinators. More (35%) honeybees, butterflies, hover flies, and ants were observed in the unsprayed vegetable patches (Table 12.3).

Table 12.3 Effect of detergent/chemical spray on native pollinators

Native pollinators	Pollinators on flowers treated with detergent spray (Average %)	Pollinators on flowers treated without detergent spray (Average %)
Apis spp.	2.35	23.75
Butterflies	1.10	7.00
Flies	0.15	3.00
Ants	0.00	2.02

Effect of lawn mowing on pollinator's abundance

It was observed that green spaces embedded within the urban matrix, especially home yards, could alleviate negative aspects of urbanization and offer habitat to pollinators. City lawns constitute a supreme green cover, and their management comprises mowing to stop the growth of dandelions and clover. As bees and other pollinators' declines are a growing concern from habitat losses, these weeds spontaneous flowers provide pollen and nectar sources throughout the year. We experimentally tested whether different lawn mowing frequencies (1, 2, or 3 weeks) influenced bee abundance and diversity in Rosehill yards by enhancing lawn floral resources. Lawns mowed every three weeks had as many lawn flowers (2.1 times more) than the other frequencies. Interestingly, lawns mowed every two weeks helped in great bee abundance yet the lowest bee strength (Table 12.4).

Table 12.4 Effect of lawn mowing on native pollinators in the domestic gardens

Insect / Pollinator	Regular mowing (Average %)	Mowing leaving grass strips (Average %)	No mowing (Average %)
Apis spp.	0.00	7.75	11.75
Butterflies	0.00	1.15	2.15
Flies	0.00	1.00	1.15
Ants	0.75	0.75	1.10

Flower resources were higher in flower density at Rosehill residential garden for efficient foraging in comparison to the rural land may be due to more flowering plants grown by the residents or the city council and grasses and weeds in the neglected areas. Sydney cities harbor inferior insect species richness, especially for Diptera and Lepidoptera than the rural regions, maybe due to interspecies competition, flower preferences, or habitat unsuitability. However, species richness by bees belonging to Hymenoptera was great including flower visitation rates in the urban areas, may positively be driven by edge density of green spaces, and associated with enhanced pollination. The phylogenetic diversity can be an important indicator and predictor of the influences of bio-diverse ecological system services. These observations are in in line with earlier studies (Muratet and Fontaine, 2015; Tanda, 2021a-i). The pollinator population density and flower visitation were critically higher in the domestic gardens than in the rural land environment, maybe due to the more flower constancy and density of the pollinator's preferences and favorable habitat as compared to the rural region due to the absence of flower resources. About 22-25% higher pollinator's population was observed in the domestic gardens in comparison to the countryside land area. Present studies established that native landscape drivers can affect local pollinators' abundance, and their conservation may be enhanced by providing food plants sources, hence sustainability of the bio-diverse system. Rural sites supported a higher overall insect species richness and biomass compared to urban sites, which remained significant when controlling for local (patch) and landscape variables. Pollinator's abundance including honeybees, butterflies, flies, and ants, was found 15% higher in floral resource areas may be due to the presence of pollinator attractant plants such as basil. Less frequent mowing also attracted more native pollinators as there were dandelion flowers and other weeds that acted as sources of nectar and pollen in the domestic gardens. Chemical sprays on mustard and radish blossoms reduced the number of pollinators, and 35% more honeybees, butterflies, hover flies, and ants were observed in the unsprayed vegetable patches, suggested not to use the chemicals at the flowering periods or determine the pollinator's abundance by monitoring before taking any action as reported by other reports (Simao et al., 2018; Muratet and Fontaine, 2015; Tanda, 2021a-i). *Apis* spp. were the most dominant flower visitors with 23.75% followed by black ant with 14.23% and small ant (*Solenopsis* sp.) with 8.02% among total flower-visiting insect pollinators in flowering plant-rich sites. These findings are in close agreement with earlier studies that reported that hymenopterans were more (Simao et al., 2018; Muratet and Fontaine, 2015;Tanda, 2021a-i). It may be due to the presence of bright yellow-colored flowers with sufficient pollen grains and nectar in mustard flowers, which attracted the various types of insect pollinators as compared to white-colored flowers radish plants grown nearby. Basil blossoms always allured honeybee foragers as compared to other flowers, maybe due to the higher nectar and pollen availability. These findings are in line with those who reported that hymenopterans were more (Simao et al., 2018; Muratet and Fontaine, 2015; Tanda, 2021a-i). At 45-50 % relative humidity, *A. mellifera* bees visit was low (1-2) in numbers. These results agree with those who reported that the flight activity of *Apis* spp. begins at 12°C-14°C and decreased with increasing wind speed in blueberry and cotton. When wind speed was higher, the insect pollinators almost did not visit the flowers (0.01%). Observations on seasonal abundance revealed that the foraging activity of *A. mellifera* was increased with an increase in temperature from 29°C-34°C. The foraging population of *A. mellifera* was highly significant and positively correlated with maximum temperature, whereas, it was negatively correlated with relative humidity during evening hours. These findings are in close agreement with flower-visiting insect pollinators and their response to the various weather parameters recorded during the peak flowering period of other crops, such as peaches, pear, okra, and cotton. To assess the effect of dish liquid detergent chemical, the homemade solution was sprayed against the aphid populations on radish and mustard in the early morning or early evening, as the cooler temperatures slow evaporation. Pollinator activity tends to be low during these hours, still impaired bees and flies foraging activity on mustard and radish blossoms and reduced (4%) the average number of pollinators. More (35%) honeybees, butterflies, hover flies, and ants were observed in the unsprayed vegetable patches (Table 12.3) are in conformity with other studies (Muratet and Fontaine, 2015; Tanda, 2021a-i). It was observed that green spaces embedded within the urban matrix, especially home yards, could alleviate negative aspects of urbanization and offer habitat to pollinators. City lawns constitute a supreme green cover, and their management comprises mowing to stop the growth of dandelions and clover. As bees and other pollinators' declines are a growing concern from habitat losses, these weeds spontaneous flowers provide pollen and nectar sources throughout the year. Lawns mowed every three weeks had as many lawn flowers as the other frequencies as reported earlier (Simao et al., 2018; Tanda, 2021a-i). Interestingly, lawns mowed every two weeks helped in great bee abundance yet the lowest bee strength. Our results showed that a lazy lawnmower approach to offer bee habitat is beneficial for keeping pollinators populations. Low frequently lawn mowing is practicable, economical, and timesaving activity management to lawn replacement or even planting pollinator gardens. Our findings provide immediate solutions for individual residents to contribute towards native pollinator's conservation (Simao et al., 2018; Muratet and Fontaine 2015; Tanda, 2021a-i).

In this study, we found that the experimentally quantified provision of pollination services was higher in urban than in rural sites. This difference was likely driven by increased visitation rates, largely of honeybees, within urban habitats and was associated with increasing Hymenopteran phylogenetic diversity and Hymenopteran species richness, which were enhanced by increasing habitat edge density but suppressed by agriculture in the surrounding landscape. In contrast, Diptera and Lepidoptera were less diverse at urban sites. A number of pollinator species were recorded from the home garden according to their ability to pollinate flowers. Some of these native pollinators have proven to be beneficial pollinators of vegetables and fruit plants in the residential gardens. Particularly noteworthy is the honeybee, *A. mellifera* L., which was characterized by a pronounced ability to control the pollination of crop plants with an efficiency comparable to that of the best wild species used in cross-pollination. In addition, the conducted urbanization studies made it possible to find the *A. mellifera* L., which significantly differs in its behavioral and morphological characteristics from all known representatives of Hymenoptera. About 15% more pollinator's abundance including honeybees, butterflies, flies, and ants, were recorded in higher floral resource areas with pollinator attractant plants such as basil. Less frequent mowing also attracted more native pollinators like dandelion, clover flowers and other weeds that acted as sources of nectar and pollen and habitat in the city domestic gardens. Furthermore, chemical sprays on mustard and radish blossoms reduced the number of pollinators, and 35% more honeybees, butterflies, hover flies, and ants were observed in the unsprayed vegetable patches. Our innovative findings, suggest to the current knowledge that more floral resources by growing vegetable and fruit plants, less mowing of lawns, and providing suitable habitat to the urban pollinators can be beneficial for the native pollinators conservation and better quality kitchen garden produce. Thus, both entpollinalogically significant species reflecting the flower-pollinators biodiversity of both areas were established (Tanda, 2021a-i). Further detailed studies on their behavior, habitat adoption, contribution to pollination, and conservation are ongoing. In conclusion, as cities expand globally, appropriate urban strategies and stakeholder engagement to offer local floral resources and to enhance the cover and connectivity of seminatural flora and another green cover could boost their value as refuges for pollinators badly impacted by crop intensification practices. Urban centers across the globe could thereby serve as food resources of pollinators and hotspots of the ecosystem service of pollination of urban crops and wildflowers. From the perspective of the rural landscape, advanced pollinators and integrated pest management technological opportunities are likely to conserve and increase pollinator's diversity for the sustainable bio-ecosystem of pollination for food security and wild native flower multiplication (Tanda, 2021a-i).

12.6 TARGETS AND GUIDELINES FOR CONSERVATION

Conservation measures within city habitats target at least bee pollinator declines greatly include the development and protection of green covers throughout the city matrix. Assuring committed sites, which may vary in flower resources based on the landscape type, adequately quality food directed towards multiple pollinating taxa is important considering the various requirements of native bees. Few reports give suggestions on how to alter spaces for plant pollinators, noticing a balance between the public and stakeholders obtaining protection target remains a complex matter [Turo and Gardiner, 2019]. The economic and political part of conservation strategies, however, may be challenging as policymakers may be unwilling to execute expensive alterations that do not manage immediate results [Howes et al., 2017]. The people have an influence in choosing plants, which are mostly selected on aesthetic values [Burr et al., 2018; Byrne, 2005]. Also, in the public landscape acceptance rely on average income of the home owner, gender, and their education [Zhao et al., 2019]. The green space preservation may not solely help pollinators; but, as positive human health benefits have been also related with green covers in the surroundings [Hunter et al., 2019; Keeler et al., 2019; Wolch et al., 2014]. Previously, landscape exposure has been reported to be negatively correlated with mortality and positively associated with physical activity and mental health aspects [Kondo et al., 2018]. Furthermore, city parks can reconnect people with similar nature and promote community activities of conservation targets [Duchemin et al., 2009]. Economic costs may hold precedence over social, ecological and recreational interests [Tibesigwa et al., 2020]. Efficient methods should be designed to ensure that urban ecological plans are not only of high-quality and fulfil desires but are also cost-effective. In spite of any hesitancy shown by potential stakeholders, the green cover inclusion and modified management process may decrease total expenses in long time period. Less frequent mowing not only helps plant pollinators by preserving floral resources but also minimizes the spending of mowing [Aguilera et al., 2018] keeping the aesthetic value. Keeping perennial plants may also lower the management costs by alleviating the frequency at which seeds require to be re-sown compared to more labor-intensive annuals [MacIvor et al., 2013]. Seed mix selections in green covers could also increase environment quality and stop ineffective costs. Seed mixes used in urban designs can undoubtedly allure more pollinators such as bumblebees and hover flies; however, the plants selected in such mixes are significant. Specific plant species can greatly increase the pollinator foraging rates within the habitats. Many species of plants used in seed mixes, sold in stores, and mostly observed in urban spaces may be of insignificant value to bee pollinators, and those that are described as pollinator alluring flowering plants lack empirical evidence [Harmon-Threatt et al., 2015]. As pollinators show a broad food preferences, explaining species-scale liking is imperative to manage many aspects of pollinator groups [Turo and Gardiner, 2019]. Setting up empirically as in bumblebee's plant-pollinator interactions could be deciding species resolution pollinator resource usage and floral visitation [Sikora et al., 2020]. Such networks are designed based on field data or pollen extractions and bee sampling from flowers directly [Tucker and Rehan, 2016]. Similar networks have been described that the inclusion and/or exclusion of plants included in seed mixes utilized in prairie restoration can significantly influence bee richness and population [Harmon-Threatt et al., 2015]. Framed networks can then show how various native pollinators rely on different food resources and depict how network attributes such as generality can move across city greener spaces. Not only can this be significant for knowing if some pollinators, especially oligolectic specialists, have adequate floral resources, but conversely, interactions can also designate whether a plant species is alluring adequate foragers. The strong management and other pollinator conservation plans, however, in part greatly rely on the passage of appropriate legislature from national and city governance. While pollinator-conserving legislature has only until recently started to gain traction, several nations have made many steps to improve attempts. From 2000 to 2017, about 110 laws related with pollinators have been passed at the US state-level [Hall and Steiner, 2019]. Measures within cities have started, as the City of Toronto government, has designed a pollinator protection plan to encourage pollinator awareness and conservation [City of Toronto, 2020]. For example, the Xerces Society, the invertebrate conservation groups have identified cities within the United States and Canada that encourage pollinator biodiversity as "Bee Cities" [Xerces Society, 2020]. Not only can the passage of such strategies offer few direct degree of ensured protection for pollinators, but they can be utilized to gather public support and promote community activities and their education [Baldock, 2020]. To accurately assess the extent to which urbanization influences pollinators, workers should modify their trends to urban study. To begin, detailed research should be established over those more stable results and approaches can be discovered. Further, research should direct more on the South American, Asia, and Australian continents as the recent report is overrepresented by North American and European findings [Buchholz and Egerer, 2020]. Research should emphasize its attention to understudied areas of pollinators as results may not be generalized to such locations easily [Stewart et al., 2018; Ayers and Rehan, 2021; Tanda, 2021a-i].

12.7 CONCLUSION

Though city landscapes may abate native bee pollinator population more broadly and help particular functional and taxonomic attributes over others, cities can still hold significance as pollinator habitats [Hamblin et al., 2017; Ayers and Rehan, 2021; Tanda, 2021a-i]. These moves in pollinators, which assist traits such as polylecty and nesting in cavities, are results of interacting local and landscape features which apply differential, non-generalizable effects on various pollinators [Cariveau and Winfree, 2015]. Extricating these urban impacts on pollinators is crucial to know how urbanization can best aid pollinator communities. Landscapes able to implement knowledgeable designs, principles and management strategies may efficiently baffle pollinator biodiversity losses. Green spaces should be designed in such a way that meets pollinator and other sensitive target group demands. Landowners should adopt protection approaches such as management intensity [Zurbuchen et al., 2010], encouraging native floral abundance and richness, and developing bare patches for nesting bees in ground [Cane et al., 2006]. Cities should aspire to green roofs and other green infrastructure to minimize fragmentation and encourage associated, pollinator resource availability and pollinator shift [Braaker, 2017]. Setting up a scientific consensus on city pollinator ecology is also crucial if implemented policy guidelines are to be successful in encouraging pollinator biodiversity and requesting to policy makers and the citizens. Pollinator health may continue to decline if a single party remains unmotivated to work. The science of communication is important to ensure meaningful alterations can happen collaboratively and collectively. This trend should, of course, not only relate to city pollinators but to all endangered bio-diverse communities. However, native pollinators, can work as a significant community to articulate conservation measures amongst the citizens and encourage the strong co-existence of ecosystem and city community (Ayers and Rehan, 2021; Tanda, 2021a-i).

Challenges to city pollinators such as intensification, habitat changes, pesticides and pollution. Additional beehives could negatively impact wild pollinators in cities. Parks and residential gardens benefit native bee pollinators. Flowering plants, nesting sites and less lawn mowing benefits pollinators. Multiple stakeholders are crucial for native pollinator conservation. Urbanization is considered to be one of the main threats to world biodiversity yet few pollinator communities, especially bees, can do excellent in cities (Ayers and Rehan, 2021; Tanda, 2021a-i). New research indicate that both local and landscape-level drivers can impact urban pollinators, with local floral resources and the number of impervious surfaces in the landscapes influencing pollinator population, richness and group composition. City intensification, pesticides, climate and enhanced honeybee colony densities all negatively influence city pollinators. Managing good areas of environment for pollinators, such as those observed in community gardens and home gardens, and improving management methods in city green covers and by increasing floral resources and nesting sites will help pollinator preservation. For pollinators, opportunities conservation work through multiple stakeholders comprising policymakers, urban citizens, planners and landscape architects (Ayers and Rehan, 2021; Tanda, 2021a-i).

REFERENCES

- Aguilera G, Ekroos J, Persson AS, Pettersson LB, Öckinger E (2018) Intensive management reduces butterfly diversity over time in urban green spaces. Urban Ecosyst., 22, 335–344.
- Aronson MFJ, Lepczyk C, Evans KL, Goddard M, Lerman SB, MacIvor JS, Nilon CH, Vargo T (2017) Biodiversity in the city: Key challenges for urban green space management. Front. Ecol. Environ., 15, 189–196.
- Ayers AC, Rehan SM (2021) Supporting bees in cities: How bees are influenced by local and landscape features. Insects, 12(2), 128. https://doi.org/10.3390/insects12020128.
- Baiser B, Lockwood JL (2010) The relationship between functional and taxonomic homogenization: Functional and taxonomic homogenization. Glob. Ecol. Biogeogr., 20, 134–144.

- Baldock KC (2020) Opportunities and threats for pollinator conservation in global towns and cities. Curr. Opin. Insect Sci., 38, 63–71.
- Baldock KCR, Goddard MA, Hicks DM, Kunin WE, Mitschunas N, Osgathorpe LM, Potts SG, Robertson KM, Scott AV, Stone GN, et al. (2015) Where is the UK's pollinator biodiversity? The importance of urban areas for flower-visiting insects. Proc. R. Soc. B Boil. Sci., 282, 20142849.
- Baldock KCR, Goddard MA, Hicks DM, Kunin WE, Mitschunas N, Morse H, Osgathorpe LM, Potts SG, Robertson KM, Scott AV, et al. (2019) A systems approach reveals urban pollinator hotspots and conservation opportunities. Nat. Ecol. Evol., 3, 363–373.
- Ballare, KM, Neff JL, Ruppel R, Jha S (2019) Multi-scalar drivers of biodiversity: Local management mediates wild bee community response to regional urbanization. Ecol. Appl., 29, e01869.
- Banaszak-Cibicka W, Twerd L, Fliszkiewicz M, Giejdasz K, Langowska A (2018) City parks vs. Natural areas – Is it possible to preserve a natural level of bee richness and abundance in a city park? Urban Ecosyst., 21, 599–613.
- Banaszak-Cibicka W, Zmihorski M (2012) Wild bees along an urban gradient: Winners and losers. J. Insect Conserv., 16, 331–343.
- Beninde J, Veith M, Hochkirch A (2015) Biodiversity in cities needs space: A meta-analysis of factors determining intra-urban biodiversity variation. Ecol. Lett., 18, 581–592.
- Bogusch P, Hlaváčková L, Gasol NR, Heneberg P (2020) Near-natural habitats near almond orchards with presence of empty gastropod shells are important for solitary shell-nesting bees and wasps. Agric. Ecosyst. Environ., 299, 106949.
- Bowler DE, Buyung-Ali L, Knight TM, Pullin AS (2010) Urban greening to cool towns and cities: A systematic review of the empirical evidence. Landsc. Urban Plan., 97, 147–155.
- Braaker S, Obrist MK, Ghazoul J, Moretti M (2017) Habitat connectivity and local conditions shape taxonomic and functional diversity of arthropods on green roofs. J. Anim. Ecol., 86, 521–531.
- Buchholz S, Egerer M (2020) Functional ecology of wild bees in cities: Towards a better understanding of trait-urbanization relationships. Biodivers. Conserv., 29, 2779–2801.
- Buchholz S, Kowarik I (2019) Urbanisation modulates plant-pollinator interactions in invasive vs. native plant species. Sci. Rep., 9, 1–9.
- Burdine JD, McCluney KE (2019) Differential sensitivity of bees to urbanization-driven changes in body temperature and water content. Sci. Rep., 9, 1–10.
- Burr A, Hall DM, Schaeg N (2018) The perfect lawn: Exploring neighborhood socio-cultural drivers for insect pollinator habitat. Urban Ecosyst., 21, 1123–1137.
- Byrne L (2005) Of Looks, Laws and Lawns: How Human Aesthetic Preferences Influence Landscape Management, Public Policies, and Urban Ecosystems. In Proceedings of the Urban Rural Interface Conference Proceedings, Atlanta, GA, USA, 13–16 March; pp. 42–46.

- Cane JH, Minckley RL, Kervin LJ, Roulston TH, Williams NM (2006) Complex responses within a desert bee guild (Hymenoptera: Apiformes) to urban habitat fragmentation. Ecol. Appl., 16, 632–644.
- Cardoso MC, Gonçalves RB (2018) Reduction by half: The impact on bees of 34 years of urbanization. Urban Ecosyst., 21, 943–949.
- Cariveau DP, Winfree R (2015) Causes of variation in wild bee responses to anthropogenic drivers. Curr. Opin. Insect Sci., 10, 104–109.
- Center for Sustainable Systems, University of Michigan. U.S. Cities Factsheet; Ann Arbor, MI, USA, 2020.
- Chang C-R, Li M-H, Chang S-D (2007) A preliminary study on the local cool-island intensity of Taipei city parks. Landsc. Urban Plan., 80, 386–395.
- City of Toronto. Toronto Pollinator Protection Strategy (2018). Available online: https://www.toronto.ca/wp-content/uploads/2018/05/9676-A1802734_pollinator-protection-strategy-booklet.pdf (accessed on 22 November 2020).
- Crum SM, Jenerette GD (2017) Microclimate variation among urban land covers: The importance of vertical and horizontal structure in air and land surface temperature Relationships. J. Appl. Meteorol. Clim., 56, 2531–2543.
- Da Rocha-Filho LC, Ferreira-Caliman MJ, Garófalo CA, Augusto SC (2018) A specialist in an urban area: Are cities suitable to harbour populations of the oligolectic bee centris (Melacentris) collaris (Apidae: Centridini)? Ann. Zool. Fenn., 55, 135–149.
- Dale AG, Frank SD (2018) Urban plants and climate drive unique arthropod interactions with unpredictable consequences. Curr. Opin. Insect Sci., 29, 27–33.
- Daniels B, Jedamski J, Ottermanns R, Ross-Nickoll MA (2020) "Plan bee" for cities: Pollinator diversity and plant-pollinator interactions in urban green spaces. PLoS One, 15, e0235492.
- Deguines N, Julliard R, De Flores M, Fontaine C (2016) Functional homogenization of flower visitor communities with urbanization. Ecol. Evol. 6, 1967–1976.
- Duchemin E, Wegmuller F, Legault A-M (2009) Urban agriculture: Multi-dimensional tools for social development in poor neighbourhood. Field Actions Sci. Rep., 2, 1–8.
- Dylewski Ł, Máckowiak Ł, Banaszak-Cibicka W (2020) Linking pollinators and city flora: How vegetation composition and environmental features shapes pollinators composition in urban environment. Urban For. Urban Green., 56, 126795.
- Dylewski Ł, Máckowiak Ł, Banaszak-Cibicka W (2019) Are all urban green spaces a favourable habitat for pollinator communities? Bees, butterflies and hover flies in different urban green areas. Ecol. Èntomol. 44, 678–689.
- Egerer M, Arel C, Otoshi MD, Quistberg RD, Bichier P, Philpott SM (2017) Urban arthropods respond variably to changes in landscape context and spatial scale. J. Urban 3, 1–10.
- Eggenberger H, Frey D, Pellissier L, Ghazoul J, Fontana S, Moretti M (2019). Urban bumblebees are smaller and more phenotypically diverse than their rural counterparts. J. Anim. Ecol., 88, 1522–1533.

- Fenoglio MS, Rossetti MR, Videla M (2020) Negative effects of urbanization on terrestrial arthropod communities: a meta-analysis. Glob. Ecol. Biogeogr. 29, 1412–1429.
- Fischer LK, Eichfeld J, Kowarik I, Buchholz S (2016) Disentangling Urban habitat and matrix effects on wild bee species. PeerJ, 4, e2729.
- Fisogni A, Hautekèete N, Piquot Y, Brun M, Vanappelghem C, Michez D, Massol F (2020) Urbanization drives an early spring for plants but not for pollinators. Oikos, 129, 1681–1691.
- Fitch G, Wilson CJ, Glaum P, Vaidya C, Simao M-C, Jamieson MA (2019) Does urbanization favour exotic bee species? Implications for the conservation of native bees in cities. Biol. Lett. 15, 20190574.
- Fortel L, Henry M, Guilbaud L, Guirao AL, Kuhlmann M, Mouret H, Rollin O, Vaissière BE (2014) Decreasing abundance, increasing diversity and changing structure of the wild bee community (Hymenoptera: Anthophila) along an urbanization gradient. PLoS One, 9, e104679.
- Frankie GW, Thorp RW, Schindler M, Hernandez J, Ertter B, Rizzardi M (2005) Ecological patterns of bees and their host ornamental flowers in two northern California cities. J. Kans. Èntomol. Soc., 78, 227–246.
- Garbuzov M, Alton K, Ratnieks FL (2017) Most ornamental plants on sale in garden centres are unattractive to flower-visiting insects. PeerJ, 5, e3066.
- Garibaldi LA, Steffan-Dewenter I, Winfree R, Aizen MA, Bommarco R, Cunningham SA, Kremen C, Carvalheiro LG, Harder LD, Afik O, et al. (2013) Wild pollinators enhance fruit set of crops regardless of honeybee abundance. Science, 339, 1608–1611.
- Geslin B, Gauzens B, Thébault E, Dajoz I (2013) Plant pollinator networks along a gradient of urbanisation. PLoS One, 8, e63421.
- Geslin B, Le Féon V, Folschweiller M, Flacher F, Carmignac D, Motard E, Perret S, Dajoz I (2016) The proportion of impervious surfaces at the landscape scale structures wild bee assemblages in a densely populated region. Ecol. Evol. 6, 6599–6615.
- Giovanetti M, Giuliani C, Boff S, Fico G, Lupi D (2020) A botanic garden as a tool to combine public perception of nature and life-science investigations on native/exotic plants interactions with local pollinators. PLoS One, 15, e0228965.
- Gonçalves A, Ornellas G, Ribeiro AC, Maia F, Rocha A, Feliciano M (2018) Urban cold and heat island in the city of bragança (Portugal). Climate, 6, 70.
- Goulson D, Nicholls E, Botías C, Rotheray EL (2015) Bee declines driven by combined stress from parasites, pesticides, and lack of flowers. Science, 347, 1255957.
- Government of Canada, Statistics Canada. Canada's Population Estimates: Subprovincial Areas, 1 July 2018. Statcan.gc.ca; 2019. Available online: https://www150.statcan.gc.ca/n1/daily-quotidien/190328/dq190328b-eng.htm (accessed on 27 November 2020.
- Grab H, Branstetter MG, Amon N, Urban-Mead KR, Park MG, Gibbs J, Blitzer EJ, Poveda K, Loeb G, Danforth BN (2019) Agriculturally dominated landscapes reduce bee phylogenetic diversity and pollination services. Science, 363, 282–284.

- Greenleaf SS, Williams NM, Winfree R, Kremen C (2007) Bee foraging ranges and their relationship to body size. Oecologia, 153, 589–596.
- Gunnarsson B, Federsel LM (2014) Bumblebees in the city: Abundance, species richness and diversity in two urban habitats. J. Insect Conserv., 18, 1185–1191.
- Hadley AS, Betts MG (2011) The effects of landscape fragmentation on pollination dynamics: Absence of evidence not evidence of absence. Biol. Rev., 87, 526–544.
- Hall DM, Steiner R (2019) Insect pollinator conservation policy innovations at subnational levels: Lessons for lawmakers. Environ. Sci. Policy, 93, 118–128.
- Hamblin AL, Youngsteadt E, Frank SD (2018) Wild bee abundance declines with urban warming, regardless of floral density. Urban Ecosyst., 21, 419–428.
- Hamblin AL, Youngsteadt E, López-Uribe MM, Frank SD (2017) Physiological thermal limits predict differential responses of bees to urban heat-island effects. Biol. Lett. 13, 20170125.
- Harmon-Threatt AN, Hendrix SD (2015) Prairie restorations and bees: The potential ability of seed mixes to foster native bee communities. Basic Appl. Ecol., 16, 64–72.
- Harrison T, Gibbs J, Winfree R (2017) Forest bees are replaced in agricultural and urban landscapes by native species with different phenologies and life-history traits. Glob. Chang. Biol., 24, 287–296.
- Harrison T, Gibbs J, Winfree R (2018) Phylogenetic homogenization of bee communities across ecoregions: XXXX. Glob. Ecol. Biogeogr., 27, 1457–1466.
- Harrison T, Winfree R (2015) Urban drivers of plant-pollinator interactions. Funct. Ecol., 29, 879–888.
- Hennig EI, Ghazoul J (2012) Pollinating animals in the urban environment. Urban Ecosyst. 15, 149–166.
- Hernandez JL, Frankie GW, Thorp RW (2009) Ecology of urban bees: A review of current knowledge and directions for future study. Cities Environ., 2, 1–15.
- Herrera CM (1995) Floral biology, microclimate, and pollination by ectothermic bees in an early-blooming herb. Ecology, 76, 218–228.
- Hostetler NE, McIntyre ME (2001) Effects of urban land use on pollinator (Hymenoptera: Apoidea) communities in a desert metropolis. Basic Appl. Ecol., 2, 209–218.
- Howes M, Wortley L, Potts R, Dedekorkut-Howes A, Serrao-Neumann S, Davidson J, Smith T, Nunn P (2017) Environmental sustainability: A case of policy implementation failure? Sustainability, 9, 165.
- Hunter P (2014) Brown is the new green: Brownfield sites often harbour a surprisingly large amount of biodiversity. EMBO Rep., 15, 1238–1242.
- Hunter R, Cleland C, Cleary A, Droomers M, Wheeler B, Sinnett D, Nieuwenhuijsen MJ, Braubach M (2019) Environmental, health, wellbeing, social and equity effects of urban green space interventions: A meta-narrative evidence synthesis. Environ. Int., 130, 104923.
- Hurd, PD (1978) An annotated catalogue of the carpenter bees (Genus *Xylocopa* Latrielle) of the western hemisphere (Hymenoptera: Anthophoridae); Smithsonian Institute Press: Washington, DC, USA.

- Jauker B, Krauss J, Jauker F, Steffan-Dewenter I (2012) Linking life history traits to pollinator loss in fragmented calcareous grasslands. Landsc. Ecol., 28, 107–120.
- Jones EL, Leather SR (2012) Invertebrates in urban areas: A review. Eur. J. Èntomol., 109, 463–478.
- Kaiser A, Merckx T, Van Dyck H (2021) The urban heat island and its spatial scale dependent impact on survival and development in butterflies of different thermal sensitivity. Ecol. Evol. 2016, 6, 4129–4140. Insects, 12, 128 15 of 18.
- Kaluza BF, Wallace H, Heard TA, Klein A, Leonhardt SD (2016). Urban gardens promote bee foraging over natural habitats and plantations. Ecol. Evol., 6, 1304–1316.
- Kearns CA, Inouye DW, Waser NM (1998) Endangered mutualisms: The conservation of plant-pollinator interactions. Annu. Rev. Ecol. Syst., 29, 83–112.
- Keeler B, Hamel P, McPhearson T, Hamann MH, Donahue ML, Prado KAM, Arkema KK, Bratman GN, Brauman KA, Finlay J, et al. (2019) Social-ecological and technological factors moderate the value of urban nature. Nat. Sustain., 2, 29–38.
- Kennedy CM, Lonsdorf E, Neel MC, Williams NM, Ricketts TH, Winfree R, Bommarco R, Brittain C, Burley AL, Cariveau D, et al. (2013) A global quantitative synthesis of local and landscape effects on wild bee pollinators in agroecosystems. Ecol. Lett., 16, 584–599.
- Klein A-M, Brittain C, Hendrix SD, Thorp R, Williams NM, Kremen C (2012) Wild pollination services to california almond rely on semi-natural habitat: Wild pollination services to California almond. J. Appl. Ecol., 49, 723–732.
- Kłysik K, Fortuniak K (1999) Temporal and spatial characteristics of the urban heat island of Lodz, Poland. Atmos. Environ., 33, 3885–3895.
- Knop E (2016) Biotic Homogenization of three insect groups due to urbanization. Glob. Chang. Biol., 22, 228–236.
- Kondo MC, Fluehr JM, McKeon TP, Branas CC (2018) Urban Green space and its impact on human health. Int. J. Environ. Res. Public Health, 15, 445.
- Konvicka M, Kadlec T (2011) How to increase the value of urban areas for butterfly conservation? A lesson from Prague nature reserves and parks. Eur. J. Èntomol., 108, 219–229.
- Kurylo JS, Threlfall CG, Parris KM, Ossola A, Williams NSG, Evans KL (2020) Butterfly richness and abundance along a gradient of imperviousness and the importance of matrix quality. Ecol. Appl., 30.
- Kuussaari M, Toivonen M, Heliölä J, Pöyry J, Mellado J, Ekroos J, Hyyryläinen V, Vähä-Piikkiö I, Tiainen J (2020) Butterfly species' responses to urbanization: Differing effects of human population density and built-up area. Urban Ecosyst.
- Lanner J, Kratschmer S, Petrović B, Gaulhofer F, Meimberg H, Pachinger B (2019) City dwelling wild bees: How communal gardens promote species richness. Urban Ecosyst., 23, 271–288.
- Lepczyk C, Aronson MFJ, Evans KL, Goddard MA, Lerman SB, MacIvor JS (2017) Biodiversity in the city: Fundamental questions for understanding the ecology of urban green spaces for biodiversity conservation. BioScience, 67, 799–807.

- Lerman SB, Contosta AR, Milam J, Bang C (2018) To mow or to mow less: Lawn mowing frequency affects bee abundance and diversity in suburban yards. Biol. Conserv., 221, 160–174.
- Leston L, Koper N (2017) Urban rights-of-way as extensive butterfly habitats: A case study from Winnipeg, Canada. Landsc. Urban Plan., 157, 56–62.
- Lovell ST. Multifunctional urban agriculture for sustainable land use planning in the United States. Sustainability 2010, 2, 2499–2522.
- Lowenstein DM, Matteson KC, Minor E (2015) Diversity of wild bees supports pollination services in an urbanized landscape. Oecologia, 179, 811–821.
- Lowenstein DM, Matteson KC, Minor E (2019) Evaluating the dependence of urban pollinators on ornamental, non-native, and 'weedy' floral resources. Urban Ecosyst., 22, 293–302.
- Mach BM, Potter DA (2018) Quantifying bee assemblages and attractiveness of flowering woody landscape plants for urban pollinator conservation. PLoS One, 13, e0208428.
- MacIvor JS, Cabral JM, Packer L (2013) Pollen specialization by solitary bees in an urban landscape. Urban Ecosyst., 17, 139–147.
- Martins KT, Gonzalez A, Lechowicz MJ (2017) Patterns of pollinator turnover and increasing diversity associated with urban habitats. Urban Ecosyst., 20, 1359–1371.
- Matteson KC, Ascher JS, Gail AL (2008) Bee richness and abundance in New York City Urban Gardens. Ann. Entomol. Soc. Am., 101, 140–150.
- Matteson KC, Langellotto GA (2010) Determinates of Inner City Butterfly and Bee Species Richness. Urban Ecosyst., 13, 333–347.
- McIntyre N, Rango J, Fagan W, Faeth S (2001) Ground arthropod community structure in a heterogeneous urban environment. Landsc. Urban Plan., 52, 257–274.
- McKinney ML (2008) Effects of urbanization on species richness: A review of plants and animals. Urban Ecosyst., 11, 161–176.
- Merckx T, Kaiser A, Van Dyck H (2018) Increased body size along urbanization gradients at both community and intraspecific level in macro-moths. Glob. Chang. Biol., 24, 3837–3848.
- Merckx T, Souffreau C, Kaiser A, Baardsen LF, Backeljau T, Bonte D, Brans KI, Cours M, Dahirel M, Debortoli, N, et al. (2018) Body-size shifts in aquatic and terrestrial urban communities. Nat. Cell Biol., 558, 113–116.
- Michener CD (2007) The bees of the world, 2nd ed, Johns Hopkins University Press: Baltimore, MD, USA.
- Middle I, Dzidic P, Buckley A, Bennett D, Tye M, Jones R (2014) Integrating community gardens into public parks: An innovative approach for providing ecosystem services in urban areas. Urban For. Urban Green., 13, 638–645.
- Morales CL, Traveset A (2009) A meta-analysis of impacts of alien vs. native plants on pollinator visitation and reproductive success of co-flowering native plants. Ecol. Lett., 12, 716–728.

- Muratet A, Fontaine B (2015) Contrasting impacts of pesticides on butterflies and bumblebees in private gardens in France. BiolConserv, 182: 148-154.
- Nagy DD, Magura T, Horváth R, Debnár Z, Tóthmérész B (2018) Arthropod assemblages and functional responses along an urbanization gradient: A trait-based multi-taxa approach. Urban For. Urban Green. 30, 157–168.
- Normandin ÉÉ, Vereecken NJ, Buddle CC, Fournier VV (2017) Taxonomic and functional trait diversity of wild bees in different urban settings. PeerJ, 5, e3051.
- Odanaka KA, Rehan SM (2019) Impact indicators: Effects of land use management on functional trait and phylogenetic diversity of wild bees. Agric. Ecosyst. Environ., 286, 106663.
- Oke TR (1973) City size and the urban heat island. Atmos. Environ., 7, 769–779.
- Ollerton J, Winfree R, Tarrant S (2011) How many flowering plants are pollinated by animals? Oikos, 120, 321–326.
- Pardee GL, Philpott SM (2014) Native plants are the bee's knees: Local and landscape predictors of bee richness and abundance in backyard gardens. Urban Ecosyst., 17, 641–659.
- Pawelek JC, Frankie GW, Thorp RW, Przybylski M (2009) Modification of a community garden to attract native bee pollinators in urban San Luis Obispo, California. Cities Environ., 2, 1–20.
- Phillips BW, Gardiner MM (2015) Use of video surveillance to measure the influences of habitat management and landscape composition on pollinator visitation and pollen deposition in pumpkin (*Cucurbita pepo*) agroecosystems. PeerJ, 3, e1342.
- Piano E, Souffreau C, Merckx T, Baardsen LF, Backeljau T, Bonte D, Brans KI, Cours M, Dahirel M, Debortoli N, et al. (2019) Urbanization drives cross-taxon declines in abundance and diversity at multiple spatial scales. Glob. Chang. Biol., 26, 1196–1211.
- Pincebourde S, Murdock CC, Vickers M, Sears MW (2016) Fine-scale microclimatic variation can shape the responses of organisms to global change in both natural and urban environments. Integr. Comp. Biol., 56, 45–61.
- Polidori C, García-Gila J, Blasco-Aróstegui J, Gil-Tapetado D (2021) Urban areas are favouring the spread of an alien mud-dauber wasp into climatically non-optimal latitudes. Acta Oecologica, 110, 103678.
- Potts SG, Biesmeijer JC, Kremen C, Neumann P, Schweiger O, Kunin WE (2010) Global pollinator declines: Trends, impacts and drivers. Trends Ecol. Evol., 25, 345–353.
- Quistberg RD, Bichier P, Philpott SM (2016) Landscape and Local correlates of bee abundance and species richness in urban gardens. Environ. Èntomol., 45, 592–601.
- Radmacher S, Strohm E (2009) Factors affecting offspring body size in the solitary bee *Osmia bicornis* (Hymenoptera, Megachilidae). Apidologie 41, 169–177.
- Richards MH, Garate AC, Shehata M, Groom D, Tattersall GJ, Welch KC (2020) Effect of nest microclimate temperatures on metabolic rates of small carpenter bees, *Ceratina calcarata* (Hymenoptera: Apidae). Can. Èntomol., 152, 772–782.

- Ricketts TH, Regetz J, Steffan-Dewenter I, Cunningham SA, Kremen C, Bogdanski A, Gemmill-Herren B, Greenleaf SS, Klein AM, Mayfield MM, et al. (2008) Landscape effects on crop pollination services: Are there general patterns? Ecol. Lett., 11, 499–515.
- Rollings R, Goulson D (2019) Quantifying the attractiveness of garden flowers for pollinators. J. Insect Conserv., 23, 803–817.
- Salisbury A, Armitage J, Bostock H, Perry J, Tatchell M, Thompson K (2015) Editor's choice: Enhancing gardens as habitats for flower-visiting aerial insects (pollinators): Should we plant native or exotic species? J. Appl. Ecol., 52, 1156–1164.
- Sattler T, Duelli P, Obrist MK, Arlettaz R, Moretti M (2010) Response of arthropod species richness and functional groups to urban habitat structure and management. Landsc. Ecol., 25, 941–954.
- Senapathi D, Goddard MA, Kunin WE, Baldock KCR (2017) Landscape impacts on pollinator communities in temperate systems: Evidence and knowledge gaps. Funct. Ecol., 31, 26–37.
- Sheffield CS, Pindar A, Packer L, Kevan PG (2013) The potential of cleptoparasitic bees as indicator taxa for assessing bee communities. Apidologie, 44, 501–510.
- Siegner AB, Sowerwine J, Acey C (2018) Does urban agriculture improve food security? Examining the nexus of food access and distribution of urban produced foods in the United States: A systematic review. Sustainability, 10, 2988.
- Sikora A, Michołap P, Sikora M (2020) What kind of flowering plants are attractive for bumblebees in urban green areas? Urban For. Urban Green., 48, 126546.
- Simao MCM, Matthijs J, Perfecto I (2018) Experimental small-scale flower patches increase species density but not abundance of small urban bees. J Appl Ecol, 55: 1759–1768.
- Sirohi MH, Jackson J, Edwards M, Ollerton J (2015) Diversity and abundance of solitary and primitively eusocial bees in an urban centre: A case study from Northampton (England). J. Insect Conserv., 19, 487–500.
- Southon GE, Jorgensen A, Dunnett N, Hoyle H, Evans KL (2017) Biodiverse perennial meadows have aesthetic value and increase residents' perceptions of site quality in urban green-space. Landsc. Urban Plan., 158, 105–118.
- Stenhouse RN (2004) Fragmentation and internal disturbance of native vegetation reserves in the Perth metropolitan area, Western Australia. Landsc. Urban Plan., 68, 389–401.
- Stewart AB, Sritongchuay T, Teartisup P, Kaewsomboon S, Bumrungsri S (2018) Habitat and landscape factors influence pollinators in a tropical megacity, Bangkok, Thailand. PeerJ. 6: 35-53.
- Tanda AS (2019) Entomofauna enhance the quality and quantity in Okra. Indian Journal Entomology 81(1): 16-17.
- Tanda AS (2020b) Entpollinatology-a strong relationship between plants and insects for crop improvement. *6th Edition of Global Conference on Plant Sciences and Molecular Biology* (*GPMB 2020*). Paris, France. (Accepted, May 26, 2020).

- Tanda AS (2021a) Chapter 10. Wild Bee Conservation Matter in Sustainable Agriculture. In: Tanda A S (eds.) Advances in Insect Pollination Technology in Sustainable Agriculture. Springer Nature Switzerland (under publication)
- Tanda AS (2021e) Wild bees and their conservation. Indian Journal Entomology (Accepted).
- Tanda AS (2021f) Insect pollinators matter in sustainable world food production. Indian Journal Entomology (Accepted).
- Tanda AS (2021i) Why insect pollinators important in crop improvement?. Indian Journal Entomology (Accepted).
- Tanda AS (2021b) Urbanization and Its Impact on Native Pollinators. The 1st International Electronic Conference on Entomology to be held in 1st–15th July 2021 virtually (Abstract accepted).
- Tanda AS (2021c) Native bees are important and need immediate conservation measures: a review. Presented at the 1st International Electronic Conference on Entomology (IECE 2021), 1–15 July 2021; Available online: https://iece.sciforum.net/ MDPI Proceedings 2021, 68, x. https://doi.org/10.3390/xxxxx): 1–19.
- Tanda AS (2021d) New ideas in pollinator and integrated pest management technology in sustainable agriculture. Journal of Integrated Pest Management (Under publication).
- Tanda AS (2020a) Entpollinatology is for sustainable agroecosystems and global agroindustry. Indian Journal of Entomology. Submitted.
- Tanda AS (2021g) Bee pollinators matter in crop improvement need immediate conservation strategies. Journal of Insect Science (Accepted).
- Tanda AS (2021h) Comparative pollination efficacies of *Apis mellifera* L. and *Tetragonula carbonaria* (Smith) on peach (*Prunus persica* L.). Indian Journal Entomology (Accepted).
- Thailand. PeerJ, 6, e5335.
- Threlfall CG, Ossola A, Hahs AK, Williams NSG, Wilson L, Livesley SJ (2016). Variation in vegetation structure and composition across urban green space types. Front. Ecol. Evol., 4.
- Tibesigwa B, Ntuli H, Lokina R (2020) Valuing recreational ecosystem services in developing cities: The case of urban parks in Dar Es Salaam, Tanzania. Cities, 106, 102853.
- Tommasi D, Miro A, Higo HA, Winston ML (2004) Bee diversity and abundance in an urban setting. Can. Èntomol., 136, 851–869.
- Tonietto R, Fant J, Ascher J, Ellis K, Larkin D (2011) A comparison of bee communities of Chicago green roofs, parks and prairies. Landsc. Urban Plan., 103, 102–108.
- Tucker EM, Rehan SM (2016) Wild bee pollination networks in northern New England. J. Insect Conserv., 20, 325–337.
- Turo KJ, Gardiner MM (2019) From potential to practical: Conserving bees in urban public green spaces. Front. Ecol. Environ., 17, 167–175.
- Turo KJ, Spring MR, Sivakoff FS, De La Flor YAD, Gardiner MM (2021) Conservation in post-industrial cities: How does vacant land management and landscape configuration influence urban bees? J. Appl. Ecol., 58, 58–69.

- Twerd L, Banaszak-Cibicka W (2019) Wastelands: Their attractiveness and importance for preserving the diversity of wild bees in urban areas. J. Insect Conserv., 23, 573–588.
- Tzortzakaki O, Kati V, Panitsa M, Tzanatos E, Giokas S (2019) Butterfly diversity along the urbanization gradient in a densely built mediterranean city: Land cover is more decisive than resources in structuring communities. Landsc. Urban Plan., 183, 79–87.
- United Nations (2020) 68% of the World Population Projected to Live in Urban Areas by 2050, Says UN. UN DESA. United Nations Department of Economic and Social Affairs. 2018. Available online: www.un.org (accessed on 15 September 2020).
- Verboven HAF, Uyttenbroeck R, Brys R, Hermy M (2014) Different responses of bees and hover flies to land use in an urban–rural gradient show the importance of the nature of the rural land use. Landsc. Urban Plan., 126, 31–41.
- Viana BF (2012) How well do we understand landscape effects on pollinators and pollination services? J. Pollinat. Ecol., 7, 31–40.
- WallisDeVries MF, Van Swaay CAM (2006) Global warming and excess nitrogen may induce butterfly decline by microclimatic cooling: Microclimate cooling and butterfly decline. Glob. Chang. Biol., 12, 1620–1626.
- Wenzel A, Grass I, Belavadi VV, Tscharntke T (2020) How urbanization is driving pollinator diversity and pollination–A systematic review. Biol. Conserv., 241, 108321.
- Werrell PA, Langellotto GA, Morath SU, Matteson KC (2009) The influence of garden size and floral cover on pollen deposition in urban community gardens. Cities Environ., 2, 1–16.
- Williams NM, Kremen C (2007) Resource distributions among habitats determine solitary bee offspring production in a mosaic landscape. Ecol. Appl., 17, 910–921.
- Williams NM, Winfree R (2013) Local habitat characteristics but not landscape urbanization drive pollinator visitation and native plant pollination in forest remnants. Biol. Conserv., 160, 10–18.
- Williams NS, Schwartz MW, Vesk PA, McCarthy MA, Hahs AK, Clemants SE, Corlett RT, Duncan RP, Norton BA, Thompson K, et al. (2009) A conceptual framework for predicting the effects of urban environments on floras. J. Ecol., 97, 4–9.
- Wilson CJ, Jamieson MA (2019) The effects of urbanization on bee communities depends on floral resource availability and bee functional traits. PLoS One, 14, e0225852.
- Wilson ES, Murphy CE, Rinehart JP, Yocum G, Bowsher JH (2020) Microclimate temperatures impact nesting preference in *Megachile rotundata* (Hymenoptera: Megachilidae). Environ. Èntomol., 49, 296–303.
- Winfree R, Aguilar R, Vázquez DP, Lebuhn G, Aizen MA (2009) A meta-analysis of bees' responses to anthropogenic disturbance. Ecology, 90, 2068–2076.
- Wolch JR, Byrne J, Newell JP (2014) Urban green space, public health, and environmental justice: The challenge of making cities 'Just Green Enough'. Landsc. Urban Plan., 125, 234–244.
- Wray JC, Neame LA, Elle E (2014) Floral resources, body size, and surrounding landscape influence bee community assemblages in oak-savannah fragments: Bee communities in oak-savannah fragments. Ecol. Èntomol., 39, 83–93.

- Wu J (2014) Urban Ecology and Sustainability: The state-of-the-science and future directions. Landsc. Urban Plan., 125, 209–221.
- Xerces Society (2020) Bee City USA [Internet]. Xerces Society. Available online: https://www.beecityusa.org/ (accessed on 20 November 2020).
- Xu J, Dang H, Wang M, Chai Y, Guo Y, Chen Y, Zhang C, Yue M (2019) Is phylogeny more useful than functional traits for assessing diversity patterns under community assembly processes? Forests, 10, 1159.
- Zhao C, Sander HA, Hendrix SD (2019) Wild bees and urban agriculture: assessing pollinator supply and demand across urban landscapes. Urban Ecosyst., 22, 455–470.
- Zurbuchen A, Landert L, Klaiber J, Müller A, Hein S, Dorn S (2010) Maximum foraging ranges in solitary bees: only few individuals have the capability to cover long foraging distances. Biol. Conserv., 143, 669–676.

13

Pollinators are Equally Important in Integrated Pest and Management Technology

Amarjit S. Tanda*

ABSTRACT

To utilise the services of bee pollinators is the main aim of pollinator integrated pest and management technology (PIPMT) in advance pest management strategies. Separating from the model of integrated pest management, we mingle bee pollinators in an action mode of pest management in vertical hierarchy. This practice is crucial in PIPMT ranking. Ingeniously, preference is given at the base of a pyramid model, which are mobilized through crop farming, landscape, pest management and integration of pollinators. In addition, the pyramid model, procedures in the shape of responsive manipulation of biotic and abiotic factors should combine together with basic program. The major target of PIPMT is to subjugate trade-offs, and to enhance co-benefits and cooperation between pest control methods using bee pollinators. We promise that PIPMT has significant potential in a sustainable agriculture, pest and bee pollination management program, as well as beneficial for the bio-ecosystem and biodiverse system.

Keywords • *Bee pollinators* • *Crops* • *Agriculture* • *Behavior* • *Decline* • *Human* • *Challenges* • *Sustainable* • *World*

13.1 INTRODUCTION

Integrated pest management technology (IPMT) is not merely a management, but rather an applied and widely used pest management technology (Tanda, 2022) that promotes the integration of various available biological, physical, chemical, biotechnological, and mechanical techniques including the use of crop pollinators (Kogan, 1998; Tanda, 2022). It aims at the minimum utilization of pesticides finalized by pest monitoring and their economic threshold levels, and other farming practices prior to the interference with responsive pest management opportunities is the application of insecticides (Zalucki et al,. 2009). Integrated crop pollination technology (ICPT), a newly designed and modified idea comparable to IPM, however associated

* School of Horticulture, University of Western Sydney, Richmond, Australia and CSIRO North Ryde NSW, Australia.

E-mail: drtanda193@gmail.com

with the pollinators and encourages the combination of several new strategies for sustainable crop pollination targets (Isaacs et al., 2017). It is newly proposed as "Pollinator and integrated pest management technology (PIPMT)", as this is an advanced technological or holistic strategy to integrate pollinators, bio-control agents, least use of pesticides, and various other IPM procedures, aiming at the management of the pests below economic injury levels for sustainable agriculture food production (Biddinger and Rajotte, 2015; Tanda, 2019, 2021, 2022). Encouraging this modern idea into practice would help not only farmers, public, and the environment from the chemical exposure, but also save the bee pollinators, natural enemies and chemical residue in food solving the insect pest resistance problem (Tanda, 1979, 2020 a,b,c; Tanda, 2022). Fundamentally, PIPMT is analogous to the IPM methodology because it suits the chemical control activities to conserve crop pollinators, but avoided more inclusive opportunities for integrating pest control activities and pollination mechanisms. Here, we have widen the PIPMT concept by promoting proactive crop pest management systems and landscape patterns as crucial components of PIPMT (Egan et al., 2020; Tanda, 2019; Tanda, 2021a,b,c; Tanda, 2022), which can be supplemented with responsive farmer focused inputs as insecticides.

Fig. 13.1 Schematic model of ecological interplay among crop pest, natural enemies, and pollinators, and effect on crop plant quality and yield.
(Schematic model designed by Dr Amarjit S. Tanda, 2022)

This idea promotes the functioning of agroecosystem directed by attacking insect pests, parasites and predators, and the crop insect pollinators, responsible for their individual services and the duties (Losey and Vaughan, 2006; Oerke, 2006; Klein et al., 2007) (Fig. 13.1). There are several grounds for pest and insect pollination management collectively. For their co-existence and functioning in farming habitat, the phylogenic and ecological laps of insect pests, predators, and bee pollinators evolve potential menaces. The key active groups, of which pests cause losses, split various environments and anthropogenic operators. Local environmental and crop complexity can aid both natural bio-agents and bee pollinators that make use of different environments and natural assets at various life phases (Shackelford et al., 2013). The significance of co-management is enhanced under a specific situation when the same species is herbivorous, predator, and working as a pollinator such as hover flies, larvae work as natural enemies but adults are crucial crop pollinators, and *Trigona spinipes* can be either crop pollinators or herbivores determined by plant species (Saunders et al., 2016). Major challenges for PIPMT are understandable, especially in the application process of insecticides that could be safe for non-targeted fauna. The programs of PIPMT comprise tactics that bring change advocating, promoting crop pollinators and control the insect pest (Egan et al., 2020; Tanda, 2019, 2021).

Pests, predators, insect pollinators, the host plants, and their environmental interaction influence the yield of crop. Mostly mutual operators and the influence of insect groups on each other functioning develop the relationship of pest and pollinators and their co-management. All these interactions can be altered by the local climatic and geographic conditions. Another reason to co-act crop pests and pollinators is that the production of field crops pollinated by insects is usually decided by both insect pollinators and pest management programs (Tamburini et al., 2019). Insect pollination benefits are mostly increased under flourishing pest control and the least pest damage to the crop (Tamburini et al., 2019). Due to this, the potential economic gains of enhanced plant production that arise either from plant pests or insect pollinators are not determined individually. These PIPMT strategies as non-autonomous and control the synergistic chances provided by efficient co-management process. The crop yield is influenced by both pest losses and pollinator population, and offers a bit coin for both systems (Saunders et al., 2016). The production is based the total benefits or expenses by measuring the small gains or misplacements of different plans and measures that help pest management and crop pollination or advancement at the cost of other.

13.2 STRATEGY OF POLLINATOR AND INTEGRATED PEST MANAGEMENT TECHNOLOGY (PIPMT) MODEL

The full PIPMT approach is generally portrayed in a pyramidic model (Hokkanen, 2015; Tanda, 2022) (Fig. 13.2). The broad model establishes as a pillar, with maximum precedence to enterprising measures at the foundation and responsive actions at the peak. The plans are executed only if proactive activities are inadequate for managing pests or retaining losses below the economic threshold levels using all possible measures.

A hierarchical schematic model can be applied for pollinators, where insect pest and crop pollinators design a PIPMT pyramid in co-management program (Egan et al., 2020). At the bottom of PIPMT model higher utilization is of environmental activities undertaken by predators and crop pollinators naturally (Dainese et al., 2019) or managed by natural components of landscapes which subdue the pests straightway (Gurr et al., 2017). Actions at the peak put back procedures associated to eco-biodiversity with non-biological operators such as chemical applications. Insect pollinator and pest control steps and possible PIPMT collaborations, multiple advantages to each other, or exchanges at each level of the PIPMT model. Influences of PIPMT elements, and their impacts on pollinators, bio-agents, and herbivores, and among each other in PIPMT model has been also presented in Table 13.1.

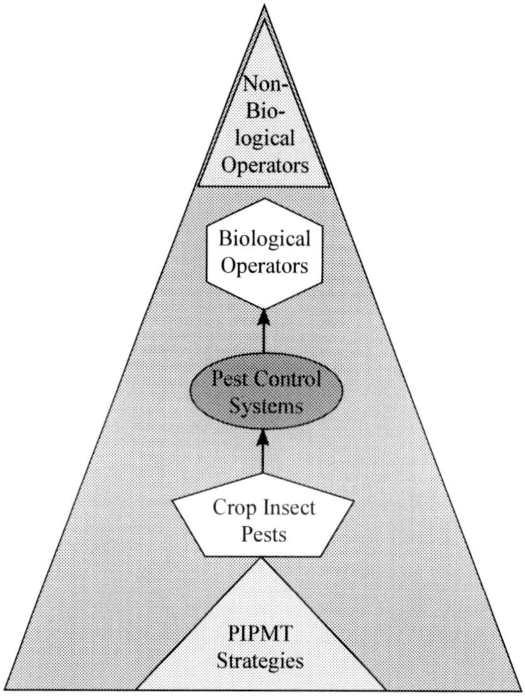

Fig 13.2 A pollinator and integrated pest management technology (PIPMT) pyramidic schematic model. (Schematic model designed by Dr Amarjit S. Tanda, 2021, 2022)

Table 13.1 The actions, reactions and influences of pollinators, bio-agents, and the herbivores in the "pollinator and integrated pest management technology (PIPMT)" model (Tanda, 2022)

PIPMT Model Components	Crop Pollinators	Bio-agents	Herbivores
Honeybees	○	○	⊕
Natural enemies	○	⊕	⊖
Pesticides	⊖	⊖	⊖
Hand pollination	○	○	○
Watering	⊕	⊖	⊕⊖
Composted nutrients	○	⊕	⊖
Flowering Plants hedges	⊕	⊕	⊕⊖
Ecological Assemblages	⊕	⊕⊖	⊕⊖
Plant cultivars	○	⊕	⊖

⊕ = Positive ⊖ = Negative ○ = Neutral (Modified after Lundin et al., 2021).

The pollinator integrated pest and management technology (PIPMT) pyramid should combine a hierarchical pest management strategic program, setting enterprising control operators at the bottom and responsive non-biological operators at the top to accomplish maximum crop pest and pollinator utilization targets by improving the fruit quality and yield. The operators at the base control environmental actions panoramically, whereas man efforts including non-biological activities work at the peak of pyramid model. To minimize the damages, monitoring of the insect pest, collection and identification and estimating the economic threshold levels kept in the middle of the management plan and direct measures required for their implementation.

13.3 POLLINATORS ARE APPROPRIATE IN INTEGRATED PEST MANAGEMENT TECHNOLOGY

PIPMT management strategy targets at direct subduing of the pests. This system also helps in assortments and plentiful groups of predators that promotes bio-diverse pest control bio-systems at the base of the pyramid model (Gurr et al., 2017). In addition, pest management activities can assist pollinators and crop pollination procedures (Isaacs et al., 2017). The PIPMT model using predators and crop pollinators to assure in conserving the natural bio-ecosystem, encouraging the bio-diversity of native wild bees and the farm eco-environment.

In an agro-landscape system, populations of semi-natural environment can assist the natural enemies and pest management applications (Chaplin-Kramer et al., 2011; Rusch et al., 2016). No doubt, inter-actions among parasites and predators differ depending on the habitat under semi-natural bio-ecosystems, besides, pests can also benefit from semi-natural environment, changing pest management processes (Karp et al., 2018). Semi natural environments look for flower pollinator's better way in advanced farming industry (Kennedy et al., 2013). The higher distinctive reaction of pest control might be due to the higher tri-trophic complexness (Snyder and Wise, 2001) and taxonomic variety of active insects assessing pest management and pollinators services (Karp et al., 2018).

Crop cultivars and cropping patterns directly influence the herbivores and bee pollinator efficiency in increasing the crop yield. Reducing the spatiotemporal efficiency of a variety could effectively decrease the pest population, in a particular plant species (Root, 1973; Delaune et al., 2019). However, enlarged plant intensity aids the bio-control through parasites and predators present in the natural environment (Redlich et al., 2018). The plant diversity on bee pollinators may be more complicated. It depends on crop identification and management strength, where mass-flower producing plants encourage crop pollinators' abundance (Rundlöf et al., 2014; Hass et al., 2018). Achieving PIPMT other advantages may therefore require more delicate selection of plant make-up rather than simply enhancing the plant variations. Host plants and other landscape plant arrangements with small-scale and erratically forms of lands and semi-natural environments benefit both cross-pollination and pest management services (Martin et al., 2019).

Crops exchange between promoting insect pollinators and suppressing the pests as red clover (*Trifolium pratense*) grown for seed production and bumblebees (*Bombus* spp) work as its pollinator, and clover seed weevils (*Protapion* spp) are the herbivores. Higher population of bumblebee queens were observed in 2 km of clover plantations and 80% less bumblebee queen abundance was found having no clover crop (Rundlöf et al., 2014). However, seed weevil population was seems to be more and reduced at situations where the cloest clover crop was <0.2 km and >0.8 km distance, correspondingly and 41% less weevils in crops lacking clover within 0.8 km area (Lundin et al., 2016). There is a scope to suppress pests by growing crops at a distance so that pest migration is limited while still encouraging crop insect pollinators.

Promoting the PIPMT at the farm and other assortment level will require a thorough understanding of the bio-ecology of herbivores, parasites and predators, and insect pollinators. The major aim at the geographic plant level should be to assure regular supply of food resources for crop pollinators and predators and eliminating pest. Among animal communities variations in movement might offer co-management chances by assuring regular food sources in spatiotemporal levels useful for beneficial insects but not for herbivory (Thies et al., 2005). In addition possibility is to take the benefit of variation in host-plant and environment between pests and pollinators, changing host plant with different non-host cultivars to disorder food resources for herbivores assuring that change in host plant food for bee pollinators and natural enemies (Schellhorn et al., 2015; Delaune et al., 2019). This strategy can be achieved for numerous crops that benefit from all pollinators and bio-control agents but get losses mainly from specific insect pests.

13.4 PIPMT STRATEGY AND FARM MANAGEMENT

Several farm measures applied have been crafted to prevent major pest eruptions, known as cultural management practices (Bajwa and Kogan, 2004). These measures involve the practices for enhancing the plant cultivars in the farm, growing pest-resistant varieties, and changing the soil texture mechanically and adding fertilizers and managing irrigation system. Many of these activities, influences the bee pollinators which are badly understood and in depth studies are demanding before the concept of PIPMT can be implemented efficiently.

The efficiency of a crop variety can be improved along its boundaries to promote pest control, using bottom-up and top-down footpaths (Root, 1973). Several examples of similar functions are there growing two or more crops in proximity (Iverson et al., 2014);

- To enhance beneficial soil biota, offering environment for natural enemies and pollinators, (Schipanski et al., 2014) planting of legumes, and other crops for erosion control, improving soil structure, and nutrient content will be helpful
- Growing flowering plants or hedgerows (Tschumi et al., 2015; Morandin et al., 2016); and minimizing weed control (DiTommaso et al., 2016).
- Flowers and hedgerows (Blaauw and Isaacs, 2014; Morandin et al., 2016) and decreased weed management (DiTommaso et al., 2016) can also encourage insect pollinators.

As the effect of cover crop and intercropping on crop pollinators are low and is well known, these practices show great potential when the agricultural design is in mass-flower production of herbal plants (Mallinger et al., 2019). Cropping system having higher crop diversity have been developed in a different way for pest control and insect pollinators. Plant selections are crucial for their influence on pests, predators, and crop pollinators (Lundin et al., 2019). Likewise, weeds would allow for diversity that help the multiplication of natural enemies and pollinators without helping herbivores or aggravating weed related issues. Advance knowledge of resource utilization is important, insects are required to completely catch the potential of crop intensification at the farm level as an efficient plan for PIPMT model.

For successful pest control program, planting of insect resistant varieties is an important method (Stenberg, 2017). A similar process is to select varieties for insect pollinator management system that could be attractive to flower pollinators for free pollination services. In PIPMT program, a major challenge for developing and variety selection would be the coeval drawing and application of many crop attributes such as insect pest resistance, pollinator management for great yield.

IPMT systems also needs modification of crop soil inputs, including fertilizers and irrigation practices. The organic soil inputs can improve total pest control through salutary influences on natural enemies and upward management by enhancing plant defense mechanisms against serious pests (Rowen et al., 2019) and increasing insect pollinator populations (Banaszak-Cibicka et al., 2019). Influence of irrigation on the crop pests, predators, and insect pollinators are generally not well documented. Overflow of watering can disarrange predators to the crops (Daane and Williams, 2003). Managing sufficient soil wetness and eluding under and over moisture can help pollinator foraging because of more nectar production (Gillespie et al., 2015).

Cultural practices and other management plans, for example, exclusion confinement and row planting, have got more awareness recently, many may require integration with bee pollinators when the crop is pollinator dependent (Minter and Bessin, 2014; Leach et al., 2016).

PIPMT measures are success at the farm level with the flower rows on crop pollination and pest management relies on the crop environment, with few observations that gains are more in straightforward than in complicated plantations (Jonsson et al., 2015; Grab et al., 2018). Management activities occasionally overlap in farm and landscape habitats, as crop practices active over a higher level to equal landscape scale system.

13.5 PIPMT MODEL AND SIGNIFICANCE OF PEST SURVEILLANCE, AND ECONOMIC THRESHOLD LEVELS

In pest control program, pest surveillance, technique of sampling, and identification of pest are the basic components that are utilized, along with economic threshold levels, to develop the control mechanism and working stratgy. The economic threshold or ET at which measure must be utilized to prevent touching the economic injury level (EIL), which is the least pest population that do monetary damages due to crop losses that excels the expenses of the control process (Stern et al., 1959). Literally, ETs for a specific insect pests and crops are unattainable, or wanting scientific data, due to the sufficient amount of facts needed for their calculations. Alternatively, the aim of bee pollination is to keep pollinator densities above a level that economic losses due to inadequate pollinators are avoided, entailing the development of our newly proposed model, "Effective Pollinator Density Level (EPDL)" a population density of pollinators level resulting in beneficial pollination for enhanced production, an analogous to the pollinator economic effective level (PEEL) or EIL (Flöhr et al., 2020; Tanda, 2022). The EPDL is a potentially beneficial density metric to assess whether pollinator management measures are asserted, however, as with EILs, calculation of a specific plant. EPDLs needs sufficient crop field–pollinators visitation data, bio-ecological population densities and economical observations. Especially, better understandable data of how to quickly decide whether crop plants are restricted for pollen by utilizing farm collection facts (Garibaldi et al., 2020) would be required to improve the EPDLs program functioning. Though many chemical control methods can be adopted quickly, management of crop pollinators may be operationally threatened to implement quickly in reply to the capability to put in more honeybee colonies for pollination. EPDL and EIL can also be integrated into one decision metric action, as we propose as the "integrated pollination effective level (IPEL) that blends crop production limited accountable activities and accessibility of herbivores and bee pollinators (Flöhr et al., 2020; Tanda, 2022).

13.6 PIPMT MODEL AND BIOLOGICAL OPERATORS

For pest management, the biological operators involve mostly cultured and field releases of bio-control agents, include insects, other invertebrates and microbes. In fact, foreign bio-control agents were generally imported with the aim of attaining classic bio-control, but additional releases to make present predators populations more powerful in the crop environment (van Lenteren et al., 2018).

During flowering period, honeybees are widely spread as biological operators for cross pollination. Alongwith natural enemies, managed bees are used to supplement natural control services provided by various invertebrates and microorganisms. Honeybee (*Apis mellifera*) is the important managed bee pollinator worldwide, while many bumblebee (*Bombus* spp) and solitary bees are also involved in different regions (Isaacs et al., 2017). The benefits of honeybees to crop pollination and yield vary much among cultivars and areas (Garibaldi et al., 2013). More research is needed to study the potential of other pollinators except honeybees for plant cross-pollination and to decide appropriate stocking population density of domesticated bee pollinators (Isaacs et al., 2017).

PIPMT systems depend upon different biological operators working independently in nature. It is unusual in our newly proposed terminology "Apipollinatology" (api=apis, polli=pollination and tology=studies), the unique technique to employ *Apis*, honeybee pollinators utilized as vectors to spread bio-control agents against insect pests and diseases alongwith the transfer of pollen grains for cross-pollination (Mommaerts and Smagghe, 2011; Tanda, 2022). Probability is there for collaboration between methods targeting to raise crops and other flora and fauna with healthy densities of native naturally friendly insects and employing other biological operators for crop pollination and pest control plans for maximum benefits. Newly freed bio-control parasites gain from complicated natural landscapes full with assets, (Perez-Alvarez et al., 2019); also, plantations with various flower resources help the honeybees and other wild bees (Smart et al., 2016; Tanda, 2021a,b,c; Tanda, 2022). Individually, releasing biological PIPMT agents into fields with insufficient food assets may result in to enhanced competitiveness between native and domesticated honeybees (Herbertsson et al., 2016) or show hostile interplay between released parasites and natural enemies present in nature (Perez-Alvarez et al., 2019; Tanda, 2021a,b,c; Tanda, 2022).

13.7 NON-BIOLOGICAL OPERATORS AND PIPMT

IPMT pest management through insecticides stays as the normal measure of pest control using farm procedures globally in six decades (Zalucki et al., 2009; Hokkanen, 2015). Methods in chemical control applications have been modified, and better powerful chemicals are being used Europe and North America regions (Douglas et al., 2020).

For artificial cross-pollination, many techniques are utilized for pollen spread to the receptive stigmas such as pollen spraying, pollination by hand, and use of other mechanical tools similar to the totally artificial inputs in the shape of insecticides applied for controlling pests (Westerkamp and Gottsberger, 2000; Potts et al., 2018). The application of artificial pollination tools has yet to get to the matching degree of pervasiveness as that of insecticides, and these techniques are unlikely to act as alternatives for honeybees on a high degree. However, hand pollination methods could serve as a habitat action for field crops sown in surroundings where management of bee pollinators in sufficient abundance is not possible (Potts et al., 2018).

Variations in life cycles, functioning patterns, and habitats of pests, predators, and crop pollinators underscore how the chemical use may be possibly less disruptive for PIPMT related benefits. The pesticide applications for pest control is associated with possibilities for exchange in pessimistic impacts on predators and plant pollinators (Bommarco et al., 2011; Rundlöf et al., 2015; Tanda, 2021a,b; Tanda, 2022). Non-selective insecticides can develop negative influences on bio-control agents that are higher than the aimed impact on the insect pest (Bommarco et al., 2011), and can also effect pollinators adversely (Rundlöf et al., 2015), which one after another can have bad consequences on crop production (Stanley et al., 2015). Individually, spraying of less hazardous pesticides on flowering time may also assist pollinators by saving their food assets from insect pest attack (Rundlöf and Lundin, 2019; Tanda, 2022).

13.8 STRATEGIES OF PIPMT MODEL

The pollinators are bumblebees and pest attacking seedlings are flea beetles (*Phyllotreta* spp) in spring rapeseed (*Brassica napus*) fields. Flea beetles were managed by seed treatments using neonicotinoids before, however as they move into the reproductive parts of the flowers with harmful impacts on bumblebee, resulting in 71% less queen cocoon production in bumblebees, due to neonicotinoid toxic effect present in the pollen grains and nectar (Rundlöf et al., 2015). Contradictorily, adoption from normal to zero tillage practices provides successful flea beetle control without any seed dressing, showing 74% less damage by flea beetle (Lundin, 2019).

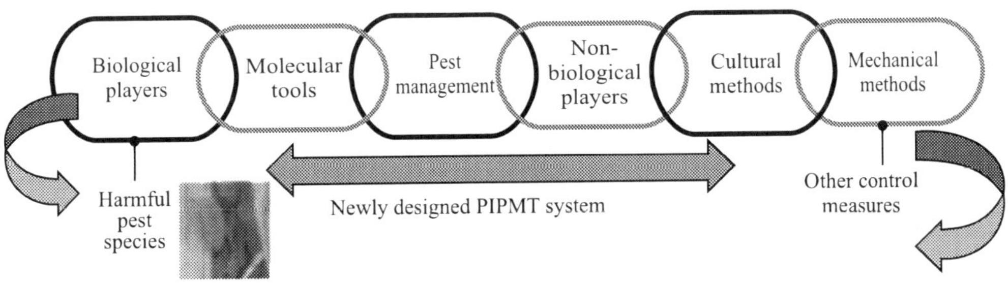

Fig. 13.3 A pollinator integrated pest management technology (PIPMT) schematic model. (Schematic model designed by Dr Amarjit S. Tanda 2022)

It is doubtful how the plants and landscapes subside harmful impacts of insecticides on predators and pollinators of crop. A complex and powerful food resourceful landscape can work as a bolster for harmful effects of insecticides on bee pollinators (Park et al., 2015), however insecticides can also disprove useful impacts of floral intricateness on bio-control agents (Ricci et al., 2019). Research on how insecticides influence pest control and crop pollination process, and how these work continuously impact the crop production, is required to regulate insecticide application to a degree that boost growers financial profits minimizing the probability of negative effects on bio-diverse environmental systems (Catarino et al., 2019). In PIPMT process the pesticide application is to decrease the maximum by analysing impacts on insect pests and crop pollinators of various management approaches from fundamental components of the PIPMT pyramidal model (Fig. 13.2), or assessing different biotechnological possibilities such as RNA interference (Zotti et al., 2018). Recent research that help to pesticide uses aiming at attacking pests over predators and bee pollinators are also main to PIPMT designing (Biddinger and Rajotte, 2015), and this could be attained by spraying insecticides at pre-blooming and where the pest attack is more as compared to the populations of pollinators and natural enemies (Fig. 13.3).

13.9 SPEEDING UP PIPMT APPLICATIONS

In the PIPMT module (Fig. 13.2), new procedure adopts the shape of an up-side down pyramidal model, with selective insecticide with minimam use of *A. mellifera* as a leader pollinator to the attacking pest, in a pollinator integrated pest management technology (Fig. 13.3). Although the necessity to integrate pollinators and pest management has developed, the up-side down view of pyramidal model has been greatly kept a thought of flexible insecticide application to minimise threat to a bee pollinator *A. mellifera* (Biddinger and Rajotte, 2015; Curtis et al., 2019). Notwithstanding, diminishing choices for insecticides and more attention on honeybee strength will probably need better plans in the farming system. To expand the evolution of PIPMT model, innovative research preferences should be concentrated on some major topics described below;

- Crafting models on the spread or movement of various crop pollinators, key crop pests, their parasites and predators in different field crops utilizing ecological systems.
- Evolving a powerful data founded bio-diverse farm managed measure technological plan using pollinator and integrated pest management (PIPMT) system (Fig. 13.2), mainly aiming at the measures' possibility to enhance sustainable agricultural food production and economic benefits to the farmers.
- Investigating how minimum insecticidal applications can be blended into PIPMT methodology that are greatly beneficial in managing key insect pests with least impacts on crop bee pollinators and predators.

13.10 DESIGNING AND ASSESSMENT OF PIPMT MODEL

Growers, entomologists, and other local experts can contribute greatly in connecting different farm pest control designing and assessment of PIPMT model studies (Lamichhane et al., 2016).

Depending on the PIPMT structure, in designing new PIPMT applied programs, the target of farm activities should be bolstered keeping in mind the crop bee pollinator services. In addition, promoting PIPMT in reality would gain from pest control strategies being included in measures offered by bee keeping specialists and bee protection educators (Sponsler et al., 2019). The important action to expedite PIPMT acceptance would be the formulating of recommendations for the field execution of components within the comprehensive PIPMT form (Fig. 13.2). Proceeding application from an inverted to an upstanding replica will need;

- methodology of understandable leading concepts,
- well-planned texting,
- communication program for counselors and professionals.

PIPMT targets will differ in different areas, farm practices, and professionals, acceptance of PIPMT will need split-level program that permit pliability in hiring different or chosen divisions, supported by understandable economic acceptance that also identifies threats and ambivalence of procedures.

For implementation of the PIPMT, a possible complex component at the landscape level is that insect pests, predators, and bee pollinators are normally influenced by various governing procedures away from the classic field of management for all insect groups. This can generate a nonlinear discrepancy whereby gains obtaining from PIPMT program at the field scale resulting away from the management division. Furthermore, field climatic plans generally aim at particular farms or environments, as plan tools promoting effective control of farm landscapes are mostly wanting (Goldman et al., 2007). An integration plan designing and optional appointments of growers, resident groups and other farm advisors (Goldman et al., 2007; Brewer and Goodell, 2012) are responsible for encouraging new farm level PIPMT procedures. These tools of strategy and associations could authorize and empower collaboration of local interferences among growers or help to field level crop rotation program that advocate crop pollinators and predators over herbivores.

13.11 CONCLUSION

For pest management and pollinators integration, there are several reasons to organize founded on:

- The frequently conjoining pest ecology,
- Parasites and predators,
- Crop pollinators,
- Their non-independent impacts on crop production.

Pollinator and integrated pest management technology (PIPMT) is a modular structure that can be applied to co-manage for ecological activities directed by herbivores, parasites and predators, and bee pollinators. The PIPMT pyramidal model performs as a resolution pillar program in hierarchy that organizes foundation-scale, operators over peak, combative reactions to attain crop pest and pollinator management targets. We spotlight here schemes for organizing pest management and crop pollination aims using PIPMT concurrently (Tanda, 2020a; Tanda, 2021a,b,c; Tanda, 2022) (Fig. 13.4).

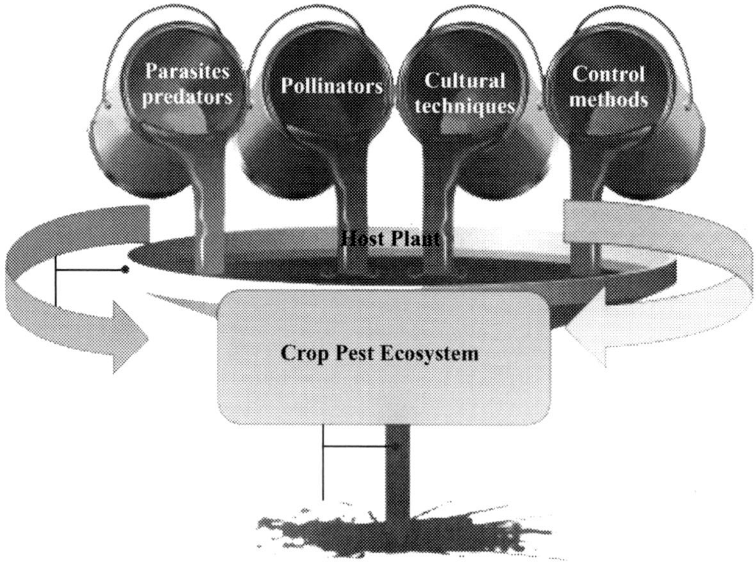

Fig. 13.4 Schematic model of pollinator and integrated pest management technology (PIPMT) schematic model (designed by Dr Amarjit S. Tanda, 2022)

This modern and new innovative proposal generates chances and moments for encouraging collaborations while restricting exchange of aggressive activities to concurrently attain pest management and crop pollination aims via highly bio-diverse founded farm and landscape measures at the formation of the PIPMT model. Many agricultural and countryside control measures within PIPMT are probably to increase extra ecological favors, as given by biological operators in soil, keeping it feasible to adjust PIPMT with the broad principal of environmental growth and development (Kleijn et al., 2019, Tanda, 2019, 2020b, c; Tanda, 2021a,b,c; Tanda, 2022). Predictive PIPMT approach would also increase ecological benefits by minimizing puts in and enhancing bio-diverse system. To conclude, executing an enlarged, well designed PIPMT approach will be helpful to the insect pollination sustainability and herbivore management process by financing on joint potentialities of the advance strategy.

REFERENCES

- Bajwa WI and Kogan M. (2004) Cultural practices: springboard to IPM. In: O Koul, GS Dhaliwal, and GW Cuperus (eds). Integrated pest management: potential, constraints and challenges. Wallingford, UK: CABI Publishing
- Banaszak-Cibicka W, Takacs V, Kesy M, et al. (2019) Manure application improves both bumblebee flower visitation and crop yield in intensive farmland. Basic Appl Ecol 36: 26–33
- Baraibar B, Westerman PR, Carrión E, and Recasens J (2009) Effects of tillage and irrigation in cereal fields on weed seed removal by seed predators. J Appl Ecol 46: 380–87

- Biddinger DJ and Rajotte EG (2015) Integrated pest and pollinator management – adding a new dimension to an accepted paradigm. Curr Opin Insect Sci 10: 204–09
- Blaauw BR and Isaacs R (2014) Flower plantings increase wild bee abundance and the pollination services provided to a pollination-dependent crop. J Appl Ecol 51: 890–98
- Bommarco R, Miranda F, Bylund H, and Björkman C (2011) Insecticides suppress natural enemies and increase pest damage in cabbage. J Econ Entomol 104: 782–91
- Brewer MJ and Goodell PB (2012) Approaches and incentives to implement integrated pest management that addresses regional and environmental issues. Annu Rev Entomol 57: 41–59
- Catarino R, Bretagnolle V, Perrot T, et al. (2019) Bee pollination outperforms pesticides for oilseed crop production and profitability. P Roy Soc B-Biol Sci 286: 20191550
- Chaplin-Kramer R, O'Rourke ME, Blitzer EJ, and Kremen C (2011) A meta-analysis of crop pest and natural enemy response to landscape complexity. Ecol Lett 14: 922–32
- Curtis R, Ludwig G, Steeger T, et al. (2019) Honeybee best management practices for California almonds. Modesto, CA: Almond Board of California
- Daane KM and Williams LE (2003) Manipulating vineyard irrigation amounts to reduce insect pest damage. Ecol Appl 13: 1650–66
- Dainese M, Martin EA, Aizen M, et al. (2019) A global synthesis reveals biodiversity-mediated benefits for crop production. Science Advances 5: eaax0121
- Delaune T, Ballot R, and Sausse C, et al. (2019) Higher surfaces of a crop in the landscape increase outbreak risks the following growing season. bioRxiv; doi.org/10.1101/641555
- DiTommaso A, Averill KM, Hoffmann MP, et al. (2016) Integrating insect, resistance, and floral resource management in weed control decision-making. Weed Sci 64: 743–56
- Douglas MR, Sponsler DB, Lonsdorf EV, and Grozinger CM (2020) County-level analysis reveals a rapidly shifting landscape of insecticide hazard to honeybees (*Apis mellifera*) on US farmland. Sci Rep-UK 10: 797
- Egan PA, Dicks LV, Hokkanen HM, and Stenberg JA (2020) Delivering integrated pest and pollinator management (IPPM). Trends Plant Sci 25: 577–89
- Flöhr A, Stenberg JA, and Egan PA (2020) The joint economic impact level (jEIL): a decision metric for integrated pest and pollinator management. In: Y Gao, H Hokkanen, and I Menzler-Hokkanen (eds). Integrative biological control: ecostacking for enhanced ecosystem services. Dordrecht, the Netherlands: Springer
- Garibaldi LA, Sáez A, Aizen MA, et al. (2020) Crop pollination management needs flower-visitor monitoring and target values. J Appl Ecol 57: 664–70
- Garibaldi LA, Steffan-Dewenter I, Winfree R, et al. (2013) Wild pollinators enhance fruit set of crops regardless of honeybee abundance. Science 339: 1608–11
- Gillespie S, Long R, and Williams N (2015) Indirect effects of field management on pollination service and seed set in hybrid onion seed production. J Econ Entomol 108: 2511–17
- Goldman RL, Thompson BH, and Daily GC (2007) Institutional incentives for managing the landscape: inducing cooperation for the production of ecosystem services. Ecol Econ 64: 333–43

- Grab H, Poveda K, Danforth B, and Loeb G (2018) Landscape context shifts the balance of costs and benefits from wildflower borders on multiple ecosystem services. P Roy Soc B-Biol Sci 285: 20181102
- Gurr GM, Wratten SD, Landis DA, and You M (2017) Habitat management to suppress pest populations: progress and prospects. Annu Rev Entomol 62: 91–109
- Hass AL, Kormann UG, Tscharntke T et al. (2018) Landscape configurational heterogeneity by small-scale agriculture, not crop diversity, maintains pollinators and plant reproduction in western Europe. P Roy Soc B-Biol Sci 285: 20172242
- Herbertsson L, Lindström SAM, Rundlöf M, et al. (2016) Competition between managed honeybees and wild bumblebees depends on landscape context. Basic Appl Ecol 17: 609–16
- Hokkanen HM (2015) Integrated pest management at the crossroads: science, politics, or business (as usual)? Arthropod-Plant Inte 6: 543–45
- Isaacs R, Williams N, Ellis J, et al. (2017) Integrated crop pollination: combining strategies to ensure stable and sustainable yields of pollination-dependent crops. Basic Applied Ecol 22: 44–60
- Iverson AL, Marín LE, Ennis KK, et al. (2014) Do polycultures promote win–wins or trade-offs in agricultural ecosystem services? A meta-analysis. J Appl Ecol 51: 1593–602
- Jonsson M, Straub CS, Didham RK, et al. (2015) Experimental evidence that the effectiveness of conservation biological control depends on landscape complexity. J Appl Ecol 52: 1274–82
- Karp DS, Chaplin-Kramer R, Meehan TD, et al. (2018) Crop pests and predators exhibit inconsistent responses to surrounding landscape composition. P Natl Acad Sci USA 115: E7863–70
- Kennedy CM, Lonsdorf E, Neel MC, et al. (2013) A global quantitative synthesis of local and landscape effects on wild bee pollinators in agroecosystems. Ecol Lett 16: 584–99
- Kleijn D, Bommarco R, Fijen TPM, et al. (2019) Ecological intensification: bridging the gap between science and practice. Trends Ecol Evol 34: 154–66
- Klein AM, Vaissiere BE, Cane JH, et al. (2007) Importance of pollinators in changing landscapes for world crops. P Roy Soc B-Biol Sci 274: 303–13
- Kogan M (1998) Integrated pest management: historical perspectives and contemporary developments. Annu Rev Entomol 43: 243–70
- Lamichhane JR, Aubertot JN, Begg G, et al. (2016) Networking of integrated pest management: a powerful approach to address common challenges in agriculture. Crop Prot 89: 139–51
- Leach H, Van Timmeren S, and Isaacs R (2016) Exclusion netting delays and reduces *Drosophila suzukii* (Diptera: Drosophilidae) infestation in raspberries. J Econ Entomol 109: 2151–58
- Losey JE and Vaughan M (2006) The economic value of ecological services provided by insects. BioScience 56: 311–23

- Lundin O (2019) No-till protects spring oilseed rape (*Brassica napus* L) against crop damage by flea beetles (*Phyllotreta* spp). Agr Ecosyst Environ 278: 1–5
- Lundin O, Rundlöf M, Smith HG, and Bommarco R (2016) Historical change and drivers of insect pest abundances in red clover seed production. Agr Ecosyst Environ 233: 318–24
- Lundin O, Ward KL, and Williams NM (2019) Identifying native plants for coordinated habitat management of arthropod pollinators, herbivores and natural enemies. J Appl Ecol 56: 665–76
- Lundin O, Rundlöf M, Mattias J, Bommarco R. and Williams NM (2021) Integrated pest and pollinator management – expanding the concept. Front Ecol Environ, 19: 283–291
- Mallinger RE, Franco JG, Prischmann-Voldseth DA, and Prasifka JR (2019) Annual cover crops for managed and wild bees: optimal plant mixtures depend on pollinator enhancement goals. Agr Ecosyst Environ 273: 107–16
- Martin EA, Dainese M, Clough Y, et al. (2019) The interplay of landscape composition and configuration: new pathways to manage functional biodiversity and agroecosystem services across Europe. Ecol Lett 22: 1083–94
- Minter LM and Bessin RT (2014) Evaluation of native bees as pollinators of cucurbit crops under floating row covers. Environ Entomol 43: 1354–63
- Mommaerts V and Smagghe G (2011) Entomovectoring in plant protection. Arthropod-Plant Inte 5: 81–95
- Morandin LA, Long RF, and Kremen C (2016) Pest control and pollination cost–benefit analysis of hedgerow restoration in a simplified agricultural landscape. J Econ Entomol 109: 1020–27
- Oerke EC (2006) Crop losses to pests. J Agr Sci 144: 31–43
- Park MG, Blitzer EJ, Gibbs J, et al. (2015) Negative effects of pesticides on wild bee communities can be buffered by landscape context. P Roy Soc B-Biol Sci 282: 20150299.
- Perez-Alvarez R, Nault BA, and Poveda K (2019) Effectiveness of augmentative biological control depends on landscape context. Sci Rep-UK 9: 8664
- Potts SG, Neumann P, Vaissière B, and Vereecken NJ (2018) Robotic bees for crop pollination: why drones cannot replace biodiversity. Sci Total Environ 642: 665–67
- Redlich S, Martin EA, and Steffan-Dewenter I (2018) Landscape-level crop diversity benefits biological pest control. J Appl Ecol 55: 2419–28
- Ricci B, Lavigne C, Alignier A, et al. (2019) Local pesticide use intensity conditions landscape effects on biological pest control. P Roy Soc B-Biol Sci 286: 20182898
- Root RB (1973) Organization of a plant–arthropod association in simple and diverse habitats: the fauna of collards (*Brassica oleracea*). Ecol Monogr 43: 95–124
- Rowen E, Tooker J, and Blubaugh C (2019) Managing fertility with animal waste to promote arthropod pest suppression. Biol Control 134: 130–40
- Rundlöf M and Lundin O (2019) Can costs of pesticide exposure for bumblebees be balanced by benefits from a mass-flowering crop? Environ Sci Technol 53: 14144–51

- Rundlöf M, Persson AS, Smith HG, and Bommarco R (2014) Late-season mass-flowering red clover increases bumblebee queen and male densities. Biol Conserv 172: 138–45
- Rundlöf M, Andersson GK, Bommarco R, et al. (2015) Seed coating with a neonicotinoid insecticide negatively affects wild bees. Nature 521: 77–80
- Rusch A, Chaplin-Kramer R, Gardiner MM, et al. (2016) Agricultural landscape simplification reduces natural pest control: a quantitative synthesis. Agr Ecosyst Environ 221: 198–204
- Saunders ME, Peisley RK, Rader R, and Luck GW. (2016) Pollinators, pests, and predators: recognizing ecological trade-offs in agroecosystems. Ambio 45: 4–14
- Schellhorn NA, Gagic V, and Bommarco R (2015) Time will tell: resource continuity bolsters ecosystem services. Trends Ecol Evol 30: 524–30
- Schipanski ME, Barbercheck M, Douglas MR, et al. (2014) A framework for evaluating ecosystem services provided by cover crops in agroecosystems. Agr Syst 125: 12–22
- Shackelford G, Steward PR, Benton TG, et al. (2013) Comparison of pollinators and natural enemies: a meta-analysis of landscape and local effects on abundance and richness in crops. Biol Rev 88: 1002–21
- Smart MD, Pettis JS, Euliss N, and Spivak MS (2016) Land use in the Northern Great Plains region of the US influences the survival and productivity of honeybee colonies. Agr Ecosyst Environ 230: 139–49
- Snyder WE and Wise DH (2001) Contrasting trophic cascades generated by a community of generalist predators. Ecology 82: 1571–83
- Sponsler DB, Grozinger CM, Hitaj C, et al. (2019) Pesticides and pollinators: a socioecological synthesis. Sci Total Environ 662: 1012–27
- Stanley DA, Garratt MPD, Wickens JB, et al. (2015) Neonicotinoid pesticide exposure impairs crop pollination services provided by bumblebees. Nature 528: 548–50
- Stenberg JA (2017) A conceptual framework for integrated pest management. Trends Plant Sci 22: 759–69
- Stern VM, Smith RF, Van den Bosch R, and Hagen KS (1959) The integration of chemical and biological control of the spotted alfalfa aphid: the integrated control concept. Hilgardia 29: 81–101
- Tamburini G, Bommarco R, Kleijn D, et al. (2019) Pollination contribution to crop yield is often context-dependent: a review of experimental evidence. Agr Ecosyst Environ 280: 16–23
- Tanda AS (2022) Chapter 11: Pest Management Technology and Bee Pollinators' Integration. In: Tanda A S (ed.) Advances in Integrated Pest Management Technology: Innovative and Applied Aspects. Springer Nature Switzerland (in press)
- Tanda AS (2021a) Urbanization and Its Impact on Native Pollinators. Presented at the 1st International Electronic Conference on Entomology (IECE 2021), 1–15 July 2021; Available online: https://iece.sciforum.net/
- Tanda AS (2021b) Wild bees and their conservation. Indian Journal Entomology Indian Journal Entomology. https://doi.org/10.55446/IJE.2021.64

- Tanda AS (2021c) Biofloral phenology, foraging behaviour and entpollinatological effect of honeybees in pomegranate (*Punica granatum*) fruit quality and yield. Journal of Horticulture 8 2: 1–3
- Tanda AS. (2020a) Biogenetic engineering in developing insect resistant crops: constraints and applications. *5th Edition of Global Congress on Plant Biology and Biotechnology* (*GPB 2020*) Valencia, Spain
- Tanda AS (2020b) An Assessment of honeybee (*Apis mellifera*) foraging activity and pollination efficacy in Australian raspberry *Rubus parvifolius* at Rosehill. Australian Journal of Entomology (Submitted)
- Tanda AS (2020c) Entpollinatology is for sustainable agroecosystems and global agroindustry. Indian Journal of Entomology (Accepted)
- Tanda AS (2019) Entomophilous crops get better fruit quality and yield: An appraisal. Indian Journal Entomology 81(2): 227–234
- Tanda AS, Goyal NP (1979) Insect pollination in Asiatic cotton (*Gossypium arboreum*). Journal of Apicultural Research 18: 64–72
- Thies C, Roschewitz I, and Tscharntke T (2005). The landscape context of cereal aphid–parasitoid interactions. P Roy Soc B-Biol Sci 272: 203–10
- Tschumi M, Albrecht M, Entling MH, and Jacot K (2015) High effectiveness of tailored flower strips in reducing pests and crop plant damage. P Roy Soc B-Biol Sci 282: 20151369
- van Lenteren JC, Bolckmans K, Köhl J, et al. (2018) Biological control using invertebrates and microorganisms: plenty of new opportunities. BioControl 63: 39–59
- Westerkamp C and Gottsberger G. (2000) Diversity pays in crop pollination. Crop Sci 40: 1209–22
- Zalucki MP, Adamson D, and Furlong MJ (2009) The future of IPM: whither or wither? Aust J Entomol 48: 85–96
- Zotti M, dos Santos EA, Cagliari D, et al. (2018) RNA interference technology in crop protection against arthropod pests, pathogens and nematodes. Pest Manag Sci 74: 1239–50

14

Challenges and Threats for Pollinator Conservation

Usman Ghani[1], Ahmad Nawaz[1], Tariq Mustafa[2], Saiqa Muneer[3], Ahmad Ghafar[1], Shanza Akbar[4], Sadia Bibi[1], Marium Aziz[1] and Aqsa Aslam[4]

ABSTRACT

Pollinators include bees, wasps, moths, butterflies, birds, flies, and small mammals. They help to pass on pollen from the male part of the flower (stamen) to the female part of the same or another bloom. Pollinators are vital in both ecological as well as agricultural terms. Recent estimates revealed that 87.5% of angiosperms use invertebrates or vertebrates for pollination. Similarly, a considerable proportion of gymnosperms are also biotically pollinated. The pollinator species rely on olfactory cues and plant architecture to recognize their host plants and a slight modification in the composition or the perception of the volatile olfactory cues can lead to the breakdown of the host recognition process. Thus, such species-specific modification of the volatile compounds emitted may impair plant-pollinators interactions. Pesticides, diseases, and loss of habitat are all contributing to reduce number of flowers, nesting sites, and egg-laying possibilities for pollinators and ultimately causing significant pollinator losses. Habitat restorations have become more crucial for rebuilding pollinator species as people continue to alter natural areas. This chapter will highlight the current and future threats and challenges for pollinator conservation.

Keywords • *Pollinators* • *Pesticides* • *Conservation* • *Ecology* • *Management* • *Natural habitat*

14.1 INTRODUCTION

Pollen migration is required for the plant to be fertilized and generate fruits, seeds, and new plants. Pollinator helps to transmit pollen from the male part of the flower (stamen) to the

[1]Department of Entomology, University of Agriculture, Faisalabad, Pakistan

[2]Department of Entomology, Sub-Campus Depalpur (Okara), University of Agriculture Faisalabad, Pakistan

[3]School of Chemistry and Molecular Biosciences, The University of Queensland, Brisbane, Australia

[4]Department of Zoology, Wildlife and Fisheries, University of Agriculture, Faisalabad, Pakistan

Corresponding author: ahmad.nawaz1793@uaf.edu.pk; nawazrajpoot@yahoo.com

female part of the same or another bloom. Some flowers are pollinated by bees, wasps, moths, butterflies, birds, flies, and small mammals, including bats, whereas others are self-pollinated. Insects and other animals visit flowers for diet, shelter, nest-building materials, and for mates (Kremen et al., 2002). Many pollinator species, including bees, collect pollen purposefully. Numerous butterflies, birds, and bats, unintentionally transfer pollen. Pollen attaches to their bodies when they suck or graze on nectar from flower blossoms and is carried from flower to flower unwittingly. Pollinators are well-known among scientists as well as the general population. They are vital in both ecological and agricultural terms, and they are on the root of developmental progress. They are major sets of ecological interlinks in terrestrial biodiversity, creating significant evolutionary diversity of pollinators and plants which treat them. Plants are immobile and are unable to move in search of mates. However, they must either self-pollinate (risking inbreeding and genetic diversity over time) or depend on carriers to move the male parts (containing pollen). This review focuses on animal pollination. Animals pollinate the majority of flowering plants. According to a recent global estimate, 87.5% (Ollerton et al., 2011) of angiosperms use invertebrates or vertebrates for pollination and a considerable proportion of gymnosperms are also biotically pollinated. However, it is obvious that the diversity and relative value of pollinators have been subjected to significant media bias and professional deception, with honeybees gaining attentiveness that is inordinate to their significance for conserving (Rios et al., 2014).

14.2 POLLINATOR DIVERSITY IN THE PRESENT

We don't have an approximate diversity of prokaryotes on the earth to the sequence of proportions at the moment, having reported predictions varying from 2-100 millions, (Costello et al., 2013) although reduced estimations are likely more plausible. Because the bulk of these organisms are insects, which are the most common honeybees, estimating the current diversity of pollinators is clearly challenging and fraught with uncertainties. However, for some taxa especially mammals and birds, as well as most likely bees (Regan et al., 2015), we can give precise evaluation, making progress towards a full account of overall evolutionary diversity and the number of species involved increasingly achievable. The Lepidoptera (especially the moths) are the most diverse group of pollinators, with over 140,000 species estimated to room flora, based on 90% organisms with operational mouthparts (Wardhaugh, 2015) while many species feed on animal/plant secretions. Coleoptera and Hymenopter are the next most diverse groups. Although, much more study has been performed and the true variety of flies as pollinators is understood. Flies is the least varied of the four major orders of pollinating insects (Larson et al., 2001; Ollerton et al., 2009, 2017; Orford et al., 2015). The remaining groupings are all quite low, while they are undoubtedly biologically important in some places and for specific plants. The pollinating groups of vertebrates are relatively low diverse. Birds are the most diversified, with over 1,000 species. Nocturnal rodents, especially in tropical forests, are possibly significant than we understand, although they have not been researched as like birds or even bats (Olesen & Valido, 2003).

Table 14.1 The important insect orders and other classes with estimated pollinator species.

Taxon	Estimated number of pollinating species
Lepidoptera	141,600
Coleoptera (beetles)	77,300
Hymenoptera	70,000
Diptera (flies)	55,000
Thysanoptera	1,500
Aves	1,089
Hemiptera	1,000
Collembola	400
Blattodea	360
Mammalia	344
Neuroptera	293
Trichoptera	144
Orthoptera	100
Mecoptera	76
Psocoptera	57
Plecoptera	37
Lacertilia	37
Dermaptera	20
Crustacea	11
Polychaeta	3

Wardhaugh (2015); van Tussenbroek et al. (2016)

14.3 SURVIVAL NEEDS OF POLLINATORS

Pesticides, diseases, and loss of habitat, which limits the number of flowers, nesting sites, and egg-laying possibilities for pollinators, are all contributing to significant pollinator losses. Habitat restorations has become more crucial for rebuilding pollinator species as people continue to alter natural areas. It has been applied in a number of landscapes, comprising agricultural areas and public parks. Similarly, the amount and diversity of bees, butterflies, and pollinating flies can be increased by adding flowering habitats such as hedgerows, crop margins, and right-of-ways. Planting gardens for insect pollinators is a habitats rehabilitation method that has gotten little research attention. Parks are areas of man managed environments in which a diversity of flowering plants are purposefully grown, usually on personal land or in community green areas like gardens in public surroundings. While yard are typically designed for esthetic purposes, they can also support significant biodiversity.

Since the 1980s, the amount of pollinators gardening guides has expanded and fauna saving efforts (National Parks) with particular gardening-for-wildlife projects. Along with popular awareness for pollinators, the number of publications and web pages on how to plant for pollinators keeps rising. Modern gardening books advise gardeners to include a wide variety of plant species that provide pollen and nectar. Butterfly, moths, bee, and wasp egg-laying places are urged to be provided by reproductive resources such as caterpillar host plants and 'bee hotels. To maintain ecologically beneficial and esthetically beautiful gardens, gardeners must make judgments about weeds eradication, pest, trimming, fertilizing, and irrigation.

Gardens are frequently considered to benefit insect pollinators since they provide a variety of resources. Despite the fact that garden areas can maintain a diverse range of pollinators due to dissimilarity in plant structure, habitat complication, regulation and landscape effects (Baldock et al., 2019). Moreover, parks may alter pollinator associations with predators and diseases, restricting pollinator immigration in the future. Knowing the conditions though gardens are favorable or harmful to bees is critical for managing the habitat's conserving capacity.

Pollinators are harmed by clearing practises such as trimming and eradicating dead plantation because certain life phases are accidentally removed. During clearance activities of dead vegetations, hibernating eggs, pupae and adults are exposed. As has been demonstrated in other areas (e.g., reduced butterfly richness on cropped fields) (Garibaldi et al., 2011) such actions may reduce pollinator abundance. The weeding, cleaning, and mulching is critical, as indicated by advice to wait until later in the growing season to remove dead plants, enabling cavity-nesting pollinators and overwintering butterfly to appear.

14.4 FACTORS THAT ARE CAUSING THE DROP

14.4.1 Degradation, Loss and Fragmentation of Habitat

- Pollinator populations are most affected by habitat degradation and fragmentation.
- Some native bees use hedgerows, field edges, embankments, and other "waste places" as breeding sites (Cane & Tapedno, 2001).
- By limiting genetic drift among demes (domestically interbreeding groups within a geographic population), rate of genetic erosion can be speed up by fragmentation and habitat destruction, increasing the likelihood of species diversity going extinct (Cane & Tapedno, 2001).
- When large habitats are divided into small isolated patches, it doesn't take long for a few of the animals that live there to drop in numbers to the level where they can't offer efficient ecosystem systems to plants (Nabhan & Buchman, 1996).
- Because wild pollinators require peaceful area for nesting, roosting, and foraging, habitat destruction and dispersion constitute a major threat to them. (Cane & Tapedno, 2001).
- Urbanization segregates most of the area that it does not destroy or absorb, in addition to destroying habitat. Food supplies are being reduced. Fewer mating, nesting, and migrating locations. Hay meadows are (over) grazed and early cut, resulting in plants not reaching flowering stage.

14.4.2 Agriculture Practices

Inappropriate pesticide, such as coating seed with a uniform or systemic insecticide (like Imidacloprid), which is transferred to all parts of plant through roots, such as pollen and nectar, poses a risk to pollinators like honeybees and other insects. Monoculture's replacement of plant communities is a problem, as most monocultures are unable to support pollinator populations. Insecticides are also used in gardens, courtyards, woodlands, and mosquito-infested marshland (Zayed & Packer, 2005). The widely employed broad spectrum insecticide are frequently hazardous to pollinators as well as the target insect. Pollinators, require protection from insecticides and different pollutants that might harm or hinder development of pollinators. Pesticides may reduce nectar, larval host plants for butterfly and moth and building stuff for pollinators (Corbet, 1995).

14.4.3 Other Factors

- Due to the emergence of resistance to acaricides, honeybee colonies have been infected with *Varroa* mites, which are viral transmitters.
- Bees are more vulnerable to extinction than other species because of specific sex site, which leaves bee especially susceptible to consequences of limited community numbers by producing infertile diploid males.
- Invasive non-native species pose a threat.
- Climate change on a global and local scale.
- Changes in surface energy and water balance are influenced by land-cover changes, which impact local climates.
- A decrease in the number of larval host plants.

14.5 COLONY COLLAPSE DISORDER (CCD)

According to Frazier et al. (2007) the causes of CCD are not yet known. The potential causes investigated by "The CCD working group" are, but are not limited to:

- **Chemical contamination/residue** in food stores, wax and bees. Because of the evolution of mite and other pathogenic resistance to chemicals, beekeepers boosting dosage amounts or experimenting with chemical mixtures, revealing honeybees to insecticidal residues that are hazardous to worker longevity.
- **Pathogens** in bees and their brood on larvae and pupae, e.g., European Foul Brood (caused by *Mellisococcus pluton*) and American Foul Brood (produced by *Paenibacillus larvae*)
- **The bees** and their offspring have a high parasitic burden, e.g, *Varroa* on honeybees.
- **Nutritional** condition of the mature bees.
- **Stress** levels in mature bees as measured by stress-inducing proteins.
- **Bees** are more sensitive to epidemic formation due to a shortage of genetic variation and ancestry. The CCD working group is not currently exploring the following topics:

- **Insecticides used in agriculture** are a necessity in agriculture. Bee colony mortality is high where insecticides are used, and where pollination is necessary, careful management is required to minimize bee losses.
- **Agricultural practices** have changed due to lower honey outputs across the country, beekeepers are looking for other ways to supplement their income (for example, almond pollination colonies, a crop that is completely reliant on bee pollination)." According to anecdotal data, colony collapse disorder is widespread on commercial level where pollinators are transported far off places and sold for pollination.
- **Genetically modified crops**, particularly, Bt (*Bacillus thurengiensis*) Corn, was implicated as a possible reason of colony collapse disorder. Whereas this option is still not excluded, the research presented here clearly suggests that Bt crop is not linked to CCD.
- **Radiation from cellular towers** Given the spread of both damaged and unaffected colony collapse disorder colonies, this seems unlikely to be the reason. In many places damaged commercial colonies found in the west, cell phone coverage is also unavailable. As a result, it is not currently a key priority.

14.6 FACTORS CONTRIBUTING TO THE DECLINE IN POLLINATORS

Pollination by different pollinators is considered one of the most important mechanisms in maintaining biodiversity, food production, and life on the earth. Currently, the pollinator's abundance and diversity are threatened by many direct drivers, including habitat fragmentation, urbanization of land, intensive agricultural practices, pesticide exposures, parasites and diseases, invasive species, and climate change.

14.6.1 Urbanization

Urbanization is a process of altering the environment with increased proportion of built surfaces including buildings, roads, sidewalks, and parking lots (Wu, 2014; Cardoso et al., 2018; Fenoglia et al., 2020; Kurylo et al., 2020; Piano et al., 2019; Sattler et al., 2010). Approximately, 90% of the world's plants and 70% of the global crop species are being pollinated by insect species (Klein et al., 2007; Ollerton et al., 2011). The reduction in the biodiversity of organisms is due to various interactive elements in the urban environment. These factors include habitat loss and fragmentation, the existence of exotic plants and animal species, and the warming effect of the urban environment (Fig. 14.1) (Harrison & Winfree, 2015).

14.6.2 Habitat Loss and Fragmentation

Habitat loss and fragmentation have been considered the most important drivers contributing to the richness and decline of insect pollinator species in the urban environment (Winfree et al., 2009). The development of roads, buildings, sidewalks, parking lots, and other impervious structures in the urban environment produced small and poor-quality habitats (Stenhouse et al., 2004). These impervious structures influence the landscape composition and alter plant composition, movement, behavior, and densities of pollinators (Greenleaf et al., 2007; Hadley et al., 2011; Hennig et al., 2012; Harrison et al., 2015). Furthermore, these impervious changes

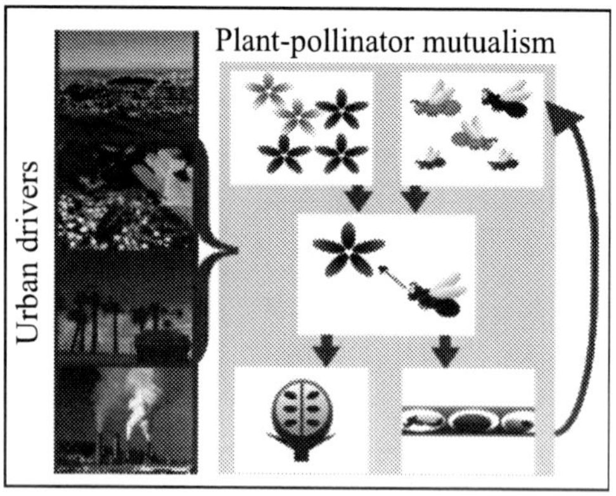

Fig. 14.1 Fragmentation of specific urban drivers that directly affect plant-pollinator interactions at different stages and plant-pollinator mutualism process (Harrison & Winfree, 2015)

result in different management practices for each fragmented area or city producing idiosyncratic habitats, which complicate the behavior of the pollinators visiting these urban landscapes.

14.6.3 Surrounding Landscape and Impervious Surface

The surrounding landscapes and impervious surfaces can impact the pollinators in areas surrounded by an agroecosystem (Ricketts et al., 2008; Kennedy et al., 2013). It has been found that higher abundance and species richness of pollinators should occur in areas existing close to natural areas from urban cores (Jones et al., 2012). Such semi-natural habitats including green spaces within an urban environment may favor pollinators (Phillips et al., 2015). These semi-natural areas can be used by different pollinators including bees to serve as nests (Bogusch et al., 2020). Urban areas surrounded by or diffused with impervious surfaces, could influence the pollinator's abundance, and species richness (Geslin et al., 2016; Egerer et al., 2017; Senapathi et al., 2017). The intensification of these impervious surfaces has a strong correlation with pollinator community shifts. The bees having greater proportions of impervious surfaces have a higher abundance of cavity-nesting bees compared to ground-nesting bees (Cane et al., 2006; Geslin et al., 2016; Kuussaari et al., 2020), which exhibits that these impervious surfaces affect the available nesting spaces which are being utilized by these ground-nesting insects.

14.6.4 Urban Vegetation and Microclimate

Ornamental and exotic plant species are found extensively within residential yards and gardens of urban areas (Threlfall et al., 2016). This increase in plant composition enhances the pollinators and other insect species' richness (Hennig, 2012; Braaker, 2017; Leston, 2017; Baldock et al., 2019; Wilson, 2019). The increasing plant species are attractive to bees and other pollinators (Martins et al., 2017), enhancing the floral abundance and frequency thus impacting and promoting the species richness (Lanner et al., 2018; Stewart et al., 2018). The high prevalence of exotic species of plants in urban areas showed that exotic plants may have a negative effect on the pollinator communities by affecting their visitations and their reproductive success compared to native plant species (Morales et al., 2009). Most recently, it was indicated that native plants were preferred over non-native plants by pollinators (Harrison et al., 2015).

Therefore, the conservation of native plant species is essential for sustaining native bees (Mach et al., 2018; Rollings et al., 2019; Giovanetti et al., 2020). The exotic plants may extend floral resource availability throughout the season thus contributing to the overall abundance of floral resources (Salisbury et al., 2015). This is especially apparent given that some evidence showed that native perennial plant species exhibit greater bee abundance compared to exotic and annual plant species (Kennedy et al., 2013; Lowenstein et al., 2019).

14.6.5 Microclimate

Microclimatic conditions of microhabitats vary across the urban environments (Pincebourde et al., 2016; Dale et al., 2018; Fisogni et al., 2020). The vegetation types and cover are considered two important key characteristics that could influence the local climate (Crum et al., 2017). The shades produced by these vegetational covers influence the microclimates of nests used by cavity-nesting bees (Richards et al., 2020), which influences the metabolic rate and development of bees (Radmacher et al., 2009; Richards et al., 2020). The deposition of nitrogen is also prevalent in urban areas and interacts synergistically with climate change and influences the local microclimates (WallisdeVries et al., 2006; Harrison & Winfree, 2015). Both these factors favor spring plant growth by reducing the abundance of pollinators (WallisdeVries et al., 2006). Such patterns have been observed in bees and other pollinators (WallisdeVries et al., 2006). Despite the knowledge that landscapes have heterogeneous microclimates, the understanding of species responding to specific habitat variables is essential as shifts in microclimate may impact pollinator populations either through extirpation or dispersal (Hamblin et al., 2018; Burdine & McCluney, 2019).

14.7 IMPACT OF CLIMATE CHANGE ON POLLINATORS

Climate change is considered as another important significant factor contributing towards the decline of pollinators and other insects' biodiversity (Parmesan 2006). The continuous warming of the environment and pattern in climate change due to drought and heat waves alters the synchrony between the flowering plants and their pollinators (Scaven and Rafferty 2013; Ockendon et al., 2014; Pandey et al., 2015; Lamaoui et al., 2018; Spinoni et al., 2018). Climate-driven changes in plant-pollinator interactions could have severe ecological and economic consequences (Potts et al., 2010; Settele et al., 2016).

14.7.1 Morphological Differences

The impact of global warming on insect body size has been found at higher temperatures. This change may occur due to the high metabolic rate, and accelerated development (Kingsolver & Huey, 2008; Oliveira et al., 2016; Gérard et al., 2018). The thermoregulatory activity of the pollinators, for example, larger bees are more capable of thermoregulating their body temperature compared to the smaller bees (Bishop & Armbruster, 1999), is highly impacted by the increased temperature. The floral morphology of the plants is linked with the natural selection used by plant pollinators (Waser et al., 1996; Nattero et al., 2010). The degree of morphological matching is high because any change to this matching could result in fitness impacts for either or both partners (Nattero et al., 2010). However, if the level of matching is low, it potentially implies a low reciprocal selection because of the high variability of the phenotypes involved in the interaction (Nattero et al., 2010). Nonetheless, in many plant–pollinator interactions, morphological matching requires precision and a consistent mechanistic fit between the two actors to ensure the contact between the sexual parts of the flower (i.e., pollen and stigma) and the body of the pollinator and therefore efficient pollination (Borrel, 2005). Floral morphology is thus linked to the pollinator taxa that they attract, notably because of the relationship between the length of nectar spurs, corolla tubes, and the proboscis of the pollinator (Borrel, 2005; Anderson et al., 2014). The relationship between proboscis length and corolla depth has been investigated in various groups like euglossine and solitary bees (Borrel, 2005) and butterflies (Corbet, 2000). In most of these studies, species with longer proboscises visit a wider range of flowers, and all the studies highlighted a positive relationship between probosci's length and the depth at which nectar is produced.

As an example of a potential morphological mismatch, two bumblebee species have evolved shorter tongue lengths over the last 40 years while their associated flowers have not become shallower (Miller-Struttmann, 2015). This tongue size matching is particularly crucial as it has been showed recently to be a strong driver of foraging efficiency in multi-species assemblages (Klumpers et al., 2019). The overall size matching will also determine the efficiency of pollen transfer: smaller pollinators (with on average shorter tongues) can have lower pollination efficiency (Howlett et al., 2011), but see through reduced pollen delivery (Adler & Irwin, 2006), therefore, having a lower single-visit pollen deposition rate and seed set while large- or middle-sized species are, most of the time, more likely to have greater contact with anthers and stigma and therefore have a greater pollination efficiency. Most of the studies have focused on pollen deposition and the pattern is largely constant. However, with regards to pollen removal from anthers, results are less consistent and some of them indicate that pollen removal efficiency is independent from body size (Zych et al., 2013). Indeed, many other morphological features from both partners are driving pollination efficiency, notably hair characteristics (e.g., size and density) (Howlett et al., 2011; Klumpers et al., 2019) which can also be impacted by temperatures during development (Adler & Irwin, 2006). Overall, morphological mismatches between plants and pollinators are most likely to occur within plant species displaying specialized flower morphologies as their morphological fit with pollinators is most restricted (Schweiger et al., 2010).

14.7.2 Phenological and Spatial Differences

Global warming has strong impact on the life cycle, and activity patterns of organisms in the ecosystem (Byers, 2017). Many phenological shifts have already been reported at the species level for the insect on emergence (Dalin, 2011; Duchenne et al., 2020) or at blooming time of flowering plants (Menzel et al., 2006). Most of the studies showed a simultaneous advance in the phenology of pollinators and plants in parallel with increasing average temperatures, as well as simultaneous delays or opposite shifts have also been described (Bartomeus et al., 2011; Duchenne et al., 2020). Mismatches between the emergence of bee species and the blooming of their main resources have been specifically recorded (Bartomeus et al., 2011). In contrast, plants that flower earlier can miss part of the activity window of their main pollinators (Kehrberger & Holzschuh, 2019) and show a decrease in seed production (Kudo & Copper,

2019). Moreover, as these phenological shifts tend to be species-specific for each species of the pollinator network, ecological communities were also disturbed. If an insect that pollinate plant species existing in slightly different climatic niches, a warming climate has the potential to cause a spatial mismatch. There are multiple examples of range shifts in pollinators that are driven by climate change (Bedford et al., 2012), including the changes in bumblebees (*Bombus* spp.). The situation is further complicated because of generalization in pollination system such as pollinators visiting multiple plant species and plant species being visited by multiple pollinator species (Waser et al., 1996). Therefore, a spatial divergence between the two is more straightforward because a reduced area of overlap has clear negative fitness implications for pollinators that cannot reproduce without their host plant (Schweiger et al., 2008; Pelini et al., 2009).

14.7.3 Disruption to Host Attraction and Foraging Behaviors

The pollinator species rely on olfactory cues, and plant architecture to recognize their host plants from a distance (Bruce et al., 2005; Beyaert & Hilker, 2014). A slight modification in the composition or the perception of the volatile olfactory cues can lead to the breakdown of the host recognition process. It has been shown that slight changes in temperature can affect volatile emissions and alters the quality of the signals produced from flowers (Farré-Armengol, 2014). The increase in temperature not only increases rate and abundance of floral volatile olfactory cues emissions, but also influences the ratios of compounds present in floral scents (Farré-Armengol, 2014). Such species-specific modification of the volatile compounds emitted may impair plant-pollinators interactions. The pollinators may become confused with more fragrant atmosphere or are no longer attracted to their host-plant floral cues. The visual stimuli displayed by plants also play an important role in host selection (Döttrel & Vereecken, 2010). Several studies have shown that the impact of high temperatures affects both vegetative and floral morphological traits (Qaderi et al., 2012), cause flower deformity (Smith & Zhao, 2016), lead to lighter flower colors (Lai et al., 2011), reduction in number of flowers and their size (Morrison & Stewart, 2002), advance the timing of the anthesis (Murcia, 1990) and limit the production of floral resources (Mu et al., 2015). The high temperatures, denature the odorant-binding proteins and olfactory receptors (Leal, 2013), resulting in a loss of biological function (Sørensen et al., 2003; Mayer, 2010). Elevated temperatures can also alter the nutritional quality for pollinators (Sato et al., 2006). Heat stress can lead to a decrease in pollen viability (Paupière et al., 2014), can decrease the concentration of soluble carbohydrates (Pressman et al., 2002) and proline (Lai et al., 2011) as well as alter the phospholipid content of pollen (Djanaguiraman et al., 2013). Moreover, plants produce a repertoire of secondary metabolites in response to heat stress (Ramakrishna & Ravishankar, 2011). Such increases in secondary metabolites as well as the accumulation of reactive oxygen species are likely to occur in pollen were detrimental to pollinators (Santiago & Sharkey, 2019). All these changes in pollen composition might decrease its suitability to pollinators, reducing their performance and health status (Vanderplanck et al., 2014; Roger et al., 2017). Finally, these changes in pollen development and viability could also affect plant fitness as they could disrupt plant fertilization process (Hedhly, 2011; Snider & Oosterhuis, 2011). Regarding all these effects on sensory cues used by pollinators for host attraction, global warming is probably already disrupting the interactions between plants and pollinators since all these modifications might alter the attractiveness of flowers to pollinators and impact their foraging behavior (Descamps et al., 2018).

14.8 CHEMICALS AND POLLUTANTS

Pesticides commonly used in agricultural, and urban environments kill invertebrate pests, diseases, and weeds. However, many pesticides—including insecticides, fungicides, and herbicides—harm pollinators and other beneficial insects (Fig. 14.2). The use of these insecticides represents a hazard to pollinator health, their diversity and abundance (Brittain & Potts, 2011; Schreinemachers & Tipraqsa, 2012; Godfray et al., 2014, 2015; van der Sluijs et al., 2015; IPBES, 2016).

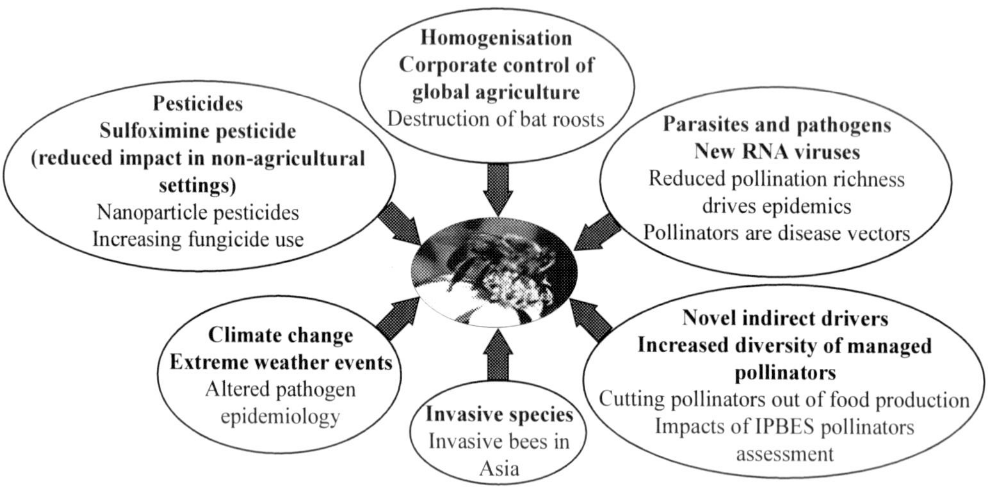

Fig. 14.2 Major drivers for pollinators and pollinator decline with positive or negative opportunities (Brown et al., 2016)

Mostly the insecticides were applied as sprays on crops, whereas the herbicides and fungicides are applied directly on to the soil before crops plantation. In both cases, the droplets, and dust from pesticide applications may fall directly on to the bees that comes in contact while fly across the treated crop fields or nearby areas (Greig-Smith et al., 1994; Craig et al., 1998; Marinelli et al., 2004). Pollinators pollinating crops may exposed to pesticides in many ways including direct contact with pesticides suspended particles in air and residue on plants, through ingestion of contaminated nectar and pollens, and through exposure to contaminated nesting sites and materials (Fig. 14.3).

Direct contact occurs when pesticides land directly on pollinators. The risk of direct contact is highest when pesticides are applied on or near flowering plants, be they crops or weeds. Residue contact occurs when pollinators visit flowers or walk on leaves that have been previously treated with pesticides. Butterflies can be at risk of exposure when laying eggs on host plants that have been treated with pesticides. Solitary bees risk exposure when collecting plant material or soil used to construct nests. Wild bees often nest in areas between row crops, in brush piles, or overgrown areas at field edges. These sites can become contaminated when pesticides are applied nearby.

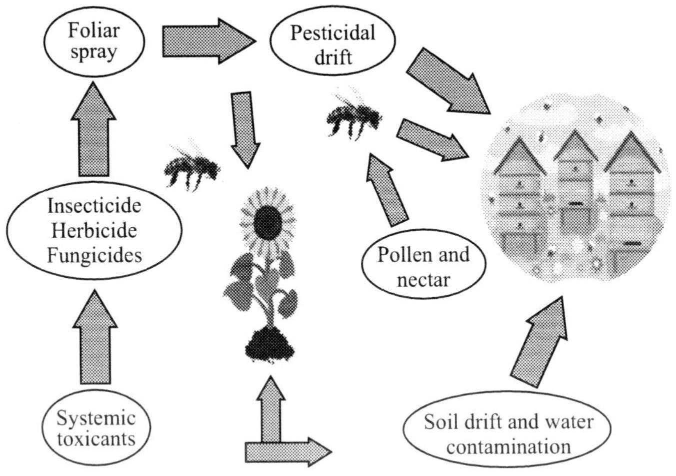

Fig. 14.3 Different routes of pesticides exposure to honeybees (Kumar et al., 2020)

The insecticides applied for seed treatments were introduced into the soil using pneumatic drilling planters. Due to the friction of the seeds during planting, the machinery produces fine insecticides dust particles that are suspended in the air. These poisonous particles cause mortality among bees if they are visiting crops in the surroundings (Georgiadis et al., 2010). The systemic insecticides applied over the field crop and ornamental plants, become the parts of plant system, and their residues present in flowers, pollen and nectar affects the pollinators visiting (Bomatin et al., 2015). In addition, some plants produce guttation drops in the morning (e.g., maize, strawberries), and elevated concentrations of systemic insecticides in such drops also kill the bee and other pollinators utilizing those contaminated drops (Tapparo et al., 2011). In most cases, the bees and other pollinators get exposure of pesticides through ingestion of residues found in the pollen and nectar of contaminated plants (Sánchez-Bayo & Goka, 2014). Here, it is important to realize that the foraging bees search the most suitable flowers that produce pollen and nectar in abundance every day. Therefore, some crops are more attractive compared to other; for example, the yellow flowers of sunflower, canola and many weeds that grow in and around the crops are more attractive to bees compared to the flowers of potato plants. The pesticide residues in pollen and nectar taken by the forager bees to their colonies are fed to the larvae and queen in the form of bee bread and honey (Kubik et al., 1999; Orantes-Bermejo et al., 2010). Thus they are also affected in similar ways as the forager bees. In addition to food, the bees also drink water to keep their body temperature under control (Schmaranzer, 2000). Pesticide residues in the soil, and drifts of insecticide sprays fall on the ground eventually move into the water and appears in the creeks, streams, and ponds of agricultural areas and beyond, which are thus contaminated with a mixture of agrochemicals (Woods et al., 2001; Siebers et al., 2003; Belden et al., 2007). Honeybees, bumblebees and wild bees like to drink from puddles, irrigation ditches, ponds and streams, and if these waters are contaminated with pesticide residues, the forager bees ingest them as well (Samson-Robert et al., 2014). Given the enormous variety of agrochemicals used in crop production, it is not surprising that, to date, residues of 173 different compounds have been found in apiaries (Sánchez-Bayo & Goka, 2014). It should be realized that through the various routes of exposure to pesticides in the environment (Fig. 14.3), bees are not threatened by one or two chemicals alone but by cocktails of many agricultural compounds.

14.9 CONCLUSION

It is evident that both considerable portion of angiosperms/gymnosperms use invertebrates or vertebrates for pollination and the bulk of these organisms are insects most commonly honeybees. Currently, the sustainability of pollinators' diversity is very challenging due to injudicious use of pesticides, diseases, and loss of habitats. The accurate measures of factors (climate change, synthetic chemicals, diseases, deforestation, urbanization, crop cultivation practices) contributing in population decline of pollinators are necessary that how these factors affect the pollinator populations. Furthermore, the detailed studies should be conducted focusing to overcome the negative impacts of such factors to improve the diversity of pollinators. Overall, habitat restoration is necessary to support significant biodiversity by putting ecological restoration at the forefront of the fight to reverse biodiversity loss and to recover ecosystem functions and services.

REFERENCES

- Adler LS & Irwin RE (2006) Comparison of pollen transfer dynamics by multiple floral visitors: experiments with pollen and fluorescent dye. Annals of Botany. 97(1): 141–150
- Anderson B, Ros P, Wiese TJ & Ellis AG (2014) Intraspecific divergence and convergence of floral tube length in specialized pollination interactions. Proceedings of the Royal Society B: Biological Sciences. 281(1795): 1414–1420
- Baldock KC, Goddard MA, Hicks DM, Kunin WE, Mitschunas N, Morse H & Memmott J (2019). A systems approach reveals urban pollinator hotspots and conservation opportunities. Nature Ecology & Evolution. 3(3): 363–373
- Bartomeus I, Ascher JS, Wagner D, Danforth BN, Colla S, Kornbluth S & Winfree R (2011) Climate-associated phenological advances in bee pollinators and bee-pollinated plants. Proceedings of the National Academy of Sciences. 108(51): 20645–20649
- Bedford FE, Whittaker RJ & Kerr JT (2012) Systemic range shift lags among a pollinator species assemblage following rapid climate change. Botany. 90(7): 587–597.
- Belden JB, Gilliom RJ, Martin JD & Lydy MJ (2007) Relative toxicity and occurrence patterns of pesticide mixtures in streams draining agricultural watersheds dominated by corn and soybean production. Integrated Environmental Assessment and Management: An International Journal. 3(1): 90–100
- Beyaert I & Hilker M (2014) Plant odor plumes as mediators of plant–insect interactions. Biological Reviews. 89(1): 68–81
- Bishop JA & Armbruster WS (1999) Thermoregulatory abilities of Alaskan bees: effects of size, phylogeny and ecology. Functional Ecology. 13(5): 711–724

- Bogusch P, Hlaváčková L, Gasol NR & Heneberg P (2020) Near-natural habitats near almond orchards with presence of empty gastropod shells are important for solitary shell-nesting bees and wasps. Agriculture, Ecosystems & Environment. 299: 106949
- Bonmatin JM, Giorio C, Girolami V, Goulson D, Kreutzweiser DP, Krupke C & Tapparo A (2015) Environmental fate and exposure; neonicotinoids and fipronil. Environmental Science and Pollution Research. 22(1): 35–67
- Borrell BJ (2005) Long tongues and loose niches: evolution of euglossine bees and their nectar flowers 1. Biotropica: The Journal of Biology and Conservation. 37(4): 664–669
- Braaker S, Obrist MK, Ghazoul J & Moretti M (2017) Habitat connectivity and local conditions shape taxonomic and functional diversity of arthropods on green roofs. Journal of Animal Ecology. 86(3): 521–531
- Brown MJ, Dicks LV, Paxton RJ, Baldock KC, Barron AB, Chauzat MP & Stout JC (2016) A horizon scan of future threats and opportunities for pollinators and pollination. PeerJ. 4: 22–49
- Bruce TJ, Wadhams LJ & Woodcock CM (2005) Insect host location: a volatile situation. Trends in Plant Science. 10(6): 269–274
- Burdine JD & McCluney KE (2019) Differential sensitivity of bees to urbanization-driven changes in body temperature and water content. Scientific Reports. 9(1): 1–10
- Byers DL (2017) Studying plant–pollinator interactions in a changing climate: a review of approaches. Applications in Plant Sciences. 5(6): 1700012
- Cane JH, Minckley RL, Kervin LJ, Roulston TAH & Williams NM (2006) Complex responses within a desert bee guild (Hymenoptera: Apiformes) to urban habitat fragmentation. Ecological Applications. 16(2): 632–644
- Cane JH & Tepedino VJ (2001) Causes and extent of declines among native North American invertebrate pollinators: detection, evidence, and consequences. Conservation Ecology. 5(1)
- Cardoso MC & Gonçalves RB (2018) Reduction by half: the impact on bees of 34 years of urbanization. Urban Ecosystems. 21(5): 943–949
- Corbet SA (1995) "Insects, plants and succession: advantages of long-term set-aside". Agriculture Economic Environment. 53: 201–217
- Corbet SA (2000) Conserving compartments in pollination webs. Conservation Biology. 14: 1229–1231
- Costello MJ, May RM & Stork NE (2013) Can we name Earth's species before they go extinct? Science 339: 413–416.
- Craig I, Woods N & Dorr G (1998) A simple guide to predicting aircraft spray drift. Crop Protection. 17(6): 475–482
- Crum SM & Jenerette GD (2017) Microclimate variation among urban land covers: the importance of vertical and horizontal structure in air and land surface temperature relationships. Journal Applied Meteorology Climate. 56: 2531–2543
- Dale AG & Frank SD (2018) Urban plants and climate drive unique arthropod interactions with unpredictable consequences. Current Opinion in Insect Science. 29: 27–33

- Dalin P (2011) Diapause induction and termination in a commonly univoltine leaf beetle (*Phratora vulgatissima*). Insect Science. 18: 443–450
- Descamps C, Quinet M, Baijot A & Jacquemart AL (2018). Temperature and water stress affect plant-pollinator interactions in *Borago officinalis* (Boraginaceae). *Ecology and Evolution*. 8(6): 3443–3456
- Djanaguiraman M, Prasad PV, Boyle DL & Schapaugh WT (2013) Soybean pollen anatomy, viability and pod set under high temperature stress. *Journal of Agronomy & Crop Science*. 199(3): 171–177.
- Dötterl S & Vereecken NJ (2010) The chemical ecology and evolution of bee–flower interactions: a review and perspectives. *Canadian Journal of Zoology*. 88(7): 668–697
- Duchenne F, Thébault E, Michez D, Elias M, Drake M, Persson M & Fontaine C (2020). Phenological shifts alter the seasonal structure of pollinator assemblages in Europe. *Nature Ecology & Evolution*. 4(1): 115–121
- Egerer MH, Arel C, Otoshi MD, Quistberg RD, Bichier P & Philpott SM (2017) Urban arthropods respond variably to changes in landscape context and spatial scale. *Journal of Urban Ecology*. 3(1): 1–6
- Farré-Armengol G, Filella I, Llusià J, Niinemets Ü & Peñuelas J (2014) Changes in floral bouquets from compound-specific responses to increasing temperatures. *Global Change Biology*. 20(12): 3660–3669
- Fenoglio MS, Rossetti MR & Videla M (2020) Negative effects of urbanization on terrestrial arthropod communities: A meta-analysis. *Global Ecology and Biogeography*. 29(8): 1412–1429
- Fisogni A, Hautekèete N, Piquot Y, Brun M, Vanappelghem C, Michez D & Massol F (2020) Urbanization drives an early spring for plants but not for pollinators. *Oikos*. 129(11): 1681–1691
- Frazier M, van Engelsdor D & Caro D (2007) Edited by the CCD working group at www.agr.state.il.us/programs/bees/CCD.pd
- Garibaldi LA, Steffan-Dewenter I, Kremen C, Morales JM, Bommarco R, Cunningham SA & Klein AM (2011) Stability of pollination services decreases with isolation from natural areas despite honeybee visits. *Ecology Letters*. 14(10): 1062–1072
- Georgiadis PT, Pistorius J & Heimbach U (2010) Gone with the wind-drift of abrasive dust from seed treatments-a risk for honeybees (*Apis mellifera* L.)?. *Julius-Kühn-Archiv*. (4): 24–33
- Gerard M, Michez D, Debat V, Fullgrabe L, Meeus I, Piot N & Vanderplanck M (2018). Stressful conditions reveal decrease in size, modification of shape but relatively stable asymmetry in bumblebee wings. Scientific Reports. 8(1): 1–10
- Geslin B, Le Féon V, Folschweiller M, Flacher F, Carmignac D, Motard E & Dajoz I (2016) The proportion of impervious surfaces at the landscape scale structures wild bee assemblages in a densely populated region. Ecology and Evolution. 6(18): 6599-6615
- Giovanetti M, Giuliani C, Boff S, Fico G & Lupi D (2020) A botanic garden as a tool to combine public perception of nature and life-science investigations on native/exotic plants interactions with local pollinators. PLoS One. 15(2): e0228965

- Greig-Smith PW, Thompson HM, Hardy AR, Bew MH, Findlay E & Stevenson JH (1994) Incidents of poisoning of honeybees (*Apis mellifera*) by agricultural pesticides in Great Britain 1981-1991. Crop Protection. 13(8): 567–581
- Hadley AS Betts MG (2012) The effects of landscape fragmentation on pollination dynamics: absence of evidence not evidence of absence. Biological Reviews. 87(3): 526–544
- Hamblin AL, Youngsteadt E & Frank SD (2018) Wild bee abundance declines with urban warming, regardless of floral density. Urban Ecosystems. 21(3): 419–428
- Harrison T & Winfree R (2015) Urban drivers of plant-pollinator interactions. Functional Ecology. 29: 879–888
- Hedhly A (2011) Sensitivity of flowering plant gametophytes to temperature fluctuations. Environmental and Experimental Botany. 74: 9–16
- Hennig EI & Ghazoul J (2012) Pollinating animals in the urban environment. Urban Ecosystems. 15(1): 149–166
- Howlett BG, Walker MK, Rader R, Butler RC, Newstrom-Lloyd LE & Teulon DAJ (2011) Can insect body pollen counts be used to estimate pollen deposition on pak choi stigmas. New Zealand Plant Protection. 64: 25–31
- Johansen CA & Mayer DF (1990) "Pollinator protection: a bee and pesticide handbook". Wicwas Press, Cheshire, Connecticut, USA
- Jones EL & Leather SR (2013) Invertebrates in urban areas: a review. EJE. 109(4): 463–478
- Kehrberger S & Holzschuh A (2019) Warmer temperatures advance flowering in a spring plant more strongly than emergence of two solitary spring bee species. PLoS One. 14(6): e0218824
- Kennedy CM, Lonsdorf E, Neel MC, Williams NM, Ricketts TH, Winfree R & Kremen C (2013) A global quantitative synthesis of local and landscape effects on wild bee pollinators in agroecosystems. Ecology Letters. 16(5): 584–599
- Kingsolver J.G. & Huey R.B. (2008) Size, temperature, and fitness: three rules. Evolutionary Ecology Research. 10(2): 251–268
- Klumpers SG, Stang M & Klinkhamer PG (2019). Foraging efficiency and size matching in a plant-pollinator community: the importance of sugar content and tongue length. Ecology Letters. 22(3): 469–479
- Kremen C, Williams NM & Thorp RW (2002) Crop pollination from native bees at risk from agricultural intensification. Proceedings of the National Academy of Sciences. 99(26): 16812–16816
- Kubik M, Nowacki J, Pidek A, Warakomska Z, Michalczuk L & Goszczynski W (1999) Pesticide residues in bee products collected from cherry trees protected during blooming period with contact and systemic fungicides. Apidologie. 30: 521–532
- Kudo G & Cooper EJ (2019) When spring ephemerals fail to meet pollinators: mechanism of phenological mismatch and its impact on plant reproduction. Proceedings of the Royal Society B. 286(1904): 20190573

- Kurylo JS, Threlfall CG, Parris KM, Ossola A, Williams NSG & Evans KL (2020) Butterfly richness and abundance along a gradient of imperviousness and the importance of matrix quality. Ecological Applications. 30(7): 21–44
- Kuussaari M, Toivonen M, Heliölä J, Pöyry J, Mellado J, Ekroos J & Tiainen J (2021) Butterfly species' responses to urbanization: differing effects of human population density and built-up area. Urban Ecosystems. 24(3): 515–527
- Kumar G, Singh S & Nagarajaiah RPK (2020) Detailed review on pesticidal toxicity to honeybees and its management. In. Ranz RER (ed) Modern beekeeping-bases for sustainable production, pp. 13-33
- Lai YS, Yamagishi M & Suzuki T (2011) Elevated temperature inhibits anthocyanin biosynthesis in the tepals of an Oriental hybrid lily via the suppression of LhMYB12 transcription. Scientia Horticulturae. 132: 59–65
- Lamaoui M, Jemo M, Datla R & Bekkaoui F (2018) Heat and drought stresses in crops and approaches for their mitigation. Frontiers in Chemistry. 6: 26–33
- Lanner J, Kratschmer S, Petrović B, Gaulhofer F, Meimberg H & Pachinger B (2020) City dwelling wild bees: how communal gardens promote species richness. Urban Ecosystems. 23(2): 271–288
- Larson BMH, Kevan PG & Inouye DW (2001) Flies and flowers: taxonomic diversity of anthophiles and pollinators. The Canadian Entomologist. 133(4): 439–465
- Leal WS (2013) Odorant reception in insects: roles of receptors, binding proteins, and degrading enzymes. Annual Review Entomology. 58(1): 373–391
- Leston L & Koper N (2017) Urban rights-of-way as extensive butterfly habitats: A case study from Winnipeg, Canada. Landscape and Urban Planning. 157: 56–62
- Lowenstein DM, Matteson KC & Minor ES (2019) Evaluating the dependence of urban pollinators on ornamental, non-native, and 'weedy' floral resources. Urban Ecosystems. 22(2): 293–302
- Mach BM & Potter DA (2018) Quantifying bee assemblages and attractiveness of flowering woody landscape plants for urban pollinator conservation. PLoS One. 13(12): e0208428
- Marinelli E, De Pace FM, Belligoli P & Oddo LP (2004) Biomonitoring by bees. Tests in Latium Region. Redia. 87: 207–10
- Martins KT, Gonzalez A & Lechowicz MJ (2017) Patterns of pollinator turnover and increasing diversity associated with urban habitats. Urban Ecosystems. 20(6): 1359–1371
- Mayer MP (2010) Gymnastics of molecular chaperones. Molecular Cell. 39(3): 321–331
- Menzel A, Sparks TH, Estrella N, Koch E, Aasa A, Ahas R & Zust ANA (2006) European phenological response to climate change matches the warming pattern. Global Change Biology. 12(10): 1969–1976
- Miller-Struttmann NE, Geib JC, Franklin JD, Kevan PG, Holdo RM, Ebert-May D & Galen C (2015). Functional mismatch in a Bumblebee pollination mutualism under climate change. Science. 349(6255): 1541–1544

- Morales CL & Traveset A (2009) A meta-analysis of impacts of alien vs. native plants on pollinator visitation and reproductive success of co-flowering native plants. Ecology Letters. 12(7): 716–728
- Morrison MJ & Stewart DW (2002) Heat stress during flowering in summer Brassica. Crop Science. 42(3): 797–803
- Mu J, Peng Y, Xi X, Wu X, Li G, Niklas KJ & Sun S (2015) Artificial asymmetric warming reduces nectar yield in a Tibetan alpine species of Asteraceae. Annals of Botany. 116(6): 899–906
- Murcia C (1990) Effect of floral morphology and temperature on pollen receipt and removal in Ipomoea trichocarpa. Ecology. 71(3): 1098–1109
- Nabhan GP & Buchmann SL (1997) Pollination services: biodiversity's direct link to world food stability. Nature's Services: Societal Dependence on Natural Ecosystems. Island Press, Washington, DC. 133–150
- Nattero J, Cocucci AA & Medel R (2010) Pollinator-mediated selection in a specialized pollination system: matches and mismatches across populations. Journal of Evolutionary Biology. 23(9): 1957–1968
- Ockendon N, Baker DJ, Carr JA, White EC, Almond RE, Amano T & Pearce-Higgins J W (2014) Mechanisms underpinning climatic impacts on natural populations: altered species interactions are more important than direct effects. Global Change Biology. 20(7): 2221–2229
- Olesen JM & Valido A (2003) Lizards as pollinators and seed dispersers: an island phenomenon. Trends in Ecology & Evolution. 18(4): 177–181
- Oliveira MO, Freitas BM, Scheper J & Kleijn D (2016) Size and sex-dependent shrinkage of Dutch bees during one-and-a-half centuries of land-use change. PLoS One. 11(2): e0148983
- Ollerton J, Dötterl S, Ghorpadé K, Heiduk A, Liede-Schumann S, Masinde S & Whittington A (2017) Diversity of Diptera families that pollinate Ceropegia (Apocynaceae) trap flowers: an update in light of new data and phylogenetic analyses. Flora. 234: 233–244
- Ollerton J, Masinde S, Meve U, Picker M & Whittington A (2009) Fly pollination in Ceropegia (Apocynaceae: Asclepiadoideae): biogeographic and phylogenetic perspectives. Annals of Botany. 103(9): 1501–1514
- Ollerton J, Winfree R & Tarrant S (2011) How many flowering plants are pollinated by animals?. Oikos. 120(3): 321–326
- Orantes-Bermejo FJ, Pajuelo AG, Megías MM & Fernández-Píñar CT (2010) Pesticide residues in beeswax and beebread samples collected from honeybee colonies (*Apis mellifera* L.) in Spain. Possible implications for bee losses. Journal of Apicultural Research. 49(3): 243–250
- Orford KA, Vaughan IP & Memmott J (2015) The forgotten flies: the importance of non-syrphid Diptera as pollinators. Proceedings of the Royal Society B: Biological Sciences. 282(1805): 20142934

- Pandey P, Ramegowda V & Senthil-Kumar M (2015) Shared and unique responses of plants to multiple individual stresses and stress combinations: physiological and molecular mechanisms. Frontiers in Plant Science. 6, 723–733
- Pardee GL & Philpott SM (2014) Native plants are the bee's knees: local and landscape predictors of bee richness and abundance in backyard gardens. Urban Ecosystems. 17(3): 641–659
- Parmesan C (2006) Ecological and evolutionary responses to recent climate change. Annual Review of Ecology, Evolution, and Systematics. 37: 637–669
- Paupière MJ, Van Heusden AW & Bovy AG (2014). The metabolic basis of pollen thermo-tolerance: perspectives for breeding. Metabolites. 4(4): 889–920
- Pelini SL, Prior KM, Parker DJ, Dzurisin JD, Lindroth RL & Hellmann JJ (2009) Climate change and temporal and spatial mismatches in insect communities. In: Climate Change pp. 215–231
- Phillips BW & Gardiner MM (2015) Use of video surveillance to measure the influences of habitat management and landscape composition on pollinator visitation and pollen deposition in pumpkin (*Cucurbita pepo*) agroecosystems. PeerJ. 3: 13–42
- Piano E, Souffreau C, Merckx T, Baardsen LF, Backeljau T, Bonte D & Hendrickx F (2020) Urbanization drives cross-taxon declines in abundance and diversity at multiple spatial scales. Global Change Biology. 26(3): 1196–1211
- Pincebourde S, Murdock CC, Vickers M & Sears MW (2016). Fine-scale microclimatic variation can shape the responses of organisms to global change in both natural and urban environments. Integrative and Comparative Biology. 56(1): 45–61
- Potts SG, Biesmeijer JC, Kremen C, Neumann P, Schweiger O & Kunin WE (2010) Global pollinator declines: trends, impacts and drivers. Trends in Ecology & Evolution. 25(6): 345–353
- Pressman E, Peet MM & Pharr DM (2002) The effect of heat stress on tomato pollen characteristics is associated with changes in carbohydrate concentration in the developing anthers. Annals of Botany. 90(5): 631–636
- Qaderi MM, Kurepin LV & Reid DM (2012) Effects of temperature and watering regime on growth, gas exchange and abscisic acid content of canola (*Brassica napus*) seedlings. Environmental and Experimental Botany. 75: 107–113
- Ramakrishna A & Ravishankar GA (2011) Influences of abiotic stress signals on secondary metabolites in plants. Plant Signal Behavior. 6: 1720–1731
- Regan EC, Santini L, Ingwall-King L, Hoffmann M, Rondinini C, Symes A & Butchart SH (2015) Global trends in the status of bird and mammal pollinators. Conservation Letters. 8(6): 397–403
- Richards MH, Garate AC, Shehata M, Groom D, Tattersall GJ & Welch KC (2020) Effect of nest microclimate temperatures on metabolic rates of small carpenter bees, Ceratina calcarata (Hymenoptera: Apidae). The Canadian Entomologist. 152(6): 772–782
- Ricketts TH, Regetz J, Steffan-Dewenter I, Cunningham SA, Kremen C, Bogdanski A & Viana BF (2008) Landscape effects on crop pollination services: are there general patterns? Ecology Letters. 11(5): 499–515

- Ríos Reyes LD, Fuchs Castillo EJ, Hodel DR & Cascante Marín A. Neither insects nor wind: ambophily in dioecious *Chamaedorea palms* (Arecaceae). Plant Biol. 16: 702–10
- Roger N, Michez D, Wattiez R, Sheridan C & Vanderplanck M (2017). Diet effects on bumblebee health. Journal of Insect Physiology. 96: 128–133
- Rollings R & Goulson D (2019) Quantifying the attractiveness of garden flowers for pollinators. Journal of Insect Conservation. 23(5): 803–817
- Salisbury A, Armitage J, Bostock H, Perry J, Tatchell M & Thompson K (2015) Editor's Choice: Enhancing gardens as habitats for flower-visiting aerial insects (pollinators): should we plant native or exotic species? Journal of Applied Ecology. 52(5): 1156–1164
- Samson-Robert O, Labrie G, Chagnon M & Fournier V (2014) Neonicotinoid-contaminated puddles of water represent a risk of intoxication for honeybees. PLoS One. 9(12): e108443
- Sanchez-Bayo, F & Goka, K (2016) Impacts of pesticides on honeybees. Beekeeping and Bee Conservation-Advances in Research. 4: 77–97
- Santiago JP & Sharkey TD (2019) Pollen development at high temperature and role of carbon and nitrogen metabolites. Plant, Cell & Environment. 42(10): 2759–2775
- Sato S, Kamiyama M, Iwata T, Makita N, Furukawa H & Ikeda H (2006) Moderate increase of mean daily temperature adversely affects fruit set of *Lycopersicon esculentum* by disrupting specific physiological processes in male reproductive development. Annals of Botany. 97(5): 731–738
- Sattler T, Duelli P, Obrist MK, Arlettaz R & Moretti M (2010) Response of arthropod species richness and functional groups to urban habitat structure and management. Landscape Ecology. 25(6). 941–954
- Scaven VL & Rafferty NE (2013) Physiological effects of climate warming on flowering plants and insect pollinators and potential consequences for their interactions. Current Zoology. 59(3): 418–426
- Schmaranzer S (2000) Thermoregulation of water collecting honeybees (Apis mellifera) Journal of Insect Physiology. 46(8): 1187–1194
- Schweiger O, Biesmeijer JC, Bommarco R, Hickler T, Hulme PE, Klotz S & Settele J. (2010) Multiple stressors on biotic interactions: how climate change and alien species interact to affect pollination. Biological Reviews. 85(4): 777–795
- Schweiger O, Settele J, Kudrna O, Klotz S & Kühn I (2008) Climate change can cause spatial mismatch of trophically interacting species. Ecology. 89(12): 3472–3479
- Senapathi D, Goddard MA, Kunin WE & Baldock KC (2017) Landscape impacts on pollinator communities in temperate systems: evidence and knowledge gaps. Functional Ecology. 31(1): 26–37
- Settele J, Bishop J & Potts SG (2016) Climate change impacts on pollination. Nature Plants.2(7): 1–3
- Siebers J, Binner R & Wittich KP (2003) Investigation on downwind short-range transport of pesticides after application in agricultural crops. Chemosphere. 51(5): 397–407

- Smith AR & Zhao D (2016) Sterility caused by floral organ degeneration and abiotic stresses in Arabidopsis and cereal grains. Frontiers in Plant Science. 7: 1503–1509
- Snider JL & Oosterhuis DM (2011) How does timing, duration, and severity of heat stress influence pollen-pistil interactions in angiosperms?. Plant Signaling & Behavior. 6(7): 930–933
- Solís-Montero L & Vallejo-Marín M (2017) Does the morphological fit between flowers and pollinators affect pollen deposition? An experimental test in a buzz-pollinated species with anther dimorphism. Ecology and Evolution. 7(8): 2706–2715
- Sørensen JG, Kristensen TN & Loeschcke V (2003) The evolutionary and ecological role of heat shock proteins. Ecology Letters. 6(11): 1025–1037
- Spinoni J, Vogt JV, Naumann G, Barbosa P & Dosio A (2018) Will drought events become more frequent and severe in Europe? International Journal of Climatology. 38(4): 1718–1736
- Stenhouse RN (2004) Fragmentation and internal disturbance of native vegetation reserves in the Perth metropolitan area, Western Australia. Landscape and Urban Planning. 68(4): 389–401
- Stewart AB, Sritongchuay T, Teartisup P, Kaewsomboon S & Bumrungsri S (2018) Habitat and landscape factors influence pollinators in a tropical megacity, Bangkok, Thailand. PeerJ. 6: 35–53
- Threlfall CG, Ossola A, Hahs AK, Williams NS, Wilson L & Livesley SJ (2016) Variation in vegetation structure and composition across urban green space types. Frontiers in Ecology and Evolution. 4: 66–70
- Van Tussenbroek BI, Villamil N, Márquez-Guzmán J, Wong R, Monroy-Velázquez LV & Solis-Weiss V (2016) Experimental evidence of pollination in marine flowers by invertebrate fauna. Nature Communications. 7(1): 1–6
- Vanderplanck M, Moerman R, Rasmont P, Lognay G, Wathelet B, Wattiez R & Michez D (2014) How does pollen chemistry impact development and feeding behaviour of polylectic bees? PLoS One. 9(1): e86209
- Wallisdevries MF & Van Swaay CA (2006) Global warming and excess nitrogen may induce butterfly decline by microclimatic cooling. Global Change Biology. 12(9): 1620–1626
- Wardhaugh CW (2015) How many species of arthropods visit flowers? Arth-Plant Interact. 9: 547–65
- Waser NM, Chittka L, Price MV, Williams NM & Ollerton J (1996) Generalization in pollination systems and why it matters. Ecology. 77: 1043–1060
- Wenzel A, Grass I, Belavadi VV & Tscharntke T (2020) How urbanization is driving pollinator diversity and pollination–A systematic review. Biological Conservation. 241: 108–120
- Wilson CJ & Jamieson MA (2019) The effects of urbanization on bee communities depends on floral resource availability and bee functional traits. PloS One. 14(12): e0225852

- Winfree R, Aguilar R, Vázquez DP, LeBuhn G & Aizen MA (2009) A meta-analysis of bees' responses to anthropogenic disturbance. Ecology. 90(8): 2068–2076
- Woods N, Craig IP, Dorr G & Young B (2001) Spray drift of pesticides arising from aerial application in cotton. Journal of Environmental Quality. 30(3): 697–701
- Wu J (2014) Urban ecology and sustainability: The state-of-the-science and future directions. Landscape and Urban Planning. 125: 209–221
- Zayed A & Packer L (2005) Complementary sex determination substantially increases extinction proneness of haplodiploid populations. Proceedings of the National Academy of Sciences. 102(30):10742-10746
- Zych M, Goldstein J, Roguz K & Stpiczyńska M (2013) The most effective pollinator revisited: pollen dynamics in a spring-flowering herb. Arthropod-Plant Interactions. 7(3): 315–322

Index

A

Abiotic pollination 51
Adult nutrition 190, 192
Agave 59
Agriculture practices 282
Agroecosystem 93
Agroindustry 6, 16
Alien plants 171
Androecium 47
Angiosperms, sexual reproduction in 11
Animal pollination, relative importance of 53
Apiculture 116
Artificial feeding 189
Artificial pollination 14
Autogamy 48

B

Basic flower anatomy 46
Bee conservation, strategies of 167
  measures in 169
Bee flower foraging 139
Bee foraging 14
Bee hotels accommodation 171
Bee pollination crisis solution 8
Bee pollination vs. non-bee pollination 142
Bee pollination, economy and 133
  opportunities and threats 141
Bee pollinators, decline of 6
Bee population, decline in 167
Bees 164, 282
  nutrition-related threat for 195
Bees and the landscape, relationship between 3
Beetles vs. bees 144
Benzenoids 208
Biotic pollination 51
Blueberry 57
Butterflies vs. bees 143
Buzz pollination 7

C

Cacao 59
Calyx 47
Canola 58
Carbohydrates feeding 191
Carbohydrates nutrition 189
Carpenter bees, role of 138
Chemical compounds, negative consequences 217
Chemical contamination/residue 282
Chemical ecology 106
Chemical self-incompatibility 108
Chemicals and pollutants 288
City landscapes 237
Cleistogamy 48
Coffee 58
Colony collapse disorder 282
Colony division 124
Colony nutrition 189,191
Commercial bee pollination 8
Commercial honeybees 18
Corolla 47

Crop diversity in India 66
Crop insect pollinators
  and biohabitats 13
  importance of 16
Crop plants,
  reproductive strategies of 67
  reproductive traits of 68
  yield enhancement of 73
Crop pollination 2, 5, 8, 13
  and bumblebees 136
Crop production and bee pollination 135
Cross fruitful varieties 50
Cross unfruitful 50
Cross-pollination 48, 49, 108
  significance of 35, 49

D

Dioecious plants 50
Disease and pest incidence 89
Dish liquid detergent chemical sprays,
  effect of 243

E

Effective pollination 5
Entpollinatological awareness 30
Entpollinatological procedure 4
Entpollinatology 2
Essential oils 209

F

Fatty acid derivatives 209
Feeding proteins 193
Fertilization 49
Floral chemistry
  evolution of 111
  functions of 109
Floral metabolite 218
Floral resources, impact of 241
Floral scents, chemical profile of 216
Floral visitors of crop plants 67
Flower pollination 6
Foraging behavior of subspecies 188
Foraging distance 188
Foraging philosophy 186
Foraging preferences 188
Foraging tasks 187
Foraging time 187
Foraging,
  controlling of 189
  factors influencing 188
  monitoring of 189
Forest ecosystem 93
Fragmentation 283

G

Green habitat, size of 237
Gynoecium 47
Gynoecious plants 50

H

Habitat loss 283
Honey crop, making space for 123
Honeybee colonies, management of 116, 119
Honeybee services 2
Honeybees and wild bees, comparative
  significance of 164
Honeybees,
  contribution of 7
  for pollination of cultivated crops 34
  role of 6, 135
  seasonal management of 115
Hover flies vs. bees 142

I

Insect pollinator populations, decline in 7
Insect pollinators 2, 204
  importance in food sustainability 10
  importance of 81
  threats to 17
Inbreeding and self-seed fertilization,
  potential genetic complications in 11

Insect pollination 30, 203
Insect pollination technology, economic significance 132
Insect pollination 203

L

Larval nutrition 190, 193
Landscape composition 238
Lawn mowing, effect of 244
Local habitats, climatic conditions of 236

M

Managed pollinators 65
Monoecious plants 50
Month-wise colony management 125
Moths vs. bees 144
Mutualistic associations 12

N

Native pollinators 234
Natural foraging 186
Nectar chemistry 109
Nectar secretion 47
Nectar, toxic metabolites of 217
Nectaries 47
Nesting sites 239

P

Parasitic mites 4
Pathogens 282
Pest management technology 265
Pest surveillance, significance of 267
Phenylpropanoids 208
PIPMT applications 270
PIPMT model and biological operators 268
PIPMT model, strategies of 263, 269
PIPMT, non-biological operators and 269
PIPMT, strategy and farm management 266
Plant biochemicals 203, 205
    for pollinator's attraction 207
Plant flowering resources 241
Plant reproduction procedures 35
Plant secondary metabolites 211
Plant variation and insect pollination mechanisms 36
Plant volatile organic compounds, metabolic engineering of 219
Plant-pollinator interactions, ecology and evolution of 107
Plant-pollinators interaction 90
Pollenizer 48
Pollinatechnology 2
Pollinating agent, role of 14
Pollinating agents 50
Pollination 4
    concepts and crop production 44,
    empirical demonstration in 32
    process of 45
    significance of 6
Pollination 48
    types of 49
Pollination biology 5
Pollination deficiency 72
Pollination services, adequacy of 56
Pollination syndromes 5
Pollinator density and crop yield 73
Pollinator dependence, degree of 65
Pollinator diversity 279
Pollinator lecty and food 239
Pollinator services and yield of crop plants 74
Pollinators abundance and foraging, effect of climate on 243
Pollinators
    and pollination limitations 55
    biology of 86
    blueprint for conservative activity of 92
    climate change and 80,
    conservation of 91-94
    conservational aspects of 80

decline rate 91
distribution of 84
diversity of 83
effect of climate changes on 83
flight and foraging activity of 87
in agriculture 51
importance of 3
morphology of 85
physiology of 85
policies regarding conservation of 91
survival needs of 280
Protein nutrition 191

S

Season-wise crop types in India 66
Secondary metabolites 207
Self fruitful 49
Self incompatible 50
Self unfruitful 50
Self-pollination 48, 49
mechanisms that prevents 108
Semi-natural environment protection 168
Sex differentiation in crops, principle of 33
Solitary bees, role of 138
Sonication 7
Stigma receptivity 48
Stingless bees, role of 137
Sunflower 57
Supering 123
Supplementary pollinator services 64
Sustainable agriculture 163
Swarm
catching and hiving 121
control of 122
prevention of 121
Swarm indications 121
Swarming 119
period and causes 120
setbacks due to 120

T

Terpenoids 208
Thrips vs. bees 145

U

Urban ecosystem 93
Urban green habitats 234
Urban landscapes 235
Urban structures, attributes of 236
Urban development, effect of 238
Urbanization and pollinator body size 240
Urbanization 283

V

Volatile organic compounds 208, 220

W

Watermelon 56
Wasps vs. bees 145
Whole plant chemistry 110
Wild bees 163
conservation of 166
Wild bee abundance, protection approaches in 172
Wild bee pollinators, conservation strategies for 173